T0214573

Communications in Computer and Information Science 859

Commenced Publication in 2007
Founding and Former Series Editors:
Phoebe Chen, Alfredo Cuzzocrea, Xiaoyong Du, Orhun Kara, Ting Liu,
Dominik Ślęzak, and Xiaokang Yang

More information about this series at http://www.springer.com/series/7899

Daniel A. Alexandrov · Alexander V. Boukhanovsky
Andrei V. Chugunov · Yury Kabanov
Olessia Koltsova (Eds.)

Digital Transformation and Global Society

Third International Conference, DTGS 2018
St. Petersburg, Russia, May 30 – June 2, 2018
Revised Selected Papers, Part II

Editors
Daniel A. Alexandrov ⓘ
National Research University Higher School
of Economics
St. Petersburg, Russia

Yury Kabanov ⓘ
National Research University Higher School
of Economics
St. Petersburg, Russia

Alexander V. Boukhanovsky
Saint Petersburg State University
of Information Technologies
St. Petersburg, Russia

Olessia Koltsova ⓘ
National Research University Higher School
of Economics
St. Petersburg, Russia

Andrei V. Chugunov
Saint Petersburg State University
of Information Technologies
St. Petersburg, Russia

ISSN 1865-0929   ISSN 1865-0937 (electronic)
Communications in Computer and Information Science
ISBN 978-3-030-02845-9   ISBN 978-3-030-02846-6 (eBook)
https://doi.org/10.1007/978-3-030-02846-6

Library of Congress Control Number: 2018958515

This Springer imprint is published by the registered company Springer Nature Switzerland AG
The registered company address is: Gewerbestrasse 11, 6330 Cham, Switzerland

Preface

The International Conference on Digital Transformation and Global Society (DTGS 2018) was held for the third time from May 30 to June 2, 2018, in St. Petersburg Russia. It is a rapidly developing academic event, addressing the interdisciplinary agenda of ICT-enabled transformations in various domains of human life. DTGS 2018 was co-organized by the ITMO University and the National Research University Higher School of Economics (St. Petersburg), two of the leading research institutions in Russia.

This year was marked by a significant rise in interest in the conference in academia. We received 222 submissions, which were carefully reviewed by at least three Program Committee members. In all, 76 papers were accepted, with an acceptance rate of 34%. More than 120 participants attended the conference and contributed to its success. We would like to emphasize the increase in the number of young scholars taking part in the event, as well as the overall improvement in the quality of the papers.

DTGS 2018 was organized as a series of research paper sessions, preceded by a poster session. The sessions corresponded to one of the following DTGS 2018 tracks:

- ESociety: Social Informatics and Virtual Communities
- EPolity: Politics and Governance in the Cyberspace
- EHumanities: Digital Culture and Education
- ECity: Smart Cities and Urban Governance
- EEconomy: Digital Economy and ICT-Driven Economic Practices
- ECommunication: Online Communication and the New Media

Two new international workshops were also held under the auspices of DTGS: the Internet Psychology Workshop, chaired by Prof. Alexander Voiskounsky (Moscow State University) and Prof. Anthony Faiola (The University of Illinois at Chicago), as well as the Computational Linguistics Workshop, led by Prof. Viktor Zakharov (St. Petersburg State University) and Prof. Anna Tilmans (Leibniz University of Hannover). The agenda of DTGS is thus becoming broader, exploring the new domains of digital transformation.

Furthermore, we would like to mention several insightful keynote lectures organized at DTGS 2018. Prof. Stephen Coleman from the University of Leeds gave a talk on the role of the Internet in restoring and promoting democracy. The lecture was partially based on his recent book *Can the Internet Strengthen Democracy?* (Polity: 2017), which was translated into Russian and published by the DTGS team before the conference. Dr. Dennis Anderson (St. Francis College, USA) shared with the participants his vision of the future of e-government and its role in society, while Dr. Christoph Glauser (Institute for Applied Argumentation Research, Switzerland) presented the tools to evaluate citizens' expectations from e-government and the ways e-services can be adjusted to serve people's needs. The keynote lecture by Prof. Anthony Faiola was devoted to e-health technologies, especially to the potential of mobile technologies to facilitate health care.

Finally, two international panel discussions were arranged. The first one – "Cybersecurity, Security and Privacy" — was chaired by Prof. Latif Ladid from the University of Luxembourg. Panel participants Dr. Antonio Skametra (University of Murcia), Dr. Sebastian Ziegler (Mandat International, IoT Forum, Switzerland), and Dr. Luca Bolognini (Italian Institute for Data Privacy and Valorization) shared their opinion on the future of privacy protection in relation to the changes of the EU personal data regulations. The second panel moderated by Dr. Yuri Misnikov (ITMO University) and Dr. Svetlana Bodrunova (St. Petersburg State University) was devoted to the online deliberative practices in the EU and Russia. Prof. Stephen Coleman, Prof. Leonid Smorgunov (St. Petersburg State University), Dr. Lyudmila Vidiasova (ITMO University), Dr. Olessia Koltsova, and Yury Kabanov (National Research University Higher School of Economics) took part in the discussion, expressing their views on the role of virtual communities in maintaining democratic practices and better governance.

Such a plentiful scientific program would have been impossible without the support and commitment from many people worldwide. We thank all those who made this event successful. We are grateful to the members of the international Steering and Program Committees, the reviewers and the conference staff, the session and workshop chairs, as well as to the authors contributing their excellent research to the volume.

We are happy to see the conference growing in importance on the global scale. We believe that DTGS will continue to attract an international expert community to discuss the issues of digital transformation.

May 2018

Daniel A. Alexandrov
Alexander V. Boukhanovsky
Andrei V. Chugunov
Yury Kabanov
Olessia Koltsova

Organization

Program Committee

Artur Afonso Sousa	Polytechnic Institute of Viseu, Portugal
Svetlana Ahlborn	Goethe University, Germany
Luis Amaral	University of Minho, Portugal
Dennis Anderson	St. Francis College, USA
Francisco Andrade	University of Minho, Portugal
Farah Arab	Université Paris 8, France
Alexander Babkin	Crimea Federal University, Russia
Maxim Bakaev	Novosibirsk State Technical University, Russia
Alexander Balthasar	Bundeskanzleramt, Austria
Luís Barbosa	University of Minho, Portugal
Vladimír Benko	Slovak Academy of Sciences, Ľ. Štúr Institute of Linguistics, Slovakia
Sandra Birzer	Innsbruck University, Austria
Svetlana Bodrunova	St. Petersburg State University, Russia
Radomir Bolgov	St. Petersburg State University, Russia
Anastasiya Bonch-Osmolovskaya	National Research University Higher School of Economics, Russia
Nikolay Borisov	St. Petersburg State University, Russia
Dietmar Brodel	Carinthia University of Applied Sciences, Austria
Mikhail Bundin	Lobachevsky State University of Nizhni Novgorod, Russia
Diana Burkaltseva	Crimea Federal University, Russia
Luis Camarinha-Matos	University of Lisbon, Portugal
Lorenzo Cantoni	University of Lugano, Italy
François Charoy	Lorraine Laboratory of Research in Computer Science and its Applications, France
Sunil Choenni	Research and Documentation Centre (WODC), Ministry of Justice, The Netherlands
Andrei Chugunov	ITMO University, Russia
Iya Churakova	St. Petersburg State University, Russia
Meghan Cook	SUNY Albany, Center for Technology in Government, USA
Esther Del Moral	University of Oviedo, Spain
Saravanan Devadoss	Addis Ababa University, Ethiopia
Subrata Kumar Dey	Independent University, Bangladesh
Alexey Dobrov	St. Petersburg State University, Russia
Irina Eliseeva	St. Petersburg State University of Economics, Russia
Anthony Faiola	The University of Illinois at Chicago, USA

Isabel Ferreira	Polytechnic Institute of Cávado and Ave, Spain
Olga Filatova	St. Petersburg State University, Russia
Enrico Francesconi	Italian National Research Council, Italy
Diego Fregolente Mendes de Oliveira	Indiana University, USA
Fernando Galindo	University of Zaragoza, Spain
Despina Garyfallidou	University of Patras, Greece
Carlos Gershenson	National Autonomous University of Mexico, Mexico
J. Paul Gibson	Mines Telecom, France
Christoph Glauser	Institute for Applied Argumentation Research, Switzerland
Tatjana Gornostaja	Tilde, Latvia
Dimitris Gouscos	University of Athens, Greece
Stefanos Gritzalis	University of the Aegean, Greece
Karim Hamza	Vrije Universiteit Brussel, Belgium
Alex Hanna	University of Toronto, Canada
Martijn Hartog	The Hague University of Applied Sciences, The Netherlands
Agnes Horvat	Northwestern University, USA
Dmitry Ilvovsky	National Research University Higher School of Economics, Russia
Marijn Janssen	Delft University of Technology, The Netherlands
Yury Kabanov	National Research University Higher School of Economics, Russia
Katerina Kabassi	TEI of Ionian Islands, Greece
Christos Kalloniatis	University of the Aegean, Greece
George Kampis	Eotvos University, Hungary
Egor Kashkin	V. V. Vinogradov Russian Language Institute of RAS, Russia
Sanjeev Katara	National Informatics Centre, Govt. of India, India
Philipp Kazin	ITMO University, Russia
Norbert Kersting	University of Muenster, Germany
Maria Khokhlova	St. Petersburg State University, Russia
Mikko Kivela	Aalto University, Finland
Bozidar Klicek	University of Zagreb, Croatia
Ralf Klischewski	German University in Cairo, Egypt
Eduard Klyshinskii	Moscow State Institute of Electronics and Mathematics, Russia
Andreas Koch	University of Salzburg, Austria
Olessia Koltsova	National Research University Higher School of Economics, Russia
Liliya Komalova	Institute of Scientific Information for Social Sciences of Russian Academy of Sciences, Moscow State Linguistic University, Russia
Mikhail Kopotev	University of Helsinki, Finland
Ah-Lian Kor	Leeds Beckett University, UK

Evgeny Kotelnikov	Vyatka State University, Russia
Artemy Kotov	National Research Center Kurchatov Institute, Russia
Sergey Kovalchuk	ITMO University, Russia
Michal Kren	Charles University, Czech Republic
Valentina Kuskova	National Research University Higher School of Economics, Russia
Valeri Labunets	Ural Federal University, Russia
Sarai Lastra	Universidad del Turabo, USA
Sandro Leuchter	Hochschule Mannheim University of Applied Sciences, Germany
Yuri Lipuntsov	Moscow State Universuty, Russia
Natalia Loukachevitch	Moscow State University, Russia
Mikhail Lugachev	Moscow State University, Russia
Olga Lyashevskaya	National Research University Higher School of Economics, Russia
Jose Machado	University of Minho, Portugal
Rosario Mantegna	Palermo University, Italy
Ignacio Marcovecchio	United Nations University Institute on Computing and Society, Macao, SAR China
João Martins	United Nations University, Portugal
Aleksei Martynov	Lobachevsky State University of Nizhny Novgorod, Russia
Ricardo Matheus	Delft University of Technology, The Netherlands
Tatiana Maximova	ITMO University, Russia
Athanasios Mazarakis	Kiel University/ZBW, Germany
Christoph Meinel	Hasso Plattner Institute, Germany
Yelena Mejova	Qatar Computing Research Institute, Qatar
András Micsik	SZTAKI, Hungary
Yuri Misnikov	ITMO University, Russia
Harekrishna Misra	Institute of Rural Management Anand, India
Olga Mitrofanova	St. Petersburg State University, Russia
Zoran Mitrovic	Mitrovic Development & Research Institute, South Africa
John Mohr	University of California, USA
José María Moreno-Jimenez	Universidad de Zaragoza, Spain
Robert Mueller-Toeroek	University of Public Administration and Finance Ludwigsburg, Germany
Ilya Musabirov	National Research University Higher School of Economics, Russia
Alexandra Nenko	ITMO University, Russia
Galina Nikiporets-Takigawa	University of Cambridge, UK; Russian State Social University, Russia
Prabir Panda	National Institute for Smart Government, India
Ilias Pappas	Norwegian University of Science and Technology, Norway
Mário Peixoto	United Nations University, Portugal

Lyudmila Vidiasova	ITMO University, Russia
Alexander Voiskounsky	Lomonosov Moscow State University, Russia
Ruprecht von Waldenfels	University of Jena, Germany
Catalin Vrabie	National University of Political Studies and Public Administration, Romania
Ingmar Weber	Qatar Computing Research Institute, Qatar
Mariëlle Wijermars	University of Helsinki, Finland
Vladimir Yakimets	Institute for Information transmission Problems of RAS, Russia
Nikolina Zajdela Hrustek	University of Zagreb, Croatia
Victor Zakharov	St. Petersburg State University, Russia
Sergej Zerr	L3S Research Center, Germany
Hans-Dieter Zimmermann	FHS St. Gallen University of Applied Sciences, Switzerland
Thomas Ågotnes	University of Bergen, Norway
Vytautas Čyras	Vilnius University, Lithuania

Additional Reviewers

Abraham, Joanna
Abrosimov, Viacheslav
Balakhontceva, Marina
Belyakova, Natalia
Bolgov, Radomir
Bolgova, Ekaterina
Borisov, Nikolay
Burkalskaya, Diana
Chin, Jessie
Churakova, Iya
Derevitskiy, Ivan
Derevitsky, Ivan
Duffecy, Jennifer
Eliseeva, Irina
Funkner, Anastasia
Gonzalez, Maria Paula
Guleva, Valentina Y.
Karyagin, Mikhail
Kaufman, David
Litvinenko, Anna
Marchenko, Alexander
Masevich, Andrey

Mavroeidi, Aikaterini-Georgia
Melnik, Mikhail
Metsker, Oleg
Mityagin, Sergey
Nagornyy, Oleg
Naumov, Victor
Nikitin, Nikolay
Papautsky, Elizabeth
Routzouni, Nancy
Semakova, Anna
Sergushichev, Alexey
Sideri, Maria
Sinyavskaya, Yadviga
Smoliarova, Anna
Steibel, Fabro
Trutnev, Dmitry
Ufimtseva, Nathalia
Vatani, Haleh
Virkar, Shefali
Visheratin, Alexander
Zhuravleva, Nina

Contents – Part II

International Workshop on Internet Psychology

International Workshop on Computational Linguistics

Contents – Part I

E-Polity: Law and Regulation

E-City: Smart Cities & Urban Planning

E-Economy: IT & New Markets

E-Society: Social Informatics

E-Society: Digital Divides

The Winner Takes *IT* All: Swedish Digital Divides in Global Internet Usage

John Magnus Roos[1,2(✉)]

[1] University of Skövde, Skövde, Sweden
`Magnus.roos@his.se`
[2] University of Gothenburg, Gothenburg, Sweden

Abstract. In the present study, we examined the influence of personality factors and demographic factors on Internet usage. Personality was defined from the Five Factor Model of personality in terms of Openness, Conscientiousness, Extraversion, Agreeableness, and Neuroticism, while demographic factors were defined as gender, age and socioeconomic status (e.g. income and educational attainment). The results from a large, representative Swedish sample ($N = 1,694$) show that global Internet usage can be explained by a high degree of Extraversion, young age and high socioeconomic status. Our findings are consistent with some previous studies, but in contrast with others. We discuss contrasting results in terms of different study designs, cultures and time periods of Internet development. The results are discussed in terms of the "rich get richer model" and digital divides, and what broader implication our findings might have for society. The study may help facilitate our understanding regarding future challenges in the Internet design.

Keywords: Personality · Gender · Age · Socio-economic status
Internet

1 Introduction and Research Questions

The concept "digital divide" usually refers to inequalities in Internet access and Internet usage within a country, between individuals at different socioeconomic levels or other demographic categories [8]. Although the digital divides seem to be decreasing in developed countries, it still seems as if certain groups of people use the Internet less than others: elderly people, people with a low income and people with low education [11, 23, 36]. The digital divide, in terms of the "rich get richer model," has also been discussed by personality theorists [26, 45, 50], proposing that the Internet foremost enhances sociable and extravert individuals and thereby strengthens their social contacts and well-being even more [26]. Therefore, it might be argued that the Internet foremost benefits people who by themselves are in a "rich" position in society regarding aspects such as monetary assets, well-being and social life. According to Norman [35], designers need to consider personality factors (such as Openness, Conscientiousness, Extraversion, Agreeableness, and Neuroticism) in order to design for the wide range of human needs.

D. A. Alexandrov et al. (Eds.): DTGS 2018, CCIS 859, pp. 3–18, 2018.

As Information Technologies (IT) advances to create more and more ways for people to become connected and to impact the world around them, IT is not only transforming the everyday life for people who use digital activities in their everyday lives but also for people who do not personally go online [11]. This means that the digital transformation will influence all individuals in a society to some extent, whether they use the Internet or not. The critical question for the present study is to examine if some individuals use the Internet more than others, depending on individual characteristics such as personality factors and demographic factors (e.g. gender, age and socioeconomic status). The main focus in the present study will be on personality factors rather than demographic factors. The reasons are twofold. Firstly, the digital divide regarding demographic factors has continuously been reported from nationally representative samples for a long period of time, such as the "Pew Internet & American Life Project" [23] and "The Swedes and the Internet" [11]. Secondly, there is a need for more knowledge regarding the relationship between personality factors and Internet usage since the previous studies have revealed more contradictive results compared to the studies of demographic factors, probably because they are based on small samples. Most personality studies related to Internet usage rely on convenience samples of homogenous groups: undergraduate students or young adults. In the present study, we will reconcile the previous conflicting results by using a large and nationally representative sample of the Swedish population ($N = 1,694$).

The purpose of the present study is to examine digital divides in Sweden, and to what extent "global Internet usage" (i.e. total Internet usage on an individual level) can be explained by individual differences in personality and demography. We formulated two research questions:

(1) Are personality factors related to global Internet usage?
(2) Are demographic factors related to global Internet usage?

2   Theoretical Grounding

2.1   Personality and the Five Factor Model

Personality contributes to individual differences in behavior, consistency of behavior over time and stability of behavior across situations [15]. The Five Factor Model (FFM) is the most popular theory of personality, positing that there are five major and universal factors of personality: Openness, Conscientiousness, Extraversion, Agreeableness, and Neuroticism [9, 18].

Openness represents the tendency to be curious and flexible. Open-minded people test new ideas and change their behavior frequently. Conscientiousness represents responsibility and dutifulness. Conscientious people actively plan, organize and carry out tasks. Agreeableness represents the tendency to be sympathetic and good-hearted. Agreeable people are trusting and forgiving. Extraversion represents sociability, optimism and well-being. Extraverts search for excitement and influence their social environments. Neuroticism represents emotional instability and low self-confidence. Highly neurotic people tend to be sad, anxious and stressed [6, 9, 28].

2.2 Demographic Factors

People's behavior is influenced by their demography in terms of age, gender and socioeconomic status (SES) [25]. SES is a measure of an individual's position within a social group based on various factors including education, income, occupation, location of residence, membership in civic or social organizations and certain amenities in the home, such as IT [6].

2.3 Theoretical Analysis

In brief, personality factors and demographic factors influence our behavior, for instance internet usage. Thus, we propose that global Internet usage is a consequence of personality factors and demographic factors (Fig. 1).

Fig. 1. Personality factors, demographic factors and global Internet usage. *Note.* Influenced by Chamorro-Premuzic [6], p. 19.

3 Literature Review

3.1 The Five Factor Model and Global Internet Usage

To the best of our knowledge, four studies have been conducted on global Internet usage and the FFM of personality (Table 1).

One study [32] shows that Openness is positively related to global Internet usage. The other studies presented in Table 1 did not show any significant relation between Openness and global Internet usage [21, 27, 31].

There are varied findings in the relation between Conscientiousness and global Internet usage. Landers and Lounsbury [27] have found that Conscientiousness is negatively related to global Internet usage, while Mark and Granzach [31] have found that Conscientiousness is positively related to global Internet usage. Other researchers [21, 32] did not find any significant relation between Conscientiousness and global Internet usage.

Table 1. Previous studies on the Five Factor Model and global Internet usage.

Study	Sample	Five Factor Personality inventory	Measurement for global Internet Usage	Control variables	Results: FFM relation to global Internet usage
Jackson et al. [21]	Adult residents of a low-income, medium-size urban community in U.K N = 117	The Big Five Personality Inventory (25 items) [22]	Two measures: (1) Time on line (minutes/day) and (2) number of Internet sessions/day	No control variables	Positively related to Extraversion (p < .05) and negatively related to Neuroticism (p < .05)
Landers and Lounsbury [27]	Undergraduate students in a lower-division psychology course in USA N = 117	Adolescent Personal Style Inventory (APSI) [30]	Single item: Eight-point scale ranging from (1) Less than 1 h per week to (2) More than 10 h per day	No control variables	Negatively related to Conscientiousness (p < .05), Extraversion (p < .05) and Agreeableness (p < .05)
McElroy, Hendrickson, Townsend & DeMarie [32]	MBA and senior undergraduate students from a variety of majors in USA N = 132	Revised NEO Personality Inventory (240 items) [9].	Average of seven items (i.e. surfing, chatting and searching information)	Gender, computer anxiety, self-efficacy	Positively related to Openness (p ≤ .05)
Mark and Ganzach [31]	Nationally. representative U.S. sample of young adults (e.g. 24–28 years of age) N = 6,921	Ten Items Personality Measure (TIPI) [19]	Single item: Six-point ranging from (1) Several times a day to (6) Less often than once every few weeks	Gender, income and education	Positively related to Conscientiousness (p < .001), Extraversion (p < .001) and Neuroticism (p < .001)

Jackson et al. [21] and Mark and Granzach [31] have found that Extraversion is positively related to global Internet usage. In contrast, Landers and Lounsbury [27] have found that Extraversion is negatively related to global Internet usage. McElroy et al. [32] did not find any significant relationship here.

The general conclusion is that Agreeableness is unrelated to global Internet usage [21, 31, 32]. The exception is a study by Landers and Lounsbury [27] who found a negative relation between Agreeableness and global Internet usage.

There are varied findings regarding the relationship between Neuroticism and global Internet usage. Jackson et al. [21] have found that Neuroticism is negatively related to global Internet usage, while Mark and Granzach [31] have found that Neuroticism is positively related to global Internet usage. Landers and Lounsbury [27] and McElroy et al. [32] on the other hand did not notice any significant relationship here.

The discrepant results from the previous studies might be explained by at least four reasons. Firstly, the sample sizes in three of the four studies are small (e.g. $N < 132$). Secondly, the four studies use different measurements and statistical analyses. Different inventories were used for measuring the FFM and different scales were used for measuring global Internet usage. Moreover, different control variables were used in the studies (Table 1). Thirdly, the time periods for data gathering differ across the studies. While the personality factors are assumed to be invariant, the Internet landscape has changed dramatically over the time periods that the four studies were conducted. For instance, the development and popularity of social media (such as Facebook, Twitter and MySpace) and blogs have affected the use of the Internet. Fourthly, there are cultural differences. The participants from the first study are from the U.K, while the participants from the other three studies are from the USA (Table 1).

3.2 Demographic Factors and Global Internet Usage

Analyzing the literature on demographic factors and global Internet usage generates a huge number of studies, but with methodological limitations. Most of the findings are based on descriptive statistics and very simple statistical analyses (e.g. correlation studies). However, there are some studies, especially regarding SES, age and global Internet usage, which have used more advanced statistical analyses.

Most studies have found that global Internet usage is negatively related to age [5, 7, 33, 36]. Internet usage seems to continuously decrease by age, but the decline is especially noticeable for people above 60 years of age [48]. In contrast, some studies have found positive relations between age and global Internet usage [1, 40, 49]. The positive relation is explained by SES as a confounding factor [1, 40].

The previous studies have either reported that men are using the Internet more frequently than women [32, 34, 39], or that there are no gender differences in global Internet usage [1, 13, 24, 49].

The present study will only consider SES in terms of education and income. There is a positive relationship between education and global Internet usage. People with higher education use the Internet more frequently than people with lower education [7, 33, 36].

According to most researchers, income is positively related to global Internet usage [1, 14, 16, 23, 29, 33, 47]. In contrast, Donat et al. [13] have found that this relationship has disappeared. Cheaper computers, Internet cafés and free Wi-Fi have opened up the Internet also for people with low income [13].

4 Methodology

4.1 Sample and Procedure

The Institute for Society, Opinion and Media (the SOM-institute) at the University of Gothenburg in Sweden annually surveys attitudes representing the Swedish population. Every year, The National SOM survey uses a systematic probability sample of 3,400 Swedish citizens aged 16–85. Each recipient of the questionnaire is selected in accordance with the goal of representing the Swedish population in regard of ethnicity, gender, age and SES. The research process of the SOM-institute is in agreement with the guidelines of the National Committee of Ethics in Sweden. The data for the present study was collected in 2014. The response rate was 53% [46]. Given attrition and missing values, the actual sample size of the present study was 1,694.

4.2 Variables and Instruments

Personality factors

The FFM of personality was measured by the BFI-10. The measure consists of ten self-reported items, two items per personality factor. The items are worded such as: "has an active imagination" (Openness), "does a thorough job" (Conscientiousness), "is outgoing, sociable" (Extraversion), "is generally trusting" (Agreeableness) and "gets nervous easily" (Neuroticism). The BFI-10 has adequate levels of convergent and discriminant validity and test-retest reliability [38]. Variables for the personality factors consisted of the mean scores based on the two items set on a five-point Likert scale ranging from 1 ("strongly disagree") to 5 ("strongly agree"). Only respondents who had answered both items received a mean score for that specific factor and were thereby included in the analyses.

Demographic factors

A question about age was included as an open-ended question. Gender was dummy-coded; male was coded as 1 and female as 0. Socioeconomic status was measured by income and education. Income was measured by "What is your current income per month?" ranging from 1 ("less than US$1,000") to 10 ("more than US$6,000"). Educational attainment was measured by "What is your highest attained level of education?" ranging from 1 ("not completed primary school") to 8 ("postgraduate education").

Global Internet usage

We designed two instruments to measure global Internet usage. The first instrument was a single-item: "How often have you used the Internet during the past twelve months?" The respondents were asked to indicate their frequency of Internet usage on a seven-point Likert scale ranging from 1 ("never") to 7 ("every day"). The second instrument was an index of ten activities: (1) Search for information/facts, (2) Sending/Receiving email, (3) Doing bank transactions, (4) Using news services, (5) Buying/Ordering goods or services, (6) Contacting the authorities, (7) Using social media, (8) Watching movies/TV series, (9) Reading blogs (10) Playing online games. Respondents were asked to indicate their frequency of usage of each activity during the past twelve months on a seven-point Likert scale, ranging from 1("never") to 7 ("every day"). The index is the mean of the ten activities. Only respondents who had indicated all ten activities received a mean score for global Internet usage and were thereby included in the analyses.

4.3   Statistical Analyses

First, we described our sample regarding gender, age and Internet usage and presented a correlation matrix between the study variables (i.e. global Internet usage, personality factors and demographic factors). Secondly, we conducted a hierarchical regression analyses for each dependent variable. Each analysis consisted of three steps. In the first step we entered the FFM, reasoning that personality was the main focus of the present study. In the second step we entered age and gender and in the third step we entered SES (i.e. income and educational attainment).

5   Empirical Analyses

5.1   Descriptive Statistics of the Study Variables

1,694 respondents (762 men, 932 women) answered the questionnaire. Ages ranged from 16 to 85 with a mean of 52.3 years. Among the respondents, 9.9% did not use the Internet anytime during the past 12 months, 2.5% used the Internet less than once a week, 5% used the Internet on a weekly basis, 11.7% used the Internet several times every week, 71% used the Internet every day. Table 2 contains a descriptive summary of means and standard deviations as well as the Pearson correlations between study variables.

5.2   Statistical Analyses

The FFM of personality has a significant ($p < 0.1$) effect on global Internet usage. In the two models, personality accounted for 1.7% and 2.7% respectively, of the total variance in global Internet usage (Table 3).

The two personality factors that explained most of the global Internet usage were a high degree of Openness and Extraversion. In step 2, when age and gender were entered, the models accounted for 22.0% and 30.4% respectively of the total variance in global Internet usage (Table 3). The effect of Extraversion and Openness was weaker, especially the effect of Openness. In step 3, when SES (i.e. education and income) was entered, the models accounted for 31.8% and 39.0% respectively of the total variance in global Internet usage. The effect of Extraversion was somewhat weaker while Openness was unrelated to global Internet usage (Table 3).

Table 2. Descriptive statistics of study variables.

	M	SD	α	1	2	3	4	5	6	7	8	9
1. Openness	3.06	0.95	.28									
2. Conscientiousness	4.00	0.75	.35	.02								
3. Extraversion	3.59	0.91	.62	.07**	.25**							
4. Agreeableness	3.70	0.68	.08	.51*	.26**	.17**						
5. Neuroticism	2.58	0.87	.56	.12**	-.24**	-.25**	-.26***					
6. Age	52.5	17.98		-.13**	.08**	-.03	.07**	-.08**				
7. Education	4.75	1.92		.19**	.00	.09**	.00	-.01	-.21**			
8. Income	4.40	2.64		-.04	.10**	.08**	.04	-.20	-.08**	.37**		
9. GIU (single-item)	6.10	1.84		.10**	-.02	.08**	-.03	-.04	-.47**	.37**	.32**	
10. GIU (index)	3.94	1.15	.82	.13**	-.06*	.10**	-.07*	.00	-.55**	.31**	.23**	.62**

Note. α = Internal consistencies for variables with more than one item. GIU = Global Internet usage (scale 1–7), education (scale 1–8), income (scale 1–10), personality factors (scale 1–5). * $p < .01$; ** $p < .05$.

Table 3. Hierarchical regression models exploring how personality and demography predict global Internet usage.

Independent variables	Step 1: Personality β	Step 2: Age, Gender, Personality β	Step 3: Age, Gender, Education, Income, Personality, β
MODEL 1 (Single-item)			
Demography			
Age		−.454***	−.396***
Gender		.072**	.053**
SES			
Education			.205***
Individual income			.190***
Personality			
Openness	.094***	.053*	.026
Conscientiousness	−.030	.004	−.022
Extraversion	.087***	.075**	.066**
Agreeableness	−.032	−.013	−.006
Neuroticism	−.047	−.061*	−.014
R^2	.017***	.220***	.318***
N	1519	1519	1451

(continued)

Table 3. (continued)

Independent variables	Step 1: Personality β	Step 2: Age, Gender, Personality β	Step 3: Age, Gender, Education, Income, Personality, β
MODEL 2 (Index)			
Demography			
Age		-.533***	-.522***
Gender		.028	.012
SES			
Education			.171***
Individual income			-.176***
Personality			
Openness	.113***	.068**	.040
Conscientiousness	-.063*	.029	-.051*
Extraversion	.115***	.097***	.087***
Agreeableness	-.065*	-.042	-.038
Neuroticism	-.012	-.052*	-.009
R^2	.029***	.304***	.390***
N	1386	1386	1332

Note. β = Standardized beta (the standardized regression coefficient). * $p < .05$; ** $p < .01$; *** $p < .001$. Global Internet usage in Model 1 is a single item. The question was "How often have you used the Internet the past twelve months?" and the response scale was a seven-point Likert scale ranging from 1 ("never") to 7 ("every day"). Global Internet usage in Model 2 is an index of ten Internet activities (see Sect. 4.2). The scale ranging from 1 (lowest level of global Internet usage) to 7 (highest level of global Internet usage).

6   Discussion and Conclusion

Regarding the first research question, we have found that personality factors are related to global Internet usage. People with a high degree of Openness and Extraversion use the Internet more frequently. It is worth noticing that the effect of Openness on global Internet usage disappears when demographic factors are taken into account, especially educational attainment. This indicates that higher education includes Openness. The Internet seems to enable Extraverts to experience their social and talkative nature and thereby enhance their already rich offline social networks rather than support Introverts to create social networks. In other words, the Internet does not seem to compensate for a weak social network in the offline world among people with a reserved and withdrawn nature. The positive relation between Extraversion and global Internet usage is consistent with most previous findings [21, 26, 31] and supports the "rich get richer model" rather than the "social compensation model" [26]. According to Kraut et al. [26], the Internet is better designed for the extraverts' need to maintain and organize weekly contacts than for the introverts' need to build new and deep social relationships.

Regarding the second research question, we have found that demographic factors are related to global Internet usage. Not very surprisingly, age is the factor that explains most of the variance in global Internet usage between different individuals (Table 3). We found a negative relation between age and global Internet usage, which corresponds to most previous findings [5, 7, 33, 36]. Regarding gender and global Internet usage, model 1 shows that men use the Internet more frequently than women, while model 2 shows that gender is unrelated to global Internet usage. The findings on gender and global Internet usage reported in the present study are consistent with the previous findings with one half reporting that men are overrepresented and the other half that there is no difference between men and women. Regarding SES, we found that both income and education are positively related to global Internet usage. Our results are similar to most previous studies [7, 23, 33, 36, 47]. However, our findings contradict what Donat et al. [13] have found, namely that the relation between income and Internet usage has disappeared due to cheaper computers, Internet cafés and free Wi-Fi. There are at least three explanations to the positive relation between SES and global Internet usage. First, higher SES is related to having more money to buy Internet services, equipment and products [1, 14, 16, 29, 33]. Secondly, higher SES is related to a life style (both work and leisure) where the Internet is used more frequently [16]. Thirdly, higher SES is related to better ability to read comprehensive texts and to navigate online [7]. According to Chaudhuri et al. [7], the Internet design (including all activities) favor people with high education.

6.1   Limitations

The results of this study must be viewed in light of its limitations. First, the participants' behaviors were self-reported. To estimate actual behavior from introspective self-reports and questionnaire ratings creates common problems in the field of social and personality psychology [4]. We recommend future researchers to pay more attention to measuring global Internet usage behaviorally, for instance through direct observations and automatic behavioral registrations [e.g. 21, 37].

Secondly, although the ambition was to provide a representative sample of the Swedish population (with the age span 16–85), the sample is slightly underrepresented regarding young adults (16–29) and people with non-Swedish citizenships [46].

Thirdly, there are some weaknesses in the operationalization of study variables. Personality factors were measured by a short scale with substantial losses and clear psychometric disadvantages in comparison to a full-length scale [38]. The internal consistency within factors were low (Table 2). Also, the two measurements of global Internet usage might be critically viewed. The first measure consists of a single-item question, while the second measure (i.e. the index) consists of ten Internet activities that have been subjectively selected by the SOM institute, rather than solidly grounded in previous theories and research. The scale might therefore suffer from low content validity. However, the internal consistency of the scale is good, $\alpha = 0.82$ (Table 2). We have noticed that different measures and statistical analyses are general problems in the field of personality and global Internet usage which prevent researchers from comparing studies across cultures and time periods (Table 1). The use of different measures for global Internet usage (including response alternatives) also hinders a shared and common understanding of demographic factors. In sum, it is difficult to know if the reported differences between the present study and some previous studies are due to actual individual differences across time periods and cultures or simply caused by methodological differences due to different study designs (e.g. samples, measures and analyses). We suggest creating a more proper platform for comprehensive studies in the field of global Internet usage and individual differences.

Fourthly, focusing solely on global Internet usage might prevent us from capturing important individual differences regarding Internet usage. We suggest more research on individual differences related to specific Internet activities (e.g. social media, e-shopping or online gaming), especially when the participants are young adults in developed countries since global Internet usage is embedded in their everyday lives, however in different ways.

Fifthly, we should note that the effect of the FFM on global Internet usage is not very large. However, the correlation and regression coefficients are comparable with the sizes reported in previous studies (see for example Landers and Lounsbury [27] p. 288; Mark and Ganzach [31] p. 279). Perhaps personality factors only have a limited predictive power of global Internet usage and therefore are not essential to include in analyses of digital divides? The present study demonstrates that demographic factors have a remarkably better predictive power (Table 3). Including both personality factors and demographic factors, the two models accounted for 31.8% and 39% respectively of the total variance in global Internet usage (Table 3). According to guidelines from two meta-analyses [17, 20], R^2 above 0.09 (i.e. 9.0%) is considered relatively large. Therefore, we suggest that the goodness of fit of both models are fairly adequate. In order to improve the R^2 in future models, we recommend researchers to include more SES variables, such as occupation and location of residence, as well as longer personality inventories.

Sixthly, according to Ybarra and Suman [48], there is an especially noticeable decline in Internet usage above the age of 60. In the present study, we have not distinguished between different age groups.

Finally, caution should be taken in generalizing the present results to other cultures. The previous studies on FFM and global Internet usage have been conducted in the U. K. and the USA, which both are similar to Sweden as developed Western and web-based societies. However, even though the FFM mostly has shown universal characteristics, and SES is one of the most established measurements in social sciences, more than half of the variance in these instruments depends on unknown factors, such as cultural effects [41]. Further, we know that IT develops differently in different cultures. Mobile Internet and Wi-Fi is nowadays very cheap in several Arabic countries, Russia and Brazil. We need to further explore if a high degree of Extraversion, high SES and young age also are adequate factors to explain global Internet usage in these countries.

6.2   Implications

The present study shows that it is primarily extraverted people who use the Internet in contrast to introverted people. Continuous technological improvements have enabled constant Internet access for most people in web-based societies like Sweden, leading to major changes in our Internet usage. It might be argued that Extraverts, compared to their more introvert counterparts, have lifestyles that involve them in Internet activities to a larger extent. With such a perspective, the digital divides can be explained by differences in internal dispositions between individuals. We would like to argue for an alternative explanation, that the Internet has been designed for certain individuals. Previous studies have suggested that the Internet primarily was designed for well-educated users and users with a low degree of Neuroticism [7, 43]. From the literature review and from the empirical analyses, we would argue that the Internet during the past decades primarily was designed for people with a high degree of Extraversion, probably because the Internet has become home to more and more social activities enabling extroverts to experience and express their social and outgoing nature. Perhaps the Internet could have been designed in another way, benefitting the lifestyle and nature of introverts to the same extent as extraverts? Early research disagreed on the relation between Extraversion and global Internet usage; some claimed that introverts preferred the Internet as a compensation model for weak social contacts in the offline world [27], while others claimed that extraverts preferred the Internet to enlarge their already rich social networks [21, 26]. If the Internet in the beginning mainly benefitted introverts, consistent with the "social compensation model" [27], the present study demonstrates that the pendulum nowadays has swung to the other side supporting the "rich get richer model." This Swedish finding is consistent with a large-scale representative study of young adults in the USA [31]. Therefore, researchers nowadays seem to agree on the relation between Extraversion and global Internet usage. This is true at least in Sweden and the USA, the third and fifth most digitalized countries in the world [3].

Perhaps Internet designers need to take one step back and think about why certain groups use the Internet less, so that we do not miss any potential users in our technological advances? Given that the Internet today is used largely by young people with a high degree of Openness (i.e. well-educated) and Extraversion who quickly perceive, understand and take advantage of new usage areas [10], a challenge in the Internet design must be to increase the global Internet usage among people who are more reserved (i.e. close-minded and introvert). Therefore, as proposed by Amichai-Hamburger [2], we suggest more co-elaborative and generative design sessions between Internet designers and psychologists which also should include all type of users. We believe that research on Internet usage and individual differences can help us create digital services which to a larger extent are inclusive to all types of needs in the future.

6.3   Conclusion

Global Internet usage in Sweden is notably related to individual differences regarding age, income, educational attainment and degree of Extraversion. Therefore, we can conclude that we have several digital divides in Sweden. Moreover, we would like to argue that the Internet will benefit the already strong groups in the society, in terms of SES, inclusiveness, health, and well-being. In general, elderly people have worse health conditions and are less included in the society than younger people. Further, people with a higher degree of Extraversion experience a higher level of well-being in their lives, compared with their more introverted counterparts [9, 12]. Women are also using the Internet less than men, which partly can be explained by their weaker positions regarding income and education (Table 3, Model 1. Men in Sweden have a better position in society than women, for instance, when it comes to salary and managerial work positions [42]. If the evolvement of the Internet continuously develops to favor these already strong groups in the society, *it* does not only contribute to digital divides. In fact, the digital divides contribute to the "rich get richer model" in a number of life outcomes far beyond the Internet, such as health, well-being, social inclusiveness, monetary assets and intellectual capability. In the end, "The Winner Takes it All" [44].

References

1. Akman, I., Mishra, A.: Gender, age and income differences in internet usage among employees in organizations. Comput. Hum. Behav. **26**(3), 482–490 (2010)
2. Amichai-Hamburger, Y.: Internet and personality. Comput. Hum. Behav. **18**(1), 1–10 (2002)
3. Baller, S., Dutta, S., Lanvin, B.: The Global Information Technology Report 2016: Innovating in the Digital Economy. World Economic Forum (2016). http://www3.weforum.org/docs/GITR2016/WEF_GITR_Full_Report.pdf
4. Baumeister, R.F., Vohs, K.D., Funder, D.C.: Psychology as the science of self-reports and finger movements. Whatever happen to actual behavior? Perspect. Psychol. Sci. **2**(4), 396–403 (2007)
5. Buselle, R., Reagan, J., Pinkleton, B., Jackson, K.: Factors affecting internet use in a saturatedacess population. Telematics Inform. **16**, 45–58 (1999)
6. Chamorro-Premuzic, T.: Personality and Individual Differences. BPS Blackwell, Glasgow (2012)
7. Chaudhuri, A.S., Flamm, K., Horrigan, J.: An analysis of the determinants of internet access. Telecommun. Policy **29**(9–10), 731–755 (2005)
8. Chayko, M.: Superconnected: the internet, digital media, & Techno-social life. Sage, Los Angeles (2017)

9. Costa, P.T., McCrae, R.R.: Revised NEO Personality Inventory (NEO-PI-R) and NEO Five-Factor Inventory (NEO-FFI): Professional Manual. Psychological Assessment Resources, Odessa (1992)
10. Costa, P.T., McCrae, R.R.: Domains and facets: hierarchical personality assessment using the revised NEO personality inventory. J. Pers. Assess. **64**(1), 21–50 (1995). https://doi.org/10.1207/s15327752jpa6401_2
11. Davidsson, P., Thoresson, A.: The Swedes and the Internet 2017. The Internet Foundation In Sweden (2017). http://www.iis.se/docs/Svenskarna_och_internet_2017.pdf
12. Diener, E., Lucas, R.: Personality and subjective well-being. In: Kahneman, D., Diener, E., Schwarz, N. (eds.) Well-Being: The Foundation of Hedonic Psychology, pp. 213–229. Russell Sage Foundation, New York (1999)
13. Donat, E., Brandweiner, R., Kerschbaum, J.: Attitudes and the digital divide: attitude measurement as instrument to predict internet usage. Informing Sci. Int. J. Emerg. Transdiscipl. **12**, 38–56 (2009)
14. Farag, S., Krizek, K.J., Dijst, M.: E-shopping and its relationship with in-store shopping: empirical evidence from the Netherlands and the USA. Transp. Rev. **26**(1), 43–61 (2006)
15. Feist, J., Feist, G.J.: Theories of Personality. Mc Graw Hill, Boston (2009)
16. Gardner, J., Oswald, A.: Internet use: the digital divide. In: Park, A., Curtice, J., Thomson, K., Jarvis, L., Bromley, C. (eds.) British Social Attitudes: The 18th Report: Public Policy, Social Ties, pp. 1–15. SAGE Books, New York (2001). http://dx.doi.org/10.4135/9781849208642.n7
17. Gignac, G.E., Szodorai, E.T.: Effect size guidelines for individual differences researchers. Personality Individ. Differ. **102**, 74–78 (2016)
18. Goldberg, L.R.: An alternative "description of personality": the Big Five factor structure. J. Pers. Soc. Psychol. **59**, 1216–1229 (1990)
19. Gosling, S.D., Rentfrow, P.J., Swann Jr., W.B.: A very brief measure of the Big-Five personality domains. J. Res. Pers. **37**(6), 504–528 (2003)
20. Hemphill, J.F.: Interpreting the magnitudes of correlation coefficients. Am. Psychol. **58**, 78–80 (2003)
21. Jackson, L.A., et al.: Personality, cognitive style, demographic characteristics and Internet use – findings from the HomeNetToo project. Swiss J. Psychol. **62**(2), 79–90 (2003)
22. John, O.P.: The "Big Five" factor taxonomy: dimensions of personality in the natural language and in questionnaires. In: Pervin, L. (ed.) Handbook of Personality Theory and Research, pp. 261–275. Guildford, New York (1990)
23. Jones, S., Fox, S.: Generations online in 2009. Pew Internet and American life project (2009). http://www.pewInternet.org/Reports/2009/Generations-Online-in-2009.aspx
24. Jonier, R., et al.: Gender, internet experience, internet identification, and internet anxiety: a ten-years follow-up. Cyberpsychology Behav. Soc. Netw. **15**(7), 370–372 (2012)
25. Kotler, P., Keller, K.L.: Marketing Management. Pearson Education Ltd., London (2009)
26. Kraut, R., Kiesler, S., Boneva, B., Cummings, J., Helegson, V., Crawford, A.: Internet paradox revisited. J. Soc. Issues **58**(1), 49–74 (2002)
27. Landers, R.N., Lounsbury, J.W.: An investigation of Big Five and narrow personality traits in relation to Internet usage. Comput. Hum. Behav. **22**, 283–293 (2006)
28. Larsen, R.J., Buss, D.M.: Personality Psychology. Domains of Knowledge About the Human Nature. McGraw-Hill, Boston (2005)
29. Lin, C.A.: Exploring personal computer adoption dynamics. J. Broadcast. Electron. Media **42**(1), 95–112 (1998)
30. Lounsbury, J.W., Loveland, J.M., Sundstrom, E.D., Gibson, L.W., Drost, A.W., Hamrick, F. L.: An investigation of personality traits in relation to career satisfaction. J. Career Assess. **11**(3), 287–307 (2003)

31. Mark, G., Ganzach, Y.: Personality and Internet usage: a large-scale representative study of young adults. Comput. Hum. Behav. **36**, 274–281 (2014)
32. McElroy, J.C., Hendrickson, A.R., Townsend, A.M., DeMarie, S.M.: Dispositional factors in Internet use: personality versus cognitive style. MIS Q. **31**(4), 809–820 (2007)
33. Mocnik, D., Sirec, K.: The determinants of Internet use controlling for income level: Cross-country empirical evidence. Inf. Econ. Policy **22**(3), 243–256 (2010)
34. Nachmias, R., Mioduser, D., Shemla, A.: Internet usage by students in an Israeli high school. J. Educ. Comput. Res. **22**(1), 55–73 (2000)
35. Norman, D.A.: Emotional Design. Why We Love (or Hate) Everyday Things. Basic Books, New York (2004)
36. NTIA. Falling through the net: Toward digital inclusion. A report on Americans' access to technology tools. National Telecommunication and Information Administration (NTIA), Economic and Statistics Administration, U.S. Department of Commerce. The Secretary of Commerce, Washington, DC (2000). http://www.ntia.doc.gov/files/ntia/publications/fttn00.pdf
37. Park, G., et al.: Automatic personality assessment through social media language. J. Pers. Soc. Psychol. **108**(6), 934–952 (2015). http://dx.doi.org/10.1037/pspp0000020
38. Rammstedt, B., John, O.P.: Measuring personality in one minute or less: a 10-items short version of the Big Five Inventory in English and German. J. Res. Pers. **41**(1), 203–212 (2007). http://dx.doi.org.ezproxy.ub.gu.se/10.1016/j.jrp.2006.02.001
39. Schumacher, P., Morahan-Martin, J.: Gender, internet, and computer attitudes and experiences. Comput. Hum. Behav. **17**, 95–110 (2001)
40. Smith, P., Smith, N., Sherman, K., Kripalani, K., Goodwin, I., Bell, A.: The internet: social and demographic impacts in Aotearoa New Zealand. Observatorio (OBS) J. **6**, 307–330 (2008)
41. Soto, C.J., John, O.P., Gosling, S.D., Potter, J.: Age differences in personality traits from 10 to 65: Big Five domains and facets in a large cross-sectional sample. J. Pers. Soc. Psychol. **100**(2), 330–348 (2011). https://doi.org/10.1037/a0021717
42. Statistics Sweden: Women and men in Sweden 2016. SCB-Tryck, Örebro (2016)
43. Tuten, T., Bosnjak, M.: Understanding differences in web usage: the role of need for cognition and the five factor model of personality. Soc. Behav. Pers. **29**(4), 391–398 (2001)
44. Ulvaeus, B., Andersson, B.: The Winner Takes It All. Abba. Polar Music International AB, Stockholm (1980)
45. Valkenburg, P.M., Schouten, A.P., Peter, J.: Adolescents' identity experiments on the Internet. New Media Soc. **7**(3), 383–402 (2005)
46. Vernersdotter, F.: Den nationella SOM-undersökningen. In: Bergström, A., Johansson, B., Oscarsson, H., Oskarson, M. (eds.) Fragment, pp. 563–588. The SOM-institute, University of Gothenburg, Gothenburg (2015)
47. Witt, E., Massman, A., Jackson, I.: Trends in youth's videogame playing, overall computer use, and communication technology use: The impact on self-esteem and the Big Five personality factors. Comput. Hum. Behav. **27**, 763–769 (2011)
48. Ybarra, M., Suman, M.: Reasons, assessments and action taken: Sex and age differences in uses of Internet health information. Health Educ. Res. **23**(3), 512–521 (2008)
49. Zhang, Y.: Age, gender, and internet attitudes among employees in the business world. Comput. Hum. Behav. **21**(1), 1–10 (2005)
50. Zywica, J., Danowski, J.: The faces of facebookers: investigating social enhancement and social compensation hypotheses; Predicting Facebook™ and offline popularity from sociability and self-esteem, and mapping the meanings of popularity with semantic networks. J. Comput. Mediat. Commun. **14**(1), 1–34 (2008)

The Relationship of ICT with Human Capital Formation in Rural and Urban Areas of Russia

Anna Aletdinova[(✉)] ⓘ and Alexey Koritsky ⓘ

Novosibirsk State Technical University, K. Marksa 20, 630073 Novosibirsk, Russia
aletdinova@corp.nstu.ru

Abstract. In the article the authors made an attempt to empirically substantiate the link between information and communication technologies and the accumulation of human capital in cities and rural areas of Russia. For that reason the Cobb Douglas model was applied. As a result, four statistically significant models were obtained, where the following two indicators served as the resultant variables: the number of personal computers in organizations per 100 workers and the number of personal computers in organizations with an Internet connection per 100 workers. The explanatory variables were human capital, measured as the average number of years of training per one employed in the region, the average monthly wage in the regions and the share of urban population in the regions. A positive effect of the average level of education on the number of personal computers used in organizations per 100 employees and the number of personal computers in organizations with Internet connection per 100 workers has been proved, and this effect decreases over time. The influence of the average monthly wage is also positive. The assumption has been confirmed that in cities where there is a higher concentration of human capital, higher population density and higher wages, the introduction of information and communication technologies into a production processes in organizations is more intensive than in rural areas. A higher level of wages of the population employed in the region's economy also acts as an incentive for organizations to use ICT more actively.

Keywords: Information communication technologies · Cobb Douglas model Human capital

1 Introduction

In the modern world there is a growing importance of Information Communication Technologies (ICT), their implementation ensures global inclusive economic growth, freedom of speech, the diffusion of knowledge, ideas and technologies; discovers and expands new markets. Digitalization of economy in different regions varies by its pace and level. According to the Internet Governance Forum (IGF) there is a significant difference in the positions of the countries in the ICT Development Index, e-Government Development Index and e-Participation Index Rankings [1]. Figure 1 shows these differences (white color indicates countries not participating in the ratings; the darker the colour is, the higher the country is ranked).

© Springer Nature Switzerland AG 2018
D. A. Alexandrov et al. (Eds.): DTGS 2018, CCIS 859, pp. 19–27, 2018.
https://doi.org/10.1007/978-3-030-02846-6_2

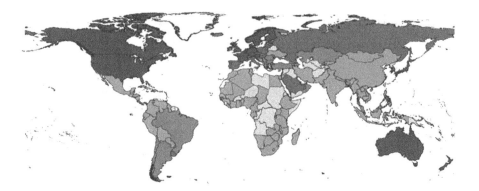

Fig. 1. Global e-Participation Index, 2016 [1].

In Russia the share of digital economy in Gross Domestic Product accounts for 2%. It is still insignificant, but the government recognizes the strategic importance of Digital Innovations to ensure prosperity, inclusive economic growth, social and cultural progress and development throughout the world. Information Communication Technologies do not only transform the economy and social sphere, but they also influence the formation of human capital. In our opinion, it is important to confirm this influence and analyze if it is the same in Russia for urban and rural population.

2   Methods

2.1   Related Work

According to research scientists, culture and technological advances are correlated with ICT, with the information society as a result of this correlation [2–5]. They are focused on technologies being the factor that eliminates differences in cultural spaces [4, 6]. The place of human capital and the role of modern employee in an organization are changing. Research scientists note that the success of an organization depends on intelligent employees, who are in command of ICT and can create innovations [7, 8].

As shown by I. Eaton and S. Kortum, technological externalities are connected with the location of economic and scientific research activity [9]. According to F. Sbergami, the diffusion of knowledge and technological externalities play an important role in long-term economic growth. This is reflected in all major models of endogenous growth [10].

Scientists, assessing the diffusion of knowledge and human capital, use the duration of fundamental education as the main criterion for its formation. For example, P. Klenow and A. Rodriguez-Clare estimated human capital taking into account the average number of years of study. The analysis of the impact of education on earnings showed differences in income per worker in terms of physical capital, human capital and total factor productivity [11]. P. Romer proved that human capital is a central prerequisite of innovation activity, it influences the capacity to adapt technological advances from other countries (insures technological adoption) [12].

The level of adoption of ICT in different countries diverges significantly, regardless of the income levels. In the paper «Cross-country differences in ICT adoption: A consequence of Culture» A. A. Erumban, S. B. De Jong tried to explain the differences in ICT adoption rates across countries by using Hofstede's cultural framework. The effect of culture on ICT adoption is explored by applying two different measures of ICT adoption, namely the average share of ICT spending in GDP across 42 countries, and per capita computer across 49 countries. The results showed that the national culture and the ICT adoption rate of a country are closely related. It appears that most of the Hofstede dimensions are important in influencing ICT adoption, in particular, the power distance and the uncertainty avoidance dimensions seem to be the most important ones. These results were robust in both datasets, even after controlling for levels of education and income [13].

Hempell examined the panel data of German service companies for the time period 1994–1998, the results indicate that ICT raise the profitability of high-skilled employees training [14]. Roztocki and Weistroffer showed that ICT impacts the individual, organizational, and country-level socioeconomic achievements, which determines the development of ICT [15]. As noted by Doring, Schnellenbach, a lot of empirical research was conducted, which was aimed at studying geographically limited knowledge spillovers, as well as the impact of knowledge transfer on the growth of innovation activity [16]. For example, Jaffe suggests that the growth of population income arises from the growth of the stock of knowledge, which promotes the use of more efficient technologies [17]. Anselin, Varga and Acs note that a more active "knowledge transfer" between universities and business firms occurs on a regional level [18].

A comparative assessment of cultures according to H. Hofstede's model shows that Russian workers are focused on avoiding uncertainty, focusing on the struggle for survival at the expense of self-expression [19]. So they might be under the influence of information and communication technologies. That's what we are going to prove in this paper.

2.2 Description of the Used Model

In our opinion, in the more densely populated regions there is a more active diffusion of knowledge and technology, as the productive and social contacts of people are not limited by far distances. It can be assumed that in cities a new technology is adopted faster and its level is higher. Consequently, there must be higher labor productivity, wages and level of the population incomes.

We have chosen two indicators characterizing the extent of information and communication technologies in Russian regions. These are the number of personal computers in organizations per 100 employees and the number of personal computers with access to the Internet per 100 employees. To assess the impact of physical and human capital as well as employment density on the income of the population of the regions, the authors use the Cobb Douglas function. The following models are proposed:

$$y_i = Ah_i^\alpha w_i^\beta, \tag{1}$$

where: y_i – the number of personal computers in organizations per 100 employees; h_i – the human capital (the average number of training years per an employee in regions of Russia);
w_i – the nominal average monthly wage in regions of Russia;

$$y_i^* = A h_i^\alpha w_i^\beta, \tag{2}$$

where: y_i^* – the number of personal computers with access to the Internet per 100 employees;

$$y_i = A h_i^\alpha n_i^\beta, \tag{3}$$

where: n_i – the share of urban population; and

$$y_i^* = A h_i^\alpha n_i^\beta \tag{4}$$

This form of production function is used, in the first place, because of the simplicity of the interpretation of the coefficients where the degrees of the corresponding variables are the coefficients of elasticity of the dependent factor with respect to the independent variable. In order to get the unknown parameters, which represent the influence of independent factors, using the known least squares formulas; the Eqs. (1)–(4) are solved in the logarithmic form. The corresponding regression equations are as follows:

$$ln y_i = ln A + \alpha ln h_i + \beta ln w_i, \tag{5}$$

$$ln y_i^* = ln A + \alpha ln h_i + \beta ln w_i, \tag{6}$$

$$ln y_i = ln A + \alpha ln h_i + \beta ln n_i, \tag{7}$$

$$ln y_i^* = ln A + \alpha ln h_i + \beta ln n_i. \tag{8}$$

Let's check the assumption that the introduction of new technologies in cities is more intensive than in rural areas.

3   Results

Table 1 shows the results of calculations of regression Eqs. (1) for the time period 2005-2016. The regression is weighted. As a weight variable "the total number of people employed in the economy of the regions" was used. All the coefficients for independent variables are statistically significant at 1% level. The coefficients of determination are quite high: at the level of 0.6–0.75, which means that the selected independent variables determine the variation of the resultant variable by 60–75%.

Table 1. Correlation between the number of personal computers in organizations per 100 employees and average monthly wage and average level of education of those employed in the economy of the regions of the Russian Federation in 2010–2016

Regression evaluation (5)	2005	2010	2014	2015	2016
LnA	−2.709***	−0.624	0.217	1.028	1.158*
	(0.681)	(0.632)	(0.755)	(0.717)	(0.663)
α	0.428***	0.300***	0.257***	0.194***	0.185***
	(0.056)	(0.050)	(0.061)	(0.057)	(0.052)
β	0.038***	0.009***	0.007***	0.008***	0.006***
	(0.007)	(0.003)	(0.002)	(0.002)	(0.001)
The coefficient of determination R^2	0.740	0.718	0.677	0.641	0.661
Fisher criterion	108.27	97.88	80.63	70.639	77.126
P- value of the criterion	0.000	0.000	0.000	0.000	0.000
Number of regions	79	80	80	82	82

Notes. *** – the significance at 1% level; (standard evaluation errors are given in brackets)

The regression analysis of statistical data in regions of Russia revealed that the level of physical capital (the volume of capital equipment per unit of labour in Russian regions) does not have any statistically significant influence on the number of personal computers in organizations in Russian regions. However, the influence of such indicators as the share of urban population and the level of average monthly wage of people employed in the regions economy was confirmed by regression analysis.

The number of regions is determined by the number of constituent entities of the Russian Federation with the exception of autonomous regions, whose statistical data are included in the relevant regions, but including data on Chukotsky Autonomous region. The data for the Chechen Republic for 2005 are not available; therefore the calculation for the year 2005 includes only 79 regions. From 2015 the data on the Republic of Crimea and the city of Sevastipol have been added.

The influence of average level of education of the population employed in regions economy on the number of personal computers used in organizations is positive. This phenomenon can be explained as follows: the higher the education level is, the easier it is to find qualified personnel for a particular organization. Consequently, it is easier to introduce and successfully implement computer technologies in the production process.

The data presented in the table show that in the course of time the influence of education level on the number of personal computers in organization per 100 employees was continually decreasing – by 2016 the coefficient for this variable decreased by more than twice.

If in 2005 the increase in average level of education of population employed in the economy of a region increased by a year, it resulted in an increase in the number of personal computers in organizations per 100 employees by almost 43%, whereas in 2016 it was only by 18% (see Table 1). Even more significantly did the average level of education influence the number of personal computers with the access to the Internet –

with a one-year increase in the average education level the number of personal computers connected to the Internet increased by 78% in 2015, and only by 20% in 2016 (see Table 2).

Table 2. Correlation between the number of personal computers connected to the internet in organizations per 100 employees and average monthly wage and average level of education of people employed in the economy of the regions of the Russian Federation in 2005–2016

Regression evaluation (6)	2005	2010	2014	2015	2016
LnA	−8.719***	−4.495***	−1.712*	−0.106	0.319
	(1.202)	(0.938)	(0.872)	(0.865)	(0.757)
α	0.783***	0.522***	0.361***	0.236***	0.206***
	(0.098)	(0.075)	(0.070)	(0.069)	(0.059)
β	0.055***	0.016***	0.009***	0.011***	0.009***
	(0.012)	(0.004)	(0.002)	(0.002)	(0.002)
The coefficient of determination R^2	0.720	0.773	0.737	0.691	0.731
Fisher criterion	97.64	131.08	107.96	88.156	107.6
P- value of the criterion	0.000	0.000	0.000	0.000	0.000
Number of regions	79	80	80	82	82

Notes. *** – the significance at 1% level; (standard evaluation errors are given in brackets)

The influence of average monthly wage is also positive. It can be assumed that high level of monthly wage makes it more profitable to use computer technologies in the production, because it helps to increase productivity and reduce staff time per unit of products (services), which stimulates the introduction of computer technology.

Besides, the assumption that in cities new technologies are implemented more intensively than in the rural areas have been checked.

Indeed, in cities the concentration of human capital, population density and wage levels are higher, consequently, it is natural to assume that there are more favourable conditions for the diffusion of advanced technologies, including computer ones. The results of the calculation of the regression Eqs. (7)–(8) with the variable "the share of urban population" are presented in the Tables 3 and 4.

The regressions are weighted. As a weight the variable "the amount of people employed in the economy of the regions of the Russian Federation" was used. All the coefficients for independent variables are statistically significant at 1% level. The coefficients of determination are quite high: at the level of 0.59–0.75, which means that the selected independent variables determine the variation of the resultant variable by 59–75%.

Table 3. Correlation between the number of personal computers in organizations per 100 employees and average monthly wage and the share of urban population employed in the economy of the regions of the Russian Federation in 2005–2016

Regression evaluation (7)	2005	2010	2014	2015	2016
LnA	−2.546***	−2.024	−0.698	−0.200	−0.128
	(0.757)	(0.610)	(0.698)	(0.679)	(0.477)
α	0.383***	0.413***	0.316***	0.288***	0.271***
	(0.067)	(0.052)	(0.060)	(0.057)	(0.038)
β	0.010***	0.001	0.005***	0.005**	0.005***
	(0.002)	(0.002)	(0.002)	(0.002)	(0.001)
The coefficient of determination R^2	0.716	0.673	0.649	0.585	0.670
Fisher criterion	95.10	79.26	71.11	55.27	80.312
P- value of the criterion	0.000	0.000	0.000	0.000	0.000
Number of regions	79	80	80	82	82

Notes. *** – the significance at 1% level; (standard evaluation errors are given in brackets)

Table 4. Correlation between the number of personal computers in organizations with connection to the Internet per 100 employees and average level of education of people employed in the economy and the share of urban population in Russian regions in 2005–2016

Regression evaluation (8)	2005	2010	2014	2015	2016
LnA	8.315****	−5.637***	−3.331***	−2.267	−1.883***
	(1.303)	(0.887)	(0.834)	(0.866)	(0.580)
α	0.701***	0.595***	0.481***	0.397***	0.361***
	(0.115)	(0.002)	(0.071)	(0.073)	(0.046)
β	0.015***	0.006***	0.004**	0.005**	0.006***
	(0.003)	(0.002)	(0.002)	(0.002)	(0.001)
The coefficient of determination R^2	0.709	0.751	0.692	0.680	0.703
Fisher criterion	92.77	116.13	86.46	59.31	93.63
P- value of the criterion	0.000	0.000	0.000	0.000	0.000
Number of regions	79	80	80	82	82

Notes. *** – the significance at 1% level; (standard evaluation errors are given in brackets)

The correlation between the human capital and the number of personal computers in organizations is positive, statistically significant at 1% level and quite high: with a one-year increase in the average education level the number of personal computers in organizations per 100 employees increased in 2005 by 38.3%, in 2010 - by 43.3%, in 2016 - by 27.1% in 2016 (see Table 3). Even more significant is this correlation between human capital and the variable "the number of personal computers connected to the Internet. In 2005 a one-year increase in the average education level of population employed in the economy of the regions was accompanied by an increase of the number of personal computers with the Internet connection in organizations by 70.1%, in 2010 the percentage fell to 59.9%, in 2014 - to 48.1% and in 2016 the percentage decreased even more to 36.1% (see Table 4). The correlation between the number of personal computers in organizations with connection to the Internet per 100 employees and the share of urban population is also positive and statistically significant at 1% level (see Table 4). Obviously, information and communication technologies are developing more successfully in cities than in rural areas.

It can be concluded that accumulation of human capital and concentration of population in cities contribute to a more effective penetration of information and communication technologies into production processes in organizations. This fact might be explained by gradual saturation with computer equipment and specialists using it in organizations in all regions of Russia.

4   Conclusions

Thus, in the article the authors show that not only information and communication technologies affect the formation of human capital, but also vice versa. High level of accumulation of human capital by the information society leads to a more intensive introduction of these technologies in organizations.

Using the example of the Russian regions the authors showed positive connections between the number of personal computers in organizations, an average number of years of training per one employed in the region, the average monthly wage in the regions and the share of urban population in the regions; as well as correlations between the number of personal computers with the Internet connection in organizations, an average number of years of training per one employed in the region, the average monthly wage in the regions and the share of urban population in the regions. Besides, information and communication technologies are developing more successfully in cities than in rural areas.

It wasn't possible to prove the influence of the level of physical capital accumulation (the volume of capital equipment per unit of labour in Russian regions) on the degree of distribution of personal computers in the organizations in regions of Russia.

It can be assumed that a higher salary or wage of employees makes the use computer technologies in production more profitable, as it helps to increase labour productivity and reduce staff time per unit of products (services), which stimulates the introduction of computer technology.

Further research is seen in modeling the relationship of other indicators of ICT use (for example, in households) with human capital.

References

1. Website of the Internet Governance Forum. http://www.intgovforum.org/multilingual/content/publications-reports. Accessed 16 Jan 2018
2. Baudrillard, J.: Simulacra and Simulation (The Body, in Theory: Histories of Cultural Materialism). University of Michigan Press, Michigan (1994). https://doi.org/10.1234/12345678

3. Morańska, D.: E-competences as a condition of the development of information society. Forum Scientiae Oeconomia **4**(2), 51–59 (2016)
4. Janssen, M., et al.: Revisiting the problem of technological and social determinism: reflections for digital government scholars. In: Electronic Government and Electronic Participation: Joint Proceedings of Ongoing Research, Workshop and Projects of IFIP EGOV, p. 254. IOS Press, no. 21 (2014)
5. Gudowsky, N., Peissl, W.: Human centred science and technology–transdisciplinary foresight and co-creation as tools for active needs-based innovation governance. Eur. J. Futur. Res. **4**(1), 8 (2016). https://doi.org/10.1007/s40309-016-0090-4
6. Obi, C.N., Leggett, C., Harris, H.: National culture, employee empowerment and advanced manufacturing technology utilisation: a study of Nigeria and New Zealand. J. Manag. Organ., 1–23 (2017). https://doi.org/10.1017/jmo.2017.70
7. Aletdinova, A.A., Bakaev, M.A.: Human capital in the information society and the wage difference factors In: Proceedings of the International Conference IMS-2017, pp. 98–101. ACM Press, New York (2017). https://doi.org/10.1145/3143699.3143744
8. Collings, D.G., Scullion, H., Vaiman, V.: Talent management: human resource. Manage. Rev. **25**(3), 233–235 (2015)
9. Eaton, I., Kortum, S.: Trade in ideas: productivity and patenting in OECD. J. Int. Econ. **40**, 251–278 (1996). https://doi.org/10.1016/0022-1996(95)01407-1
10. Sbergami, F.: Agglomeration and economic growth: same puzzles. HEI Working Paper, 02/2002, pp. 1–34 (2002)
11. Klenow, P.J., Rodriguez-Clare, A.: Economic growth: a review essay. J. Monet. Econ. **40**(3), 597–617 (1997). https://doi.org/10.1016/S0304-3932(97)00050-0
12. Romer, P.M.: The origins of endogenous growth. J. Econ. Perspect. **8**(1), 3–22 (1994). https://doi.org/10.1257/jep.8.1.3
13. Erumban, A.A., De Jong, S.B.: Cross-country differences in ICT adoption: a consequence of culture. J. World Bus. **41**(4), 302–314 (2006). https://doi.org/10.1016/j.jwb.2006.08.005
14. Hempell, T.: Do computers call for training? Firm-level evidence on complementarities between ICT and human capital investments. ZEW Discussion Paper No 03-20 (2003) https://doi.org/10.2139/ssrn.416440
15. Roztocki, N., Weistroffer, H.R.: Conceptualizing and researching the adoption of ICT and the impact on socioeconomic development. Inf. Technol. Dev. **22**(4), 541–549 (2016). https://doi.org/10.1080/02681102.2016.1196097
16. Doring, T., Schnellenbach, J.: What do we know about geographical knowledge spillovers and regional growth? A survey of the literature. Reg. Stud. **40**(03), 375–395 (2006). https://doi.org/10.1080/00343400600632739
17. Jaffe, A.B.: Patents, Patent Citations, and the Dynamics of Technological Change. NBER Reports, 8–11 (1998)
18. Anselin, L., Varga, A., Acs, Z.J.: Local geograpfic spillovers between University Research and High Technology Innovations. J. Urban Econ. **106**, 407–443 (1997). https://doi.org/10.1006/juec.1997.2032
19. Minkov, M., Hofstede, G.: A replication of Hofstede's uncertainty avoidance dimension across nationally representative samples from Europe. Int. J. Cross Cult. Manage. **14**(2), 161–171 (2014). https://doi.org/10.1177/1470595814521600

Toward an Inclusive Digital Information Access: Full Keyboard Access & Direct Navigation

Sami Rojbi[1(✉)], Anis Rojbi[2], and Mohamed Salah Gouider[1]

[1] SMART Laboratory, University of Tunis, Tunis, Tunisia
`sami.rojbi@isggb.rnu.tn, ms.gouider@yahoo.fr`
[2] CHART Laboratory, University of Paris 8, Paris, France
`anis.rojbi@univ-paris8.fr`

Abstract. The laws prohibit the discrimination of people with special needs. Accessibility has become a legal obligation for the State, which must ensure equal opportunities for access to services and knowledge. Many people have difficulty in accessing graphical interfaces or controlling the mouse. To promote a high degree of web usability, w3c guidelines emphasize the need to allow the user to interact with web pages not only through a pointing device, but through the keyboard as well. Among their appearance, access keys implementations were criticized. This article gives an overview about access keys drawbacks and presents perspectives on how to support web app interaction through a keyboard.

Keywords: Web · Accessibility · Usability · Access keys · Keyboard shortcuts
Disabilities · W3c guidelines · Human computer interaction

1 Introduction

Some countries have laws to ensure digital accessibility for people with disabilities. Web accessibility has become a legal obligation and there are acts relating to a prohibition against discrimination on the basis of disability. W3C published accessibility guidelines for web content WCAG [19] and user agent UAAG [22]. Both emphasize the need to allow the user to interact with web pages not only through a pointing device but through the keyboard as well. The specification of HTML4 [13] came with the access keys with the aim of providing a direct navigation by using a keyboard and without any need for a mouse. At the beginning, access keys were used exclusively to activate a link or give the focus to a control of a form. Later, XHTML2 and HTML5 generalized their use with any element of a web page. The desired goal is to provide users with a direct navigation and a full keyboard access similar to that offered by assistive technologies, which are based on the use of keyboard shortcuts to improve the accessibility of people with disabilities.

A keyboard shortcut is a combination of keys, usually activated simultaneously to invoke some functionality. By such simplicity of use, the keyboard shortcuts provide a very attractive mechanism for accessibility offered by almost any application.

For web apps, two conditions are necessary to achieve this accessibility goal: First, web designers must associate access keys with the elements of each page. Second, the

© Springer Nature Switzerland AG 2018
D. A. Alexandrov et al. (Eds.): DTGS 2018, CCIS 859, pp. 28–39, 2018.
https://doi.org/10.1007/978-3-030-02846-6_3

user must choose a user agent that recognizes access keys. W3C specifications are intended to be implemented by user agents. Some have not implemented the specification of access keys and therefore they do not handle them. The others have implemented access keys in the same way as their own keyboard shortcuts. The diversity of implementations has led to conflict and lack of usability.

Despite consensus on the utility of access keys, accessibility experts criticized the existent implementations. We attempt to analyze access keys drawbacks and give perspectives to approve web app interaction through a keyboard.

2 Statement of the Problem

In the beginning, access keys were introduced in 1997 by the HTML 4.0 specification [13]. Then, they were quickly implemented by user agents and deployed in Web pages. In 1999, the guidelines for web content accessibility WCAG 1.0 [20] gave them the level AAA of technical accessibility. In 2000, w3c published HTML Techniques for Web Content Accessibility [16] including samples on how to use the link element and the *accesskey* attribute as well as to skip navigation. The Section 508 standards [12] said: *"When software is designed to run on a system that has a keyboard, product functions shall be executable from a keyboard..."* and *"A method shall be provided that permits users to skip repetitive navigation links"*. Similarly to desktop applications that have procured their users the performance of keyboard shortcuts, web apps are supposed to offer them ease when using access keys.

Access keys aimed to facilitate navigation for motor-disabled, blind people and even sighted Keyboard fetishists. The web designers who got tired to approve the accessibility of their websites through html gymnastics using *"table hack"* and *"skip extensive navigation"* techniques [6] have expected this solution to offer them an easy web development.

Contrary to what was expected, access keys have fallen in conflicts. Clark [6] describes the access key as *"the delinquent teenager of accessible HTML"* and says that it is difficult to make access keys work. In 2002, a study [23] conducted to research the available access keys which had not already been reserved by various other assistive technologies demonstrated that *"the keystroke combination encoded within the web page may conflict with a reserved keystroke combination in an assistive technology or future user agent"*. This study leads to the unavailability of access keys to all users and pushed the Canadian Access Working Group, who had previously suggested the use of access keys, to recommend *"not to use access keys on Web sites of Government of Canada"*.

Likewise, accessibility experts discussed the conflicts [24] caused by access keys and suggested limiting their use to number keys 0–9. This restriction to the use of number keys is equally recommended by both the accessibility guideline of public websites of British Government and the repository of Internet services accessibility for the French administration. Of course, this recommendation was insufficient to provide the expected accessibility and did not succeed to avoid all conflicts with the most used assistive technologies which use number keys as keyboard shortcuts.

Quickly, the accessibility issue has had an international legal framework. The book of Thatcher and al. [5] discussed how to build accessible websites and evaluate their regulatory compliance. Law asked for the use of the attributes *title* and *alt* but not *accesskey* (Priority 3). To improve keyboard access, it required the use of the event manager *onkeyDawn* instead of *onDoubleclick* and it has forbidden the use of *onChange* with a select menu unless if *size* is not greater than one.

Later, XHTML2 [1] comes to deprecate the *accesskey* attribute in favor of the XHTML Access Module [21]. XHTML modules describe the elements and attributes of the language as well as their content model. The XHTML Access Module [21] gave the element *access* with its attributes *common, activate, key, targetid* and *targetrole*. Although the difference between the specifications of HTML4 and XHTML2, they had a common point. Both used a single character from the document character set as an access shortcut. But, it is assigned to the *accesskey* attribute in the case of HTML4 or to the *key* attribute in the case of XHTML2. Although the XHTML Access Module has the advantage of generalizing the use of access keys after being limited to links or controls of a form, it failed to bring the expected accessibility.

Finally, HTML5 [15] has preferred to allow the use of the *accesskey* attribute with any element but without providing a new solution for access keys drawbacks.

3   Discussion About Access Keys Conflicts

Access keys specification provides no solution to avoid conflict. Indeed, each user agent implementation has tried to develop its own solution. This approach has pushed Joe Clark to declare that "*it is difficult to operate access keys*" [6].

3.1   Access Keys Relay on Software Keyboard Shortcuts

At the beginning there were only desktop applications along static websites. Then, web applications came with a performance and functionalities comparable to desktop applications and sought to offer their users the ease of use provided by keyboard shortcuts. Access keys imitated keyboard shortcuts, on the point of view of their target (allow to activate a functionality without using a pointing device) as well as on the point of view of their operating mechanism (type simultaneously a combination of keys on the keyboard) [6]. This complete imitation [25] caused serious conflicts [26].

For a disabled person, web browsing requires a harmony between three applications: web application, user agent and assistive technology. At the appearance of access keys, most of the keys were already reserved for operating systems, user agents or assistive technologies [27]. User agents have been limited to implement access keys, each one in its particular way, and without worrying about their safety of use. Web designers, therefore, should choose the good access keys for their web applications taking into account the diversity of user agents and their different mechanism to implement access keys. At the HTML level, defining an access key consists in assigning a single letter as the value of the *accesskey* attribute of the concerned element. In case of conflict, activating an access key can disable the equivalent menu of a user agent/screen reader. In other cases, the application keyboard shortcut can run instead of the web page access key.

3.2   Access Keys Conflict with User Agents and Assistive Technologies

There is a difference in the context of use between access keys and keyboard shortcuts of applications. Access keys do not have their own keyboard event listener, but they depend on the user agent keyboard shortcuts listener (Fig. 1). For an application, it only needs to avoid operating system keyboard shortcuts. But, an access key has to be distinguished from all the operating system, the user agent and the assistive technology keyboard shortcuts. By reusing the same principle of applications' keyboard shortcuts, access keys easily fall in conflict.

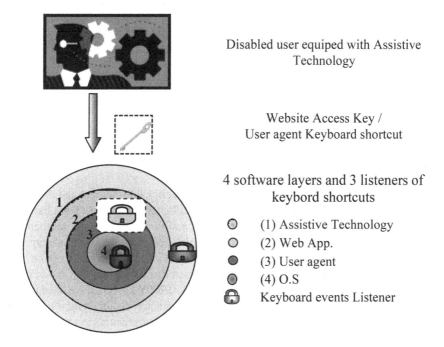

Disabled user equiped with Assistive Technology

Website Access Key / User agent Keyboard shortcut

4 software layers and 3 listeners of keybord shortcuts

○ (1) Assistive Technology
○ (2) Web App.
● (3) User agent
◐ (4) O.S
🔒 Keyboard events Listener

Fig. 1. Both access keys and keyboard shortcuts of the user agent are received by the same keyboard events listener.

In most cases of conflict with a graphical user agent, the web page access key wins. Thus, the user will lack his/her user agent keyboard shortcut. To illustrate this negative result, we take the example of an Internet Explorer browser displaying a web page which contains an access key d. The execution of this access key will lock the address bar making keyboard navigation impossible [24]. This is annoying for a motor-disabled person. What is worse, blind people who use a screen reader like JAWS are concerned by this conflict.

Screen readers have built-in keystroke shortcuts for going into different modes. IBM Home Page Reader does not allow using access keys. It cannot distinguish between its own keyboard shortcuts and access keys of web pages which are invoked in the same manner [23]. It has a "Links reading mode" that can be started by the access keystrokes Alt + L and a "Controls reading mode" that starts by the keystrokes Alt + O. With these two modes along others which we have not mentioned, IBM HPR seems not to need access keys which came to facilitate access to links and form controls.

To use access keys on IBM HPR, a visually impaired user must first click with the mouse on the graphical interface of the application. Similarly, the graphical user agent iCab which runs only under Mac OS, needs sometimes to click on page before using access keys. We recall here that the aim of access keys is to provide an alternative of navigation that does not use the mouse.

Advanced computer users are usually the most expected to use keyboard shortcuts. By contrast, surveys [28] showed a decrease of use of access keys, especially among advanced assistive technology users. This is explained by the fact that these users are aware of potential conflicts and thus they prefer to use commands offered by their familiar assistive technology.

3.3   There is no Safe Key

By looking for access keys that do not conflict with the various assistive technologies, a study [23] has led to a disappointing result. Although theoretically any alphanumeric character can be an access key, only three characters are still free. There are slash /, backslash \ and right square bracket].

Actually, Jaws [29] uses Slash key to move to next clickable element and it uses the keystroke SHIFT + Slash to move to the previous clickable element. The characters backslash and right square bracket would be inaccessible on a keyboard another than the North American Standard (QWERTY) keyboard. For example, if we use Opera 7 for Windows as a user agent and we would like to activate a web page access key, we have to type simultaneously *Esc* + *Shift* + the character of the *access key*. Now, if our access key is a right square bracket and we have an AZERTY keyboard we have to type simultaneously the two keys *alt gr* + *]* to obtain our character. Asking a motor disabled user to type simultaneously four keys at once to facilitate his/her access to a web resource is not an accessibility solution. We are stating here that it is no need to enhance the accessibility relying on inaccessible keys.

Several recommendations [24] suggested to restrict the use of access keys to numeric keys with a view to avoid conflicts with applications that have frequently used alphabetic characters as keyboard shortcuts. This orientation to the use of numeric keys has been exploited under the current access keys specification that allows only the use of a single character and has resulted in some recommendations that could not unify [25]. Among these recommendations, we have chosen to present the one adopted by the accessibility guide for public sites of the British government and by the repository Accessibility of Internet Services of the French administration (Table 1).

Table 1. Restriction of access keys use to numeric keys

0	List of used keyboard shortcuts
1	Home page
2	News page
3	Site map
4	Search form
5	FAQ, Glossary, thematic index
6	Navigation Help
7	Mail contact
8	Copyright, terms of use, license…
9	Guestbook, Feedback…

It can be stated that this proposition is limited since it does not preserve the goal of access keys to allow keyboard navigation without the need for a pointing device. Access keys appeared to enhance the accessibility of links and form controls. In general, ten access keys are not sufficient either to allow a keyboard activation of all links on a web interface or to give focus to all its form controls.

If the aim of this proposal is to avoid unsafe alphabetic characters allocated as keyboard shortcuts by various applications, it becomes also beneficial to avoid numeric access keys which are used by assistive technologies. Windows Eyes uses the numbers 0 to 9 for reading by area. Jaws [29] uses numbers 1 through 6 for Heading at level and numbers 7 through 0 for Place Marker 1 through 4 (Table 2). Thus, there are no safe keys. From our point of view, the solution is not to look for free keys that will avoid conflicts, but it is to rethink the implementation of access keys.

Table 2. Jaws keystrokes using numeric keys

Description	Jaws command
Next heading at level	1 through 6
Prior heading at level	SHIFT+ 1 through 6
First heading at level	INSERT+ALT+CTRL+ 1 through 6
Last heading at level	INSERT+ALT+CTRL+SHIFT+1 through 6
Read place marker 1 through 4	Numbers row 7 through 0
Move place marker 1 through 4	SHIFT+ Numbers row 7 through 0
Switch to a specific tab number	CTRL+ n (where n is a digit from 1 to 8)
Switch to last tab	CTRL+ 9

4   Discussion About Access Keys Usability

The usability of access keys concerns their effectiveness, efficiency and the satisfaction of their users about their ease of use. First, effectiveness means that users have done exactly what they wanted to do. Due to conflicts and lack of unification of the activation methods of access keys, users may not achieve their goals. Second, efficiency indicates that the effort required to complete the task is adequate which is not always guaranteed especially for disabled people. The decrease of use of access keys, especially among advanced assistive technology users shows a lack of satisfaction.

To activate an access key, there is no unified or universal method [24]. Convenient combination of keys depends on the used machine (Table 3). Therefore, at work, at home or in a cyber space, a user may need different keystrokes of keys to activate the same link (or to give focus to the same form control) on the same web page. Worse, he/she has to guess the needed access key which is defined on the source code level but not rendered on the user interface. Disabled users hence find additional difficulties.

Table 3. Various keystrokes to activate an access key

Access key keystrokes	Example of user agents
Alt + [key]	Chrome, K-Meleon, Amaya, Safari 4+ for Windows Firefox 1&1.5 for Windows and Linux Opera15+, IE4 and Netscape7 for PC Galeon and Epiphany for Linux
Alt + [key], then Enter	MS Internet Explorer 5+ for Windows
Alt + Shift + [key]	Early versions of Firefox2 and Opera 12 for windows Firefox 2 for Linux
Alt + Shift + [key], then Enter	Firefox 3 for Windows and Linux
Alt + Ctrl + [key]	Firefox 14+, Chrome, Opera 15+, Safari for Mac
Ctrl + [key]	Mozilla 13 and earlier for Mac IE 5+, Netscape, Safari 1.2 and 4+ for Mac Netscape 6+ for windows
Ctrl + [key], then Enter Ctrl + Cmd + [key]	MS Internet Explorer for Mac (depending on the version)
Ctrl, then [key]	Konqueror 3.3+ for Mac
Ctrl + OPT + [key]	Safari 3 and Chrome 3+ for Mac
Esc + Shift, then [key]	Opera 7+ for Windows, Mac and Linux
Don't handle access keys	Netscape 6 and earlier for PC and Mac, Camino, Galeon, Konqueror before version 3.3.0, Omniweb, Safari before version 1.2, Opera before version 7 for Windows and Linux, Internet Explorer 4.5 for Mac, Lunascape, Lynx, Mosaic, NetSurf, etc.

4.1 Lack of Access Keys Rendering Standard

Applications indicate the character of keyboard shortcut, in most cases by underlining it and recently by displaying it in an info-bubble after pressing a specific key as Alt. By contrast, most graphical user agents do not provide any indication about the character assigned to the *accesskey* attribute. Users have to guess the character to use as an access key.

Until yet, the *accesskey* attribute is not mapped to WAI-ARIA. It is not exposed via an accessibility API and it is not displayed as part of the user interface [17]. In addition to screen readers, only iCab and Konqueror report access keys. The user agent iCab runs until yet, only on Mac OS. For the Konqueror user agent, there is recently a KDE on windows in experimental state (windows.kde.org). The number of users having these two user agents which defer access keys is still very limited around the world and particularly in centers of disabled education and monitoring in poor countries which look for the least expensive technology. A recent survey [30] shows that the most used user agent are Firefox, Internet Explorer, Chrome and Safari and only 1.3% of assistive technology users use other browsers.

Up to the present time, there has been no standard [24] in the area of access keys rendering although Web Content Accessibility Guidelines [16] said *"User agents may include features that allow users to bind keyboard strokes to certain actions"*.

Graphical user agents neither render the value of the access key, nor the modifier key to be accompanied with. This resulted in a lack of usability. In fact, first-time learnability is a determinant quality component of usability. In our case, it measures how easy it is for users to activate access keys the first time they encounter and without referring to a documentation/help page. The matter is essentially with discoverability which concerns the degree of ease with which the user can find all new access keys.

4.2   Lack of Unification of Access Keys

If we want to use access keys, we have to choose a convenient user agent because many of them don't handle access keys [23]. Table 3 indicates some of these user agents and shows various implementations of access keys.

First, access keys depend on used operating system. For using MS Internet Explorer navigator to activate an access key, we have to press the Alt key if our operating system is Windows but we need the Ctrl key if we have a Macintosh operating system. By contrast, Apple systems generally ask for using the "cmd" key in addition to the access key. Second, access keys depend also on the user agent. Considering for example the Windows operating system, some user agents as Mozilla, Netscape, K-Meleon use the Alt key, but Opera makes an exception by using the combination Shift + Esc to activate an access key. Third, we usually have to tape the access key character simultaneously with a modifier key as Alt, Ctrl or Esc but sometimes we have to tape them successively like in the case of the konqueror user agent.

Finally, after taping the suitable combination of keys, we often must press successively the enter key to invoke the access key functionality. In most cases, Internet explorer needs to use the enter key but on Mac OS, it doesn't need.

The lack of unification of access keys has led to a lack of usability. The matter goes well beyond the issue of learnability to the access keys memorability. In the context of user interface/experience, memorability measures how easily users can reestablish proficiency when they return to the design after a period of not using it. In general, it is hard to remember how a visited link can be activated through a keyboard. First, the access key value is not rendered explicitly in the user interface. Second, the needed modifier keys depend on the used OS as well as on the user agent and sometimes on its version. Thus, only users having good memory skills and a visual learning style can remember both the entered access key keystrokes and the software configuration details of the platform they have last used. Particularly, the elderly or cognitively impaired person will find an additional difficulty.

4.3   Hard Activation of Access Keys for Disabled People

Usually, a user can directly see the main characteristics of his/her work environment. His visual experience is enough for him to know easily how equipped the PC is on the level of operating system and user agent. In contrast, a visually impaired user depends on the assistive technology which he uses to browse the web. He/she has to guess the convenient keystrokes to his/her environment. These keystrokes depend on the user agent and on the operating system too. It is difficult to a visually impaired to predict the combination of keys that must be used. In fact, screen readers render the character of the access key without specifying the modifier key that must be accompanied with.

Similarly, an old person, a person with Alzheimer's or with cognitive disability usually needs an assistance offering him a cognitive reward. We recall that there are many different keystrokes to activate an access key (Table 3). For only Internet Explorer, we find three different ways depending on the user agent version and the used operating system. So we do not expect from such a disabled person either to remember all the different ways of activating access keys, or to determine the one that suits his environment. This shows a deep lack of access keys learnability and discoverability.

If a disabled person succeeds to guess the combination of keys to activate an access key on a web page, a second challenge lies ahead. Depending on the used user agent, we have to type simultaneously at least two or three keys. This is a difficult task especially for a motor disabled user for whose access keys came to facilitate his/her access to web pages. In addition, to reduce conflicts with applications, several recommendations [24] encourage the use of numeric access keys [25]. Given that the keypad is ignored in actual implementations by almost all desktop user agents, it becomes necessary to simultaneously use the Shift key. Then, we will have to use at least three keys at once. This leads to the inefficiency of access keys.

5   Toward Adaptive Access Keys

Web browsing platforms are remarkably diverse but not standardized. They are varied from the implementation point of view as well as capabilities. It is in this context of platform heterogeneity that access keys' conflicts have been studied. By developing for a specific audience, it becomes possible to avoid conflict. In fact, sites intended for intranet use, can operate safe access keys. We encourage such practices whose effectiveness has been proven empirically for the used equipment. In particular, pedagogic sites accessed only in training centers for disabled people can provide this solution while they explicitly render the convenient keystrokes to activate access keys. Apart from this particular situation, user modeling and adaptation techniques can provide new perspectives.

The user modeling provides an interdisciplinary analysis of how humans act in specific computing environments. It makes it possible to address several sources of heterogeneity: quite different end users, highly diversified and un-standardized platforms, distinct interaction modalities and a very varied working environment. Thus, it allows the design of a single implementation of software (for millions of users) which works as if it was designed for every user only known in the time of use. The purposes are: personalization, increased usability and improved accessibility.

We need to generate a user model (profile) that includes all relevant information associated with a specific user. The profile data strongly depend on the context of the application that will use them [10]. Since access keys depend on user agents and operating systems, it becomes interesting to detect the characteristics of the user's platform. This allows knowing if the used user agent can handle access keys and what combination of keys should be used to activate them.

Technically, the detection can be done on the client side or server side. Javascript provides the *window.navigator.userAgent* property which returns the user agent string. Likewise, PHP offers the environment variable *$_SERVER['HTTP_USER_AGENT']* and the function *get_browser()* which attempts to determine the capabilities of the user's browser. There are also free packages for user agent detection, which are different in terms of speed and coverage. This solution remains unreliable since some browsers send non standard HTTP headers.

The best alternative is to use a Device Description Repository (such as WURFL, DetectRight, DeviceAtlas and UAProf) which offers a detection service based on an updated database of mobile devices coupled with an API to query the database with different programming languages. Detected device type includes: Device (mobile), PDA, PMP, Tablet, STB and Game/Handheld Console, but not assistive technology. The w3c IndieUI user context [18] provides relevant information about user's settings and preferences (color, type/font, media and screen reader) which can be accessed help to a user settings listener.

Once we have the user profile, we can use it to generate adaptive access keys whose values and rendering are defined at runtime. Access keys rendering can be provided through an adaptive annotation [2]. Thanks to adaptation techniques, we can avoid the access keys values that may cause a conflict with the used browser and select those that are most accessible on the user keyboard.

In fact, keyboards can be AZERTY/QUERTY, with/without a NUMPAD and can vary in number of keys. Mobile tiny keyboards also vary depending on their OS (iOS, Android, Windows phone). A usability study [11] shows that links containing 9 characters can be easier to tape than others with only 7 characters in terms of number of taps and required time.

Recently, a w3c HTML 5.1 Working Draft [14] discussed the use of two possible access keys and proposed an element's assigned access key processing model. Full keyboard devices can pick the first value as a shortcut key, while devices with small numeric keypad might pick the second. Instead of adapting access keys values, another approach seeks to rethink their implementation [9] in accordance with UAAG 2.0 [22] which requested ensuring full keyboard access (level A) as well as providing direct navigation and activation (level AA).

6 Conclusion and Perspectives

More than a billion people live with a disability and the number is increasing. At least 110 million people experience very severe functional difficulties. About 314 million persons are visually impaired among whom 45 million are blind and the number of elderly people is estimated to reach 2 billion worldwide in 2020 [7]. As a result, accessibility needs are growing on the level of law as well as technology. The version 3 2016 of the French government's General Accessibility Reference for Administrations (RGAA) has been revised as version 3 2017 which added checkpoints to improve the contrast of a page and the relevance of off-form buttons [3]. Similarly, the Section 508 standards, has been recently refreshed by a final corrected rule to consider TTY devices access [12].

The aim of this study is not to discourage the use of access keys but rather it seeks to show that web applications can offer all their users this accessibility process. Accessibility rules are not restricted to the case of handicapped persons [4]. Access keys are not intended only for users equipped with assistive technologies. Even with well-designed interfaces, menus and toolbars fail to provide to an expert user the performance and productivity obtained through the use of keyboard shortcuts. The values of access keys must be deferred in an explicit manner to all, without exception.

It has become possible to define access keys for any element of a web page, but their actual use is still restricted to links and form controls. Until now, Web designers haven't seen any significance/benefit of associating an access key with an IMG element despite being very aware of the difficulties faced by the blind/visually impaired people to access images: (1) Existing assistive technologies are limited to replacing the image with an alternative text. (2) The content of the images could become accessible only after a manual transcription which is provided by Tactile Graphics Specialists.

It is only in recent years that research [8] has shown that it is possible to automate the transformation of iconographies (images, illustrations) into touch-haptic images that can be explored by touch. The automatic transcription of an image contained in a web page through the activation of its access key will soon be possible and certainly, it will be one of the big accessibility challenges of Web 4.0.

References

1. W3C: XHTML 2.0. W3C Working Draft, 26 July 2006
2. Raufi, B., Ferati, M., Zenuni, X., Ajdari, J., Ismaili, F.: Methods and techniques of adaptive web accessibility for the blind and visually impaired. Procedia Soc. Behav. Sci. **195**, 1999–2007 (2015). https://doi.org/10.1016/j.sbspro.2015.06.214
3. SGMAP: RGAA guidelines version 3 2017, 28 July 2017. https://goo.gl/MDoHbw
4. Sloïm, E.: Improve the quality and accessibility of your websites. Améliorez la qualité et l'accessibilité de vos sites web. J. Arch. **239**, November 2010
5. Thatcher, J., et al.: Web Accessibility Web Standards and Regulatory Compliance. Springer-Verlag, New York (2006). ISBN 1590596382
6. Clark, J.: Building Accessible Websites. New Riders (2002). ISBN 0-7357-1150-X

7. Okoye, K., Jahankhani, H., Tawil, A.H.: Accessibility of dynamic web applications with emphasis on visually impaired users. J. Eng. **2014**(9), 531–537 (2014). https://doi.org/10.1049/joe.2014.0136

8. Bouhlel, N., Rojbi, A.: New tools for automating tactile geographic map translation. In: 16th international ACM SIGACCESS Conference on Computers & Accessibility (ASSETS 2014), pp. 313–314, Rochester, New York, USA (2014). https://doi.org/10.1145/2661334.2661335

9. Rojbi, S., Rojbi, A.: KeybNav: a new system for web navigation through a keyboard. In: 6th International Conference on Information and Communication Technology and Accessibility (ICTA 2017), Muscat, Oman. IEEE (2017). https://doi.org/10.1109/ICTA.2017.8336057

10. Rojbi, S., Soui, M.: User modeling and web-based customization techniques: an examination of the published literature. In: 4th International Conference on Logistics (LOGISTIQUA), pp. 83–90, Hammamet, Tunisia. IEEE (2011). https://doi.org/10.1109/LOGISTIQUA.2011.5939407

11. Gould, S.J.J., Cox, A.L., Brumby, D.P., Wiseman, S.: Short links and tiny keyboards: a systematic exploration of design trade-offs in link shortening services. Int. J. Hum Comput Stud. **96**, 38–53 (2016). https://doi.org/10.1016/j.ijhcs.2016.07.009

12. US Access Board: Guide to the Section 508 Standards: Web-based Intranet and Internet Information and Applications (1194.22) (2001). https://goo.gl/QnPEBe. Accessed 22 Jan 2018

13. W3C: HTML 4.0 Specification. W3C Recommendation, revised on 24 April 1998. https://www.w3.org/TR/1998/REC-html40-19980424/

14. W3C: HTML 5.1. W3C Working Draft 07 October 2015

15. W3C: HTML 5: A vocabulary and associated APIs for HTML and XHTML. W3C Candidate Recommendation, 17 December 2012

16. W3C: HTML Techniques for Web Content Accessibility Guidelines 1.0. W3C Note 15 September 2000

17. W3C: HTML Accessibility API Mappings 1.0. W3C Working Draft 31 January 2018

18. W3C: IndieUI: User Context 1.0, Contextual Information for User Interface Independence. W3C Working Draft 30 April 2015

19. W3C: Web Content Accessibility Guidelines (WCAG) 2.1. W3C Working Draft 07 December 2017

20. W3C: Web Content Accessibility Guidelines 1.0. W3C Recommendation 5 May 1999

21. W3C: XHTML Access Module. Module to enable generic document accessibility. W3C Working Draft 07 January 2008

22. W3C: User Agent Accessibility Guidelines (UAAG) 2.0. W3C Working Group Note 15 December 2015

23. WATS.ca - Web Accessibility Technical Service: Using Accesskeys - Is it worth it? 08 May 2003. http://john.foliot.ca/using-accesskeys-is-it-worth-it/

24. Denis, L.: Accesskey, the unprocessed accessibility test. "Accesskey, l'essai non transformé de l'accessibilité". Open Web group, 19 May 2008. https://openweb.eu.org/articles/accesskey_essai_non_transforme

25. Alsacréations: Accesskey the great failure of web accessibility. "Accesskey le grand échec de l'accessibilité du web", 29 October 2012. https://goo.gl/p64gnJ

26. Sax, E.: The accesskey attribute – do we still need it? (2010). https://goo.gl/ML2PF2

27. WATS.ca: Accesskeys and Reserved Keystroke Combinations, 04 December 2003. http://john.foliot.ca/accesskeys-and-reserved-keystroke-combinations/

28. WebAIM: Screen Reader User Survey #3. https://goo.gl/M7ytP1

29. Freedomscientific: JAWS Keystrokes. https://goo.gl/5SCGeB. Last revised 02 Sept 2014

30. WebAIM: Screen Reader User Survey #7. https://goo.gl/GSvmPb

E-Communication: Discussions and Perceptions on the Social Media

Social Network Sites as Digital Heterotopias: Textual Content and Speech Behavior Perception

Liliya Komalova[1,2](✉) ⓘD

[1] Institute of Scientific Information for Social Sciences
(Russian Academy of Sciences),
Nakhimovsky prospect 51/21, 117997 Moscow, Russia
[2] Moscow State Linguistic University,
Ostozhenka str., 38, 119034 Moscow, Russia
GenuinePR@yandex.ru

Abstract. The relevance of this study relies on the M. Foucault's concept of heterotopia and anthropocentric paradigm of semiosociopsychology introduced by T.M. Dridze. The goal of the pilot research described in this contribution is to examine how people not involved in social network websites' (SNS) communication are influenced by it, what kind of emotional effect SNS-discussions produce on bystanders, and what are the grounds of this effect. We analyzed subjects' (N = 7) emotional shift towards negative and positive emotional reactions in response to discussions on hot home and international politics (n = 20). Discussions in Russian language took place on "VKontakte" social network platform. The research used an experiment that utilized a three-condition (current emotional state of subjects; subjects' gender; type of the stimuli) between-subjects design. The findings suggest that negative emotional reactions of not involved participants are more like to those who take active part in SNS-communication. The findings suggest that discussions on hot home and international politics provoke a variety of emotions. Textual content of discussions was mentioned as the main ground for subjects' emotional reactions. No gender differences in perception of communicants' speech behavior and textual content of discussions were found.

Keywords: Heterotopia · Social network · Speech behavior · Emotion
Feeling · Text · Harmful behavior

1 Introduction and Literature Review

The underlying relevance of this study relies on the M. Foucault's concept of heterotopia, defined as a place of otherness combining several, sometimes not matching, spaces (as, for example, is cemetery that unites live and death), functioning in heterochrony (like libraries providing access to 9th and 20th century literature in 2018), supporting a system of openness and localization, and juxtaposing other cultural places (9). Digital heterotopia was defined as "cultural memory spaces, which juxtapose many otherwise incompatible spaces, online and offline, experts and amateurs, science and popular culture, which make endless knowledge claims, but which do so with a rational belief in the power of consensus" (Haider and Sundin 10).

Digital heterotopia is a kind of 20th century invention which transferred online (Haider and Sundin 10) and came in change of such hetero-chronic entities as museums capturing place and time, and revitalizing every event in user-friendly space and time within digital reality. Social network websites have become, in these surroundings, the most appropriate containers (archives) achieving the aim of digital heterotopias (Rymarczuk and Derksen 25). "Alternatively referred to as a virtual community or profile site, a social network is a website that brings people together to talk, share ideas and interests, or make new friends" (3).

Social network website can be considered a sovereign separated entity, "another place in relation to ordinary cultural spaces", and at the same time, "this place is connected with the set of all locations" (Foucault 8, p. 198) of the society and the wider world community, as for the events being discussed there are somehow related to every member of this community. In this sense social networks are becoming alternative media, designing "other" (changed) reality, virtual image of reality built upon sharing and commenting information on a wide variety of issues. McKenzie Wark pointed out that Foucault's description of the ship, the "heterotopia par excellence" (Foucault 9), as a "placeless place" applies to cyberspace as well, "particularly when it is a network, linking terminals in different places and times into a unified environment" (Wark 32, p. 140) .

Social network websites (SNS) have the ability to "juxtapose in a single place several spaces, several locations that are in themselves incompatible" (Foucault 8, p. 200]. SNS-users have the opportunity to simultaneously explore a multitude of diverse events, moving synchronically within virtual places in the real time (for example, reading, liking, sharing, commenting on today current events), as well as navigate retrospectively. Thus, when interacting with a social network website individuals "find themselves in a sort of absolute break with their traditional time" (Foucault 8, p. 200): there is no longer the beginning or the end, no more step-by-step motion from zero to infinity, but there is a possibility to shift from one given point of the so called "space and time continuum" to another, a possibility to get access to multiple locations at a particular moment at the same time, and the opportunity to gather this "time and space continuum" within one web resource. In this regard, along with museums and libraries social network websites can be called "heterotopias of indefinitely accumulating time" (Foucault 8, p. 201), providing a user a unique entrance to a given space.

Many articles have been devoted to findings of basic characteristics of digital heterotopias as private stance-taking (Chen and Lu 2; Johansoon 11; Latif 16), as well as digital heterotopias mobilizing political activities (7; de Sá Medeiros et al. 4; Salgueiro Marques et al. 26; de Vries et al. 5), motivating people to speak about events disputed in digital heterotopias in face-to-face settings (12), digital heterotopias as disturbing spaces (Pertierra 19; Rymarczuk and Derksen 25) lacking of privacy and fostering users' intentions to bullying and aggression (Rachoene and Oyedemi 23; Chen and Lu 2; Song and Oh 30). Seeking to contribute to this agenda, this study analyzes SNS-discussions on hot issues of home and international politics perceptions by male and female bystanders to fill the gap in the literature by addressing the following overarching questions. In what extent people not involved in SNS-driven communication are influenced by it? What kind of influence (emotional effect) SNS-discussions produce on bystanders? What are the grounds of this effect (what bystanders react on)?

2 Methodology

Besides M. Foucault's concept of heterotopia, the research is relied on anthropocentric paradigm of semiosociopsychology introduced by T.M. Dridze. The theory suggests that we live mostly upon discourse and texts based experiences; people lack "crude" reality, they are used to build their lives on the grounds of visual and speech perceptions of what is broadcasted in cultural resources by social media (6). "The modern world, where we work, sleep or take our leisure depends more on the created spaces we have manufactured (…) than on the natural or inherent characteristics of different locations" (27, p. 222).

Nowadays we are sure that "modern technologies (…) bring about changes in the inner-world of their users" (18) that have significant social and cultural consequences", easily fitting into traditional schemes (19, p. 175). As Mark Zuckerberg declared: "When we started Facebook, we built it around a few simple ideas [schemes]. People want to share and stay connected with their friends and the people around them" (34). But the problem is that nowadays social network websites have overgrown their basic principles, and now people are forced to contribute their social network "lives". One who does not do this, is suggested non-standard. "Public press would have us believe that anyone not willing to share their mundane day to-day business on a prime social media platform is a heinous monster likely to commit mass homicide" (13, p. 131). Thus, the SNS' "inhabitants" run specific risks: if before people considered virtuality as an escape from the painful reality, now they need to cope with both real and virtual negative impacts. Internet communications became "a venue for a variety of harmful behaviors (such as cyberbullying, trolling, cyberstalking, etc.)" (1, p. 151).

Emotional sphere is the first to suffer from such interventions. The aim of the present research is to analyze emotional reactions of bystanders under harmful speech behaviors and textual contents in social network website-discussions on hot home and international political issues.

The pilot research described in this contribution used an experiment that utilized a three-condition (current emotional state of subjects; subjects' gender; type of the stimuli) between-subjects design. Subjects (N = 7, four females and three males) were assigned to reflect their reactions towards social network written speech communication content (n = 20), discussions on hot home and international politics. Discussions took place on "VKontakte" social network platform (vk.com) in Russian language. Those discussions were used because pre-test respondents rated they correlated with the destructive discourse criteria (21, 2014; Komalova 14), so they consisted of harmful speech behaviors and textual contents.

2.1 Analysis Strategy

To answer research questions, four different strategies were employed. The Friedman test for repeated measures (Social Science Statistics) was used to examine whether the distribution of three types of subjects' emotional responses (negative, neutral and positive emotions) to the analyzed social-network contents was reliable. Wilcoxon signed-rank one-tailed test [Social Science Statistics] was conducted to evaluate differences between emotional reactions to textual contents and speech behaviors. Mann-Whitney U test (28) was used to assess differences between males and females responses to the analyzed discussions. And Spearman's Rho (correlation) (Social Science Statistics) was used to find associations between subjects' emotional reactions to the stimuli and three variables "current emotional state of subjects", "communicants' speech behavior", "textual content of discussions". The set of statistical instruments belongs to nonparametric methods, and can be implemented on non-classified data far from distribution requirements and more resistant to different interferences. However, they allow estimating only average trends.

2.2 Recruitment of Participants

A pre-test was used to create the stimuli for the experiment. Participants for the pre-test and experiment subjects were recruited within students of the Moscow State Linguistic University.

We worked under the notion that young people are more flexible to social transformations and are more willing to exposure their political activity through social network platforms, being a place of socialization (17; 13) and identification (auto-esteem and self-image) (7, p. 50, 70). Brady Robards assumes that "for many young people, participation [in social network sites] is now mandatory for inclusion amongst peer groups. For some of these young people, large parts of their social lives have been played out on these sites" (Robards 24).

Table 1. Features of pre-test and experiment participants.

Pre-test	Male participants	Female participants
Number of participants	9	24
Mean age (years)	20	20
Linguaculture	Russians	Russians
Education status	Third grade high school students	Third grade high school students
Experiment	Male subjects	Female subjects
Number of participants	3	4
Mean age (years)	20	20
Linguaculture	Russians	Russians
Education status	Third grade high school students	Third grade high school students

That's why 20-years old people were welcomed in experiment; all of them were native Russian speakers studying applied linguistics. Different people participated in the pre-test and experiment. Table 1 provides description of participants and subjects.

2.3   Stimuli Construction

A multi-step procedure was used to create the stimuli. 33 students collected real comments from such social network websites as vk.com, LiveJournal.com, utube.com, twitter.com, politforums.net, nastej.ru, debatepolitics.ru, politikforum.ru, bolshoyforum. com.

Respondents in the pre-test rated the full-text discussions, using special criteria from our prior research (21; 14). Respondents were to markup time period a discussion line was realized, to rate whether each discussion line concerned "home politics" or "international politics", whether those discussions were "emotionally positive", "emotionally neutral" or "emotionally negative". Respondents rated discussions, using perceptual emotional typology multiscale (Potapova and Potapov 22), providing detailed descriptions of emotions experienced by communicants. A total number of 937 discussion lines corresponding with the data selection criteria were obtained. To carry out the pilot experiment, 20 discussions were selected out of the collected data. Table 2 provides characteristics of the selected dataset.

Table 2. Stimuli dataset characteristics.

Characteristics	Discussion lines:	
	From 1 to 10	From 11 to 20
Resource	vk.com	vk.com
Main topic	International politics	Home politics
Subtopics	Relations between Russia, Europe and USA around situation in Syria and Ukraine	Activities of ruling party and opposition
Time of publication	November, 2015 – February, 2016	December, 2015 – February, 2016
Emotional coloring of content	Emotionally mixed and balanced: negative, positive, neutral	Emotionally mixed and balanced: negative, positive, neutral

2.4   Experiment Procedure

All the subjects gave written consent to participate in experiment. The stimuli were cleared out of videos and avatars. Gender and racially neutral anonymous nick-names were created for communicants, so that those factors would not confound results. Subjects read discussions and were prompted to think deeply about the content, make reflections about their emotional state before and after reading. Before reading stimuli discussions, a shortened A.E. Wessman & D.F. Ricks personal feeling scale (PFS) (33; 31) was used to measure the mood or current affective state of the subjects. The battery consists of four ten-point self-rating scales, which an individual subject could use to give reports of his/her experience on a number of important aspects of mood. They cover such affective dimensions as "tranquility – anxiety", "energy – fatigue", "elation – depression", "self-confidence – feeling of inadequacy". Subjects answered demographic and personal questions. Once they finished with reading-task, they were to write what emotional state just read discussion line provoked on him/her (name it and explain how he/she fixed this emotion). 140 experimental matrixes were obtained. As the data was heterogeneous (sometimes subjects reported mixed emotions in regard to different parts of discussion under investigation, sometimes they were confused and do not mention any emotional reaction – for raw data see Table 3), normalization was implemented. All measures are made using normalized data (see Sect. 3).

Table 3. Raw data characteristics.

Subjects' gender	PFS scores	Subjects emotional reactions to stimuli			The ground of subjects' emotional reactions	
		Negative	Neutral	Positive	Speech behavior	Textual content
A female	21	16	7	2	13	9
B female	26	11	2	6	3	14
C female	14	16	1	5	6	16
D female	25	15	3	4	2	16
E male	35	8	9	3	6	12
F male	31	6	15	0	5	19
G male	35	9	0	3	2	8
Sums:		81	37	23	37	94
	Total sums:			141		131

3   Results

As shown on Table 4, subjects tend to react emotionally to the social network sites discussions on hot home and international political issues. Most of the analyzed reported emotional reactions are negative.

Table 4. Emotional reactions of subjects.

Subjects	Subjects' emotional reaction		
	Negative	Neutral	Positive
A female	*.113*	.049	.014
B female	*.078*	.014	.042
C female	*.113*	.007	.035
D female	.016	.021	*.028*
E male	.056	*.063*	.021
F male	*.042*	.016	0
G male	*.063*	0	.021

Answering the first research question (RQ), we can say that none of the subjects left untouched, all of them demonstrated emotional involvement after reading the stimuli discussion lines. Each of 20 discussions provoked emotional shift in subjects' current mood.

For the first RQ the Friedman test for repeated measures was used. No statistically significant differences between negative/neutral/positive subjects' emotional reactions was found (the χ^2_r statistic = 3,4286 (ρ = 0,18009), the result is not statistically significant at ρ < 0,1).

The second RQ dealt with what kind of influence (emotional effect) discussions produce on passive participants of SNS. The findings suggest that discussions on hot home and international politics provoke a variety of emotions. The most frequently among them are: negative – irritation (7), mockery (7), sadness (6), fatigue (5), fear (5); positive – joy (5), interest (4). According to subjects' reports, positive emotions motivated them look for more discussions on the topic, learn special literature, and investigate the issue. Oppositely, experiencing negative emotions, subjects were to stop reading any other discussions, wanted to get read of the experimental task.

For the third RQ results showed that textual content of discussions was mentioned as the main ground for subjects' emotional reactions (Table 5). Wilcoxon signed-rank one-tailed test showed the data supported this research question (mean difference = 23.86, sum of positive ranks = 3, sum of negative ranks = 25, Z-value = –1.8593 (nb. N too small), W-value = 3, the result is statistically significant at ρ = 0,05).

Table 5. To what subjects react on in discussion lines.

Subjects	The ground of subjects' emotional reactions	
	Communicants' speech behavior	Textual content of discussion
A female	*.099*	.068
B female	*.022*	.016
C female	.045	*.122*
D female	.015	*.122*
E male	.045	*.091*
F male	.038	*.145*
G male	.015	*.061*

Mann-Whitney U test for male and female groups did not support our hypothesis towards gender differences in perception of communicants' speech behavior and textual content of discussions (Table 5). Nor the Friedman test for repeated measures gave statistically significant results on male or female subjects' emotional reactions on the stimuli (Table 4). Men and women experienced similar emotions stimulated by harmful speech behavior and textual content of the analyzed SNS-discussions.

To find statistical dependence between subjects' emotional reaction and three variables (1) Wessman & Ricks PFS scores, (2) communicants' speech behavior, and (3) textual content of discussions, Spearman's rank correlation coefficient was used (Table 6). Grey filling marks negative correlations between parameters.

Table 6. Correlation between subjects' emotional reactions and measured variables.

Subjects' emotional reaction	Wessman & Ricks PFS scores	The ground of subjects' emotional reactions	
		Communicants' speech behavior	Textual content of discussion
Negative	.509	.596	.445
Neutral	.036	.490	.198
Positive	.309	.293	.381

We supposed that lower PFS scores would correspond with high level of negative emotional reactions on the stimuli. As seen on Table 6, the increase of PFS scores correlates with the decrease of negative and positive emotional reactions that supported our hypothesis. The explanation of such effect can be that people in a bad or good mood are more resistant to negative emotions, as though they are, in case of bad mood already are experiencing negative emotions, and do not feel any emotional changes, that is why do not fix emotional shift; those in a really god mood can compensate negative emotions.

Decrease of both harmful speech behavior and harmful textual contents correspond with the growth of positive emotional reactions on discussions. More facts of communicants' harmful speech behavior and a decrease of harmful textual discussions' content correlate with the growth of negative emotions experienced by subjects. These correlations explain the more powerful effect of speech behavior on subjects' perception of the analyzed stimuli. It means speech behavior would more likely provoke emotional reactions in comparison with textual content.

However, all correlations turned to be weak and the data did not support them: by normal standards, the associations between measured variables would not be considered statistically significant.

4　Discussion

As one can see nowadays digital heterotopias win popularity in ordinary lives of many people intervening in their emotional sphere, being a symbol of modernity, demanding permanent online status. At the same time heterotopic places possess the same characteristics and realize the same social processes as real physical places do. That is why, to our opinion, "digital citizenship" requires more hard-working, so that a SNS-user needs to serf between and upon discursive spaces reconstructing his/ her own image of the reality, opposing negative effects of "hard" communications. In this framework digital heterotopia is considered a semiotic system melting the worldview of people drawn into the web, being in permanent search, constructing and deconstructing their own place in the society.

The research was based on a well-known presupposition that political issues' discussion is a potentially conflictive topic that would provoke not only destructive interactions between interlocutors, but also provide harmful psychological effects on not involved participants (bystanders). That is why we used destructive discourse criteria (21; 14) for data collection. However, for more concrete and complex representation of research questions emotionally neutral and positive discussions are to be included in experimental dataset.

Our findings argued that bystanders emotionally react more on textual content in comparison with speech behavior. This postulate is to be compared with the results (20) saying that SNS' communicants are more sensitive to speech behavior.

Results of present experiment corresponds with findings (15) supporting that in real live interaction bystanders mostly pay attention and consequently react on the form of communication (communicants' speech behavior) and content of communication. Even being not involved in communication, people do not take into account the intention with which communicants deliver their speeches (for example: to influence, to provoke, to inform etc.).

Our findings support for extending semiosociopsychological paradigm (6) to the social network websites' interaction and M. Foucault concept of heterotopia (9, 8) to online multi-user platforms. Future researches should attempt to refine upon these findings by enlarging the number of subjects and specifying the stimuli to a concrete topic.

5 Conclusion

The goal of this research was to examine how people not involved in SNS-generated communication are influenced by it, what kind of emotional effect SNS-discussions produce on bystanders, and what are the grounds of this effect. We analyzed subjects' emotional shift towards negative and positive emotional reactions in response to discussions on hot home and international politics. The findings suggest that negative emotional reactions of not involved participants are more like to those who take active part in SNS-communication. In order to prevent harmful effects on SNS-participants, the role of bystanders who witness destructive speech behaviors and see harmful textual contents is regarded as especially crucial, as for bystanders (passive participants) possess numerical superiority and have potential to restrain online perpetrators, and to return SNS as digital heterotopias the power of consensus.

Acknowledgements. The author thanks DSn, professor Rodmonga Potapova for consultations and the students of the Moscow State Linguistic University participated in the pre-test for collecting the experimental data that became the stimuli for this research project.

Funding. This research was supported by a grant from Russian Science Foundation (RSF) according to the research project № 18-18-00477.

References

Bogolyubova, O., Panicheva, P., Tikhonov, R., Ivanov, V., Ledovaya, Y.A.: Dark personality on Facebook: harmful online behaviors and language. Comput. Hum. Behav. **78**, 151–159 (2017). https://doi.org/10.1016/j.chb.2017.09.032. https://www.sciencedirect.com/science/article/pii/S0747563217305587

Chen, G.M., Lu, Sh.: Online political discourse: Exploring differences in effects of civil and uncivil disagreement in news website comments. J. Broadcast. Electron. Media **61**(1), 108–125 (2017). https://doi.org/10.1080/08838151.2016.1273922

Computer Hope (2017): https://www.computerhope.com/jargon/s/socinetw.htm

de Sá Medeiros, H., Araújo Diniz, J.M., de Oliveira Arruda, D.M.: Difusión de acciones no éticas de partidos políticos en Brasil y reacciones de los usuarios de Facebook. Intercom Rev. Bras. Ciênc. Comun. **39**(3), 79–97 (2016). http://dx.doi.org/10.1590/rbcc.v39i3.2559, http://portcom.intercom.org.br/revistas/index.php/revistaintercom/article/view/2559

de Vries, M., Kligler-Vilenchik, N., Allyan, E., Ma'oz, M., Maoz, I.: Digital contestation in protracted conflict: The online struggle over al-Aqsa Mosque. Commun. Rev. **20**(3): Power and sovereignty in hypermedia space: Middle East case studies, 189–211 (2017)

Dridze, T.M.: Text activity in the structure of social communication [Текстовая деятельность в структуре социальной коммуникации] (in Russian). Nauka, Moscow (1984). http://www.isras.ru/files/File/Publication/Dridze.pdf

Durán Sánchez, C.A.: Aspectos interventores en la participación política y electoral de jóvenes. Una reflexión sobre la información, interacción y difusión de contenidos en redes sociales para futuras investigaciones en Santander. Desafíos **27**(1), 47–81 (2015). https://doi.org/10.12804/desafios27.01.2015.02, https://revistas.urosario.edu.co/index.php/desafios/article/view/3630/2652

Foucault, M.: Intellectuals and power: Selected political articles, speeches and interviews [Интеллектуалы и власть: Избранные политические статьи, выступления и интервью] (in Russian). Part 3. Praksis, Moscow (2006)

Foucault, M.: Of other spaces. Diacritics **16**, 22–27 (1986)

Haider, J., Sundin, O.: Beyond the legacy of the enlightenment? online encyclopedias as digital heterotopias. First Monday **15**(1–4) (2010). http://dx.doi.org/10.5210/fm.v15i1.2744, http://firstmonday.org/ojs/index.php/fm/article/view/2744

Johansson, M.: Everyday opinion in news discussion forums: public vernacular discourse. Discourse Context Media **19**, 5–12 (2017)

Jöuet, J.: The Internet as a new civic form: the hybridisation of popular and civic web uses in France. Javn. Public J. Eur. Inst. Commun. Cult. **16**(1), 59–72 (2009). https://doi.org/10.1080/13183222.2009.11008998

Kennedy, J.: Rhetorics of sharing: data, imagination, and desire. In: Lovink, G., Rasch, M. (eds.) Unlike Us Reader: Social Media Monopolies and Their Alternatives, 2013, pp. 127–136. Institute of Network Cultures, Amsterdam (2013). https://monoskop.org/images/7/7b/Lovink_Geert_Rasch_Miriam_eds_Unlike_Us_Reader_Social_Media_Monopolies_and_Their_Alternatives.pdf

Komalova, L.R.: Aggressogen discourse: Multilingual typology of verbalized aggression [Агрессогенный дискурс: Мультилингвальная типология вербализации агрессии] (in Russian). Sputnik + , Moscow (2017). http://elibrary.ru/item.asp?id=28993951

Komalova, L.R.: Interpersonal communication: From conflict to consensus [Межличностная коммуникация: От конфликта к консенсусу] (in Russian). INION RAS, Moscow (2016). https://elibrary.ru/item.asp?id=26901304

Latif, E.A.: The oralization of writing argumentation, profanity and literacy in cyberspace. In: Høigilt, J., Mejdell, G. (eds.) The Politics of Written Language in the Arab World. Writing change, pp. 290–308. BRILL, Leiden (2017)

Miconi, A.: Under the skin of the networks: How concentration affects social practices in web 2.0 environments. In: Lovink, G., Rasch, M. (eds.) Unlike Us Reader: Social Media Monopolies and Their Alternatives, 2013, pp. 89–102. Institute of Network Cultures, Amsterdam (2013). https://monoskop.org/images/7/7b/Lovink_Geert_Rasch_Miriam_eds_Unlike_Us_Reader_Social_Media_Monopolies_and_Their_Alternatives.pdf

Pertierra, R., Ugarte, E., Pingol, A., Hernandez, J., Dacanay, N.: Txt-ing Selves: Cellphones and Philippine Modernity. De La Salle University Press, Manila (2002)

Pertierra, R.: The new media and heterotopic technologies in the Philippines. Hum. Technol. Interdiscip. J. Hum. ICT Environ. **4**(2), 169–185 (2008). https://jyx.jyu.fi/dspace/handle/123456789/20226

Potapova, R., Komalova, L.: Lexico-semantical indices of "deprivation – aggression" modality correlation in social network discourse. In: Karpov, A., Potapova, R., Mporas, I. (eds.) SPECOM 2017. LNCS (LNAI), vol. 10458, pp. 493–502. Springer, Cham (2017). https://doi.org/10.1007/978-3-319-66429-3_49

Potapova, R., Komalova, L.: Lingua-cognitive survey of the semantic field "aggression" in multicultural communication: typed text. In: Železný, M., Habernal, I., Ronzhin, A. (eds.) SPECOM 2013. LNCS (LNAI), vol. 8113, pp. 227–232. Springer, Cham (2013). https://doi.org/10.1007/978-3-319-01931-4_30

Potapova, R.K., Potapov, V.V.: Kommunikative Sprechtätigkeit. Russland und Deutschland im Vergleich. Böhlau Verlag Köln, Weimar (2011). https://www.degruyter.com/view/product/210018

Rachoene, M., Oyedemi, T.: From self-expression to social aggression: cyberbulling culture among South African youth on Facebook. Communicatio: South African Journal for Communication Theory and Research **41**(3): The participatory turn and self-expression, 302–319 (2015). https://doi.org/10.1080/02500167.2015.1093325

Robards, B.: Leaving MySpace, joining Facebook: 'Growing up' on social network sites. Contin. J. Media Cult. Stud. **26**(3): Mediated Youth Cultures, 385–398 (2012). https://doi.org/10.1080/10304312.2012.665836, http://www.tandfonline.com/doi/full/10.1080/10304312.2012.665836

Rymarczuk, R., Derksen, M.: Different spaces: exploring Facebook as heterotopia. First Monday **19**(6–2) (2014). http://dx.doi.org/10.5210/fm.v19i6.5006, http://firstmonday.org/ojs/index.php/fm/article/view/5006/4091

Salgueiro Marques, Â.C., Mauro Sá Martino, L., Ferreira Coêlho, T.: Alterity, social suffering and political power in Facebook accounts of oneself on the SP invisível project. Intercom Rev. Bras. Ciênc. Comun. **39**(3), 55–78 (2016). http://dx.doi.org/10.1590/rbcc.v39i3.2558, http://portcom.intercom.org.br/revistas/index.php/revistaintercom/article/view/2558

Saunders, P.: Space, urbanism and the created environment. In: Held, D., Thompson, J. (eds.) Social Theory of Modern Societies: Anthony Giddens and his Critics, pp. 68–82. Cambridge University Press, Sydney (1989)

Sidorenko, E.V.: Mathematical methods in psychology [Методы математической обработки в психологии]. Social-Psychological Center, Saint-Petersburg (1996)

Social Science Statistics. http://www.socscistatistics.com/tests/Default.aspx

Song, J., Oh, I.: Factors influencing bystanders; behavioral reactions in cyberbulling situations. Comput. Hum. Behav. **78**, 273–282 (2018). https://doi.org/10.1016/j.chb.2017.10.008, https://www.sciencedirect.com/science/article/pii/S0747563217305824

Tanski, T.S.: The personal feeling scales as related to the DRAW-A-group projective techniques: Master degree thesis. North Texas State University, Denton (1968). https://digital.library.unt.edu/ark:/67531/metadc130967/m2/1/high_res_d/n_03702.pdf

Wark, M.K.: Lost in space: into the digital image labyrinth. Contin. J. Media Cult. Stud. **7**(1), 140–160 (1993). http://dx.doi.org/10.1080/10304319309365594

Wessman, A.E., Ricks, D.F.: Mood and Personality. Holt, Rinehart & Winston, New York (1966)

Zuckerberg, M.: 'Making Control Simple'. In: The Facebook blog, 27 May 2010. http://blog.facebook.com/blog.php?post=391922327130,   https://www.facebook.com/facebook/videos/10150203264730484/

The Influence of Emoji on the Internet Text Perception

Aleksandra Vatian$^{(\boxtimes)}$, Antonina Shapovalova, Natalia Dobrenko,
Nikolay Vedernikov, Niyaz Nigmatullin, Artem Vasilev, Andrei Stankevich,
and Natalia Gusarova

ITMO University, 49 Kronverkskiy Prosp., Saint-Petersburg 197101, Russia
alexvatyan@gmail.com

Abstract. Subject of Research. The paper deals with emoji - a small
digital image or icon used to express an idea or emotion in electronic communication. The aim of the work is to find the dependencies between
the use of emojis in text messages and the extent to which the messages attract users' attention while viewing a page, especially in Russian-
speaking Internet community. **Method.** Social network "Vkontakte" was
chosen for the basis of the study, and four most extensive and popular
communities were selected within it. The structure of a typical post in
the VKontakte group was studied to identify the most obvious ways of
expressing reactions to a post. Using the linear regression algorithm,
graphs were constructed for the relationship between the frequency of
use of emojis in the post and main indicators of attitude toward the
post. **Main Results.** For all types of communities there is a clear tendency to reduce any type of reaction to a post with the increase in the
frequency of emojis in it. Most responded posts contain no emojis at all,
and such reports constitute the majority of the analyzed posts. The only
exceptions have become fan communities. They also feature this trend,
but the "attenuation" of interest is slower. Entertaining and motivational
communities also reflect the phenomenon of slow fading of interest, but
not so clearly and only in special cases.

Keywords: Emoji · Social network · Internet community
Internet text perception

1 Introduction

Emoji is a quite new phenomenon - as a way of social interaction in the Internet
community. Emoji is a small digital image or icon used to express an idea or emotion in electronic communication [1]. They originate from emoticons, that are
combinations of graphic signs, representing a pictogram with a certain emotional
coloring. The main goal of emoticons was never an expression of an emotion -
their creator, Scott Falman, reckoned on the use of pictograms in order to save
symbols in messages (the length of an SMS message should not exceed 160 characters). Considering an emoticon to be an object for research is rather difficult,

© Springer Nature Switzerland AG 2018
D. A. Alexandrov et al. (Eds.): DTGS 2018, CCIS 859, pp. 55–66, 2018.
https://doi.org/10.1007/978-3-030-02846-6_5

primarily because the variety of their forms is limited only by the imagination of authors and is absolutely not documented.

A different situation is developing with emojis. Invented in 1999 by Sigetaka Kurita for the economy of symbols in a text message, they were the "direct descendants" of emoticons. According to the creator, their primary goal was to express emotions. The question for which specific tasks emoji are applied remains open due to the rapid development of mass Internet culture. Currently, emojis are so popular that they have already received a full-fledged emoji dictionary (Emojipedia) and their own unique Unicode notations, and this makes the area even more attractive and accessible for various studies.

The relevance of research on this phenomenon in various areas of Internet culture and, in particular, in social networks, is especially clear if one assesses the overall impact of emotions on modern society. Applied creativity, pop culture, architecture, cinematography and animation, and medicine have found application to such a phenomenon as emoji.

Since communication culture in different regions can vary greatly, so it is important to investigate this question in Russian-speaking Internet community. The social network "VK" seems to be the best for these purposes. It is necessary to compare the results of similar studies conducted with another target audience participation in order to receive the complete conclusions.

The aim of the work is to find the dependencies between the use of emojis in text messages and the extent to which the messages attract users' attention while viewing a page. The study of this issue appears to help to reveal whether the use of emojis affects the Internet text perception by network users, and if it does, how the meaning of the text is distorted.

2   Related Works

The impact of emojis on network communication is under consideration from different points of view. Originally influence of emojis was being studied mainly in aspect of the sentiment-analysis. For example, the article [12] details the issue of the distribution and application of emojis in messages (tweets). It deals with such issues as determining the emotional color of a tweet, relating the emotionality of a tweet to the emojis used, searching for the most popular emojis and associated emotions. The paper [11] considers the question of determining the emotional color of tweets containing and not containing emojis. The paper [14] studies the predecessor emoji - emotikon, its connection with emotional coloring of a message, differences in application depending on the country and mentality. In [3] there is a way to analyze the emotional coloring of users' posts in web blogs by means of emotional vectors on the basis of emojis.

But in recent years the interest of researchers has moved to applied problems of use of emojis, and these problems are considered in different aspects.

The researchers [13] examine emojis not only as historical, social, and cultural objects, but as examples of technical standardization. But their research is too generalized and doesn't contain quantitative data. Researchers in other works variously structure the audience being studied. For example, [17] investigate such groups as rural, small town, and urban Chinese adults. The authors [9] compare smartphone user preferences concerning emojis across 212 countries and cultures. They showed that users from different countries present significantly different preferences on emojis. The authors [15] conducted their experiments through six groups of participants, in some Southeast Asia countries. They discuss peculiarities in using emojis caused by agriculture-oriented population and diverse minority languages.

The authors [5] consider gender aspects of application of emojis. According to authors [5], their paper is the first effort to explore the emoji usage through a gender point of view. Their results demonstrate that emoji usage significantly varies between males and females. But their research is very generalized and doesn't consider other characteristics of users. Authors [4] argue that it is necessary to emphasize gender differences in application of emojis, and they propose a number of measures for this purpose.

Application of emojis in such means of network communication as WeChat in Southern China [17], WhatsApp in Switzerland [6], Tencent Weibo in China [16] is investigated. But the most part of researches is executed on material of English-language Twitter.

Mutual influence of the text and emojis was studied in such works as [6,10,16, 17]. The authors [17] select scenarios of communication where purely text-based communication may not suffice. Dürscheid et al. [6] argue that emojis cannot be considered the basis of a new universal language and may be used only as additional graphic signs. The authors [10] discuss the interplay between emoji and textual context. Their research experimentally disproves a widespread hypothesis that using emojis in textual contexts reduces the potential for miscommunication. The authors [16] examine how emojis influence two types of interactions enabled by microblogging, namely commenting and retweeting. Their result is as follows: messages with more emojis receive more comments but less retweets.

The analysis of network content of the last years shows that emojis become part of social media marketing and business communications in general. According to [7], five leading brands have already included emojis in their advertizing campaign. For example, instead of personalising its product with consumer's names, PepsiCo included various moods and country-specific emojis to encourage people to share images on social media. The recent researches answer this challenge. Namely, the researchers [13] formulate their task as "conduits for affective labor in the social networks of informational capitalism". The authors argue that emoji are of enormous interest to businesses in the digital economy. At the same time there are works which show also the reverse tendency, i.e. a negative role of an emojis in business communications. For example, the findings of [8] show that emojis in work-related contexts do not increase perceptions of author's warmth but actually decrease perceptions of his competence.

The review allowed us to formulate the goals and features of our study.

– The aim of the work is to find the dependencies between the use of emojis in text messages and the extent to which the messages attract users' attention while viewing a page.
– Since communication culture in different regions can vary greatly, so it is important to investigate this question in Russian-speaking Internet community. The social network "VK" seems to be the best for these purposes.
– It is necessary to compare the results of similar studies conducted with another target audience participation in order to receive the complete conclusions.

The study of this issue appears to help to reveal whether the use of emojis affects the Internet text perception by network users, and if it does, how the meaning of the text is distorted.

3   Methods of Research Organisation

3.1   Selecting and Preparing a Dataset

For the basis of the study the authors assume that in modern social networks emojis primarily designed to draw a person's attention to a message have become a distracting object preventing to focus on the main sense of a message. Their frequent use dissipates attention and interferes with the perception of the information stated in the message causing a negative reaction among social network users.

Social network "Vkontakte" was chosen to test the hypothesis as it has a number of advantages for the study compared with other social networks:

– poor investigated platform;
– possibility of unhindered extraction of the information required for research;
– relative familiarity with the target audience.

In order to confirm or refute this assumption, the structure of a typical post in the VKontakte group was studied to identify the most obvious ways of expressing reactions to a post:

– likes and reposts - positive attitude expressions to the information offered being the most significant factors in the subsequent analysis;
– comments - interest expressions to the information provided without a clearly expressed emotional color.

Additionally, the groups studied were classified basing on their orientation and purpose of creation according to four most extensive and popular categories:

– educational communities (https://vk.com/ege100ballov);
– motivational communities (https://vk.com/club29970330);
– entertainment communities (https://vk.com/leprazo);
– fan communities (https://vk.com/fcbayern).

In total, 55 groups were studied, with 11 groups in each category. In each group we selected the last 100 posts posted in the group at the time of collection of information. This number of retrieved posts from each group allows to create a relatively fast fillable dataset and rapid display of dynamics without low accuracy that may occur in the case of a larger sample. The latter is due to the fact that "young" communities without a sufficient number of posts can get caught in the analysis, and then older communities become the main factor greatly influencing on the result with a larger sample. If the total number of posts in the study group was less than 100, then all the information placed on the community wall in text form would have been analyzed. Thus, the material for testing the hypothesis has been 3631 posts.

The information was collected using an algorithm written in Python 3.6, and in the final form it was an .xls file of the form (Fig. 1).

	A	B	C	D	E	F	G
1	Number	Post_len	Smiles	Frequency	Likes	Reposts	Comments
2	1	118	0	0	2190	202	135

Fig. 1. An example of information representation

Number is the serial number of the post, Post_len is the length of the post in the characters, Smiles is the number of emotions in the post, Frequency is the frequency of use of emojis in the post, expressed as a Emojis/Post_len relationship, Likes is the number of likes, Reposts is the number of reposts, Comments - number of comments.

3.2 Choosing an Evaluation Algorithm

After data collection with the use of the programming language R, a linear regression algorithm was implemented to reveal hidden and implicit regularities. This choice was due to the fact that in this case the work with this algorithm helps to visualize the data in a simple form and draw the appropriate conclusions by comparing the graphs visually. The formulas used in linear regression are simple to understand. Having one parameter (Frequency) explained and a set of some attributes of the object studied (Reposts, Comments, Likes), you can successfully plot dependency graphs.

Using the linear regression algorithm, graphs were constructed for the relationship between the Frequency under investigation and the three main indicators of attitude toward the post: Likes, Reposts, and Comments. The objects for the study were common datasets for groups of the same type and one dataset of an arbitrary community for each class of communities studied.

3.3 Results

The plots of the ratio of Frequency to the values of the indicators Likes, Reposts and Comments are given in the figures below. Common datasets are given in Figs. 2, 3, 4 and 5. An arbitrary communities are given in Figs. 6, 7 and 8. Finally, fan communities are in Fig. 9.

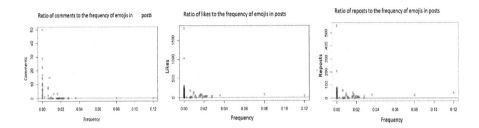

Fig. 2. Frequency to the values of the indicators Comments (a), Likes (b) and Reposts (c) in educational communities

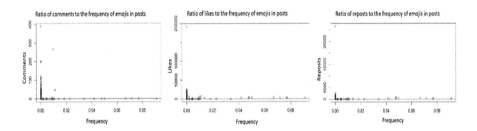

Fig. 3. Frequency to the values of the indicators Comments (a), Likes (b) and Reposts (c) in entertainment communities

4 Results and Discussion

The most complete picture has been obtained by analyzing all the assembled posts (Figs. 2, 3, 4 and 5), however, special cases (Figs. 6, 7, 8 and 9) help approach the problem more precisely by reducing data on one measured parameter from two graphs.

On the graphs in Fig. 2(a) and Fig. 6(a), one can see that in educational communities there is a clear tendency to decrease the comments to a post with the growth of emojis in it. This applies, as a general case, to all data collected in educational groups, as well as a particular case with an arbitrarily chosen group of this sample. The same dynamics can be traced in the case of likes (Figs. 2(b) and 6(b)) and reposts (Figs. 2(c) and 6(c)). These two parameters reflect a faster decline in a post interest; if we consider the total sample, in a particular case

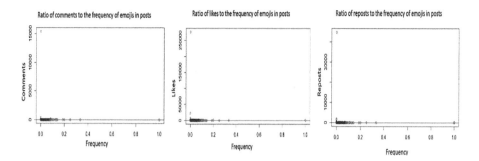

Fig. 4. Frequency to the values of the indicators Comments (a), Likes (b) and Reposts (c) in motivational communities

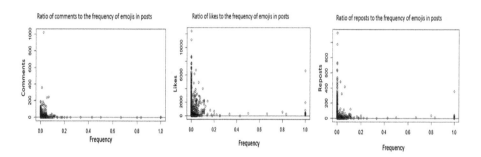

Fig. 5. Frequency to the values of the indicators Comments (a), Likes (b) and Reposts (c) in fan communities

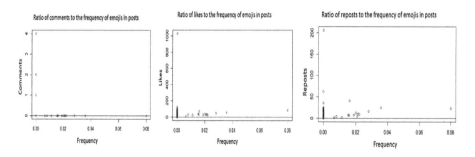

Fig. 6. Frequency to the values of the indicators Comments (a), Likes (b) and Reposts (c) in educational community

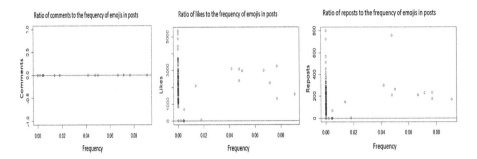

Fig. 7. Frequency to the values of the indicators Comments (a), Likes (b) and Reposts (c) in entertainment community

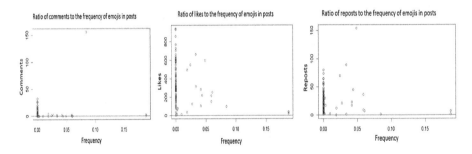

Fig. 8. Frequency to the values of the indicators Comments (a), Likes (b) and Reposts (c) in motivational community

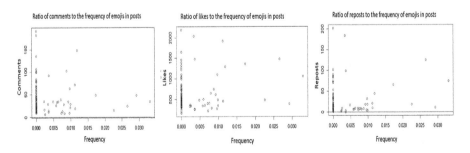

Fig. 9. Frequency to the values of the indicators Comments (a), Likes (b) and Reposts (c) in fan community

the trend will be almost the same for all three parameters. This may indicate a uniform activity of users in this public.

In entertainment communities it is not possible to trace anything from the comments, since in most communities they are closed (Fig. 7(a)). The general schedule for the comments shown in Fig. 3(a), reflects an incomplete picture, leading to low accuracy associated with close comment groups. The other two parameters operate identically in both the general sample and the control group. Moreover, the dynamics of falling interests to posts with the increase in the frequency of emojis in the text in these cases is almost the same. If we look at the generalized graph on the likes (Fig. 3(b)) and the graph on the likes of a separate group (Fig. 7(b)), they will differ little from each other as well as from the corresponding repost graphs (Fig. 3(c) – general case, Fig. 7(c) – private).

The general sample of motivational communities shows very weak response rates of participants for all three factors. This picture can be traced even in the absence of emojis in a post. If to superimpose the graphs one on the other, taking into account the differences in unit lengths along the ordinate axis, then these graphs substantially coincide (Fig. 4(a) – comments, Fig. 4(b) – likes, Fig. 4(c) – reposts). This may indicate that in this group of communities there are "super popular" motivational posts that smooth out their appearance at the bottom of the chart. Consider a special case of analysis of the motivational community. The graph in Fig. 8(a) shows that people who visit this public are not inclined to leave comments except in rare cases. In the case of Likes and Reposts (Fig. 8(b) and Fig. 8(c)), depending on the number of emojis the attenuation of interest in posts occurs much more slowly than in examples from educational and entertainment communities. This type of activity may indicate that users visiting motivational communities have a high level of personal emotional involvement, but in the majority they do not consider themselves to be part of a large social group.

Fan communities at all levels of research differ in that the attenuation of interest in posts in each of the cases is smoother than in the other three study groups. If we evaluate the comments (Fig. 5(a)), the likes (Fig. 5(b)) and the reposts (Fig. 5(c)) regardless of a community subject (football, pets, music group, etc.), we will see here the same situation. Unlike motivational communities, attenuation occurs more slowly not only in the case of the likes (Fig. 9(b)) and reposts (Fig. 9(c)), but also in the case of commenting posts (Fig. 9(a)). Active communication between community members makes it clear that in fan communities, not only personal, but general emotional involvement is also strong. Members of the community consider themselves involved in something common.

Thus for all types of communities, regardless of directivity, there is a clear tendency to reduce any type of reaction to a post with the increase in the frequency of emojis in it. Most responded posts contain no emojis at all. Such reports constitute the majority of the analyzed posts, which is completely identical to the result obtained in [11] for English-language tweets. Moreover, it confirms once again the justification of research the post emojis solely on the basis of a text [4]. The only exceptions have become fan communities. They also feature this trend, but the "attenuation" of interest is slower. Entertaining and motivational communities also reflect the phenomenon of slow fading of interest, but not so clearly and only in special cases. This is due to the emotional involvement of readers in the topic of posted messages, which confirms the correctness of the assumption about changing the emotional color of personal messages taken as the basis of articles [1, 7].

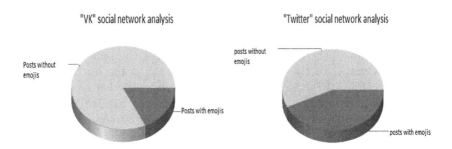

Fig. 10. Quantity of posts with emojis and without them in Russian-language (VK, a) and English-language (Twitter, b) social networks

In Fig. 10(a, b) the comparison of quantity of posts with emojis and without them in Russian-language (VK) and English-language (Twitter) social networks is provided. It proves that posts without emojis are getting more widespread, which indicates their greater informativity. The similar result was received in [6]. This conclusion correlates with the results [12, 16] but in our research we have revealed some language-dependent differences: in Twitter (43%) - Fig. 10(b) - the number of posts with emojis is higher than in "VK" (18%) - Fig. 10(a). This difference can be due to a number of reasons:

- Limit the number of characters in the Twitter messenger.
  The social network "VK" does not impose any restrictions on the length of the post. This allows users to express their thoughts without resorting to abbreviations from emojis, while Twitter has a rather strict 140 character limit, which causes users to strive for brevity, reducing the priority of emojis as a way of expressing emotions. In this case, they perform the tasks originally set before emoticons.
- Differences in mentality.
  In western pop culture, emojis occupy a weighty place. Some famous books are rewritten in Emoji language, and in the Dutch city Vathorst they decorate a building facade. In Russia, the culture of emojis is not at that peak of popularity, which is partly caused by a small percentage of posts with emojis.

5   Conclusion

Our analysis has confirmed that the hypothesis set forth at the beginning of the article has fully justified itself. Emojis interfere with the perception of information in social networks, as in the modern world they are seen more as an object for attracting attention to advertising information. An exception is appeared to be the communities with increased emotional involvement (groups of sport fans, fans of the series, music group, etc.). In this case, emojis expand the level of acceptable frequency of their use, since they act as an emotional catalyst and serve for expressing feelings, not for attracting attention.

Acknowledgements. This work was financially supported by the Government of Russian Federation, (Grant 08-08).

References

1. emoji. https://en.oxforddictionaries.com/definition/emoji
2. Adams, P.: Emoji trends in business messaging 2015–2016. https://blog.intercomassets.com/wp-content/uploads/2016/12/09135612/Intercom-Emoji_Trends_in_Business_Messaging_2015-2016.pdf
3. Aoki, S., Uchida, O.: A method for automatically generating the emotional vectors of emoticons using weblog articles. In: Proceedings of the 10th WSEAS International Conference on Applied Computer and Applied Computational Science, Stevens Point, Wisconsin, USA, pp. 132–136 (2011)
4. Been, R., Bleuel, N., Fonts, A., Davis, M.: Expanding emoji professions: reducing gender inequality. https://unicode.org/L2/L2016/16160-emoji-professions.pdf
5. Chen, Z., Lu, X., Shen, S., Ai, W., Liu, X., Mei, Q.: Through a gender lens: an empirical study of emoji usage over large-scale android users. arXiv preprint arXiv:1705.05546 (2017)
6. Duerscheid, C., Siever, C.M.: Beyond the alphabet-communcataion of emojis. Kurzfassung eines (auf Deutsch) zur Publikation eingereichten Manuskripts (2017)
7. Gilliland, N.: Five examples of brands using emojis in marketing campaigns (2017). https://econsultancy.com/blog/68745-five-examples-of-brands-using-emojis-in-marketing-campaigns
8. Glikson, E., Cheshin, A., van Kleef, G.A.: The dark side of a smiley: effects of smiling emoticons on virtual first impressions. Soc. Psychol. Pers. Sci. 1948550617720269 (2017). https://doi.org/10.1177/1948550617720269
9. Lu, X., et al.: Learning from the ubiquitous language: an empirical analysis of emoji usage of smartphone users. In: Proceedings of the 2016 ACM International Joint Conference on Pervasive and Ubiquitous Computing, pp. 770–780. ACM (2016). https://doi.org/10.1145/2971648.2971724
10. Miller, H.J., Kluver, D., Thebault-Spieker, J., Terveen, L.G., Hecht, B.J.: Understanding emoji ambiguity in context: the role of text in emoji-related miscommunication. In: ICWSM, pp. 152–161 (2017)
11. Pak, A., Paroubek, P.: Twitter as a corpus for sentiment analysis and opinion mining. In: LREc, vol. 10, pp. 1320–1326 (2010)
12. Roberts, K., Roach, M.A., Johnson, J., Guthrie, J., Harabagiu, S.M.: Empatweet: annotating and detecting emotions on Twitter. In: LREC. vol. 12, pp. 3806–3813. Citeseer (2012)
13. Stark, L., Crawford, K.: The conservatism of emoji: work, affect, and communication. Soc. Media+ Soc. **1**(2), 2056305115604853 (2015). https://doi.org/10.1177/2056305115604853

14. Yamamoto, Y., Kumamoto, T., Nadamoto, A.: Role of emoticons for multidimensional sentiment analysis of Twitter. In: Proceedings of the 16th International Conference on Information Integration and Web-based Applications and Services, pp. 107–115. ACM (2014). https://doi.org/10.1145/2684200.2684283

15. Yuhui, F., Menlan, Q., Yan, L., Haiyang, Y.: Cross-culture business communication by emoji in GMS. In: The First International Symposium on Business Cooperation and Development in South-East and South Asia under B&R Initiative (ISBCD 2016), pp. 0181–0186. Atlantis Press (2016)

16. Zhang, Z., Zhang, Y.: How do explicitly expressed emotions influence interpersonal communication and information dissemination? a field study of emoji's effects on commenting and retweeting on a microblog platform. In: Proceedings of the Pacific Asia Conference on Information Systems, PACIS 2016, p. 1 (2016). http://aisel.aisnet.org/pacis2016/124

17. Zhou, R., Hentschel, J., Kumar, N.: Goodbye text, hello emoji: mobile communication on wechat in China. In: Proceedings of the 2017 CHI Conference on Human Factors in Computing Systems, pp. 748–759. ACM (2017). https://doi.org/10.1145/3025453.3025800

Power Laws in *Ad Hoc* Conflictual Discussions on Twitter

Svetlana S. Bodrunova$^{(\boxtimes)}$ ⓘ and Ivan S. Blekanov ⓘ

St. Petersburg State University, 7-9 Universitetskaya nab., St. Petersburg
199034, Russia
s.bodrunova@spbu.ru

Abstract. *Ad hoc* discussions have been gaining a growing amount of attention in scholarly discourse. But earlier research has raised doubts in comparability of *ad hoc* discussions in social media, as they are formed by unstable, affective, and hardly predictable issue publics. We have chosen inter-ethnic conflicts in the USA, Germany, France, and Russia (six cases altogether, from Ferguson riots to the attack against *Charlie Hebdo*) to see whether similar patterns are found in the discussion structure across countries, cases, and vocabulary sets. Choosing degree distribution as the structural proxy for differentiating discussion types, we show that exponents change in the same manner across cases if the discussion density changes, this being true for neutral vs. affective hashtags, as well as hashtags vs. hashtag conglomerates. This adds to our knowledge on comparability of *ad hoc* discussions online, as well as on structural differences between core and periphery in them.

Keywords: *Ad hoc* discussion · *Ad hoc* publics · Network structure
Twitter · Degree distribution · Power law

1 Introduction

Public discussions online, and, of these, the ones on social networking sites, have become a growing area of scholarly attention, as they have been perceived as a manifestation of the public sphere, a crucial condition for efficient democratic deliberation [28]. Thus, understanding how networked discussions form and evolve is necessary for elaboration of proper criteria for their efficiency evaluation.

One of the issues recently raised by social science and communication scholars is that of comparability of the online discussions formed by the so-called issue publics [23, 28: 422]. Such discussions form quickly (or even burst out) around events or burning issues. Due to their *ad hoc* nature [16], they may dissolve just as quickly, involve various actors, and are affective [40] and, thus, are shaped by emotions rather than by rational argumentation. The question remains whether we have grounds for comparing such discussions, as the differences in discussion substance may lead to critical differences in the discussion structure and connectivity, which would make, e.g., cross-cultural and cross-language comparisons of same-topic discussions impossible. Also, conclusions made for one discussion in terms of actor roles, dynamics, or

© Springer Nature Switzerland AG 2018
D. A. Alexandrov et al. (Eds.): DTGS 2018, CCIS 859, pp. 67–82, 2018.

other constitutive parameters would not allow for predicting them for other discussions if the network structures are not assessed and recognized as similar.

This question has not yet been properly addressed in the social network analysis literature, as structural similarities of networked discussions, despite all the attention given to them, remain understudied in comparative perspective, especially in the view of the public sphere theory. The latter has elaborated its own vision of how to assess the efficacy of public discussions. Linking the two research areas for elaborating the SNA parameters for such assessment of the discussion networks would address one of the existing gaps in social network studies. For example, a number of more recent works have juxtaposed calculated and *ad hoc* publics [17, 35] in search of networking patterns of both types of discussions, but these studies were single-case, and we still lack the knowledge whether structural patterns vary across cases, cultures, and types of vocabularies used for data collection.

Another input for social network analysis from the public sphere theory would be addressing the difference between users who are key for the discussion outburst and random discussion participants. Traditional view would imply various manifestations of power distance between the two user groups, and one needs to know whether online discussions show stable patterns of differences between key and random users.

We address these research gaps by collecting data and analyzing web graph structures for five discussion outbursts about inter-ethnic (inter-national and inter-race) conflicts in the USA, Germany, France, and Russia of the 2010 s, reviewing altogether six discussion cases, as we split one into neutral and affective parts. We collect data from Twitter based on single hashtags and keyword conglomerates and show that there are repeatable structural patterns of the discussions across these cases.

The remainder of the paper is organized as follows. Section 2 discusses today's literature on various aspects of discussion structure for comparative tasks. In Sect. 3, we describe the cases and formulate the hypotheses for their comparison. Section 4 shows and discusses the discovered results. In conclusion, we discuss applicability of our findings for further studies of online conflictual discussions.

2   Comparing Discussion Structure on Social Networks: A Literature Review

Ad hoc *discussions: a plea for grounds for comparison.* Public discussions on social networking sites have drawn scholarly attention to their various aspects, including the connection between the deliberative power of 'ordinary citizens' and the platform and network features that might either empower or disempower user groups in their opinion expression, shape the discussion outcomes, and impact the respective decision-making, as well as inspire political mobilization via the 'logic of connective action' [7]. Thus, the discussions in online networks have been thoroughly studied in their substantial aspects. But one of the key issues in this research area is the principal possibility of comparisons between discussions on similar topics happening in various parts of the world and in varying times.

So far, comparative studies of cross-country social network discussions have been rare [12]; one of the reasons for that is the scholarly argument of low (or, rather, unknown) comparability of the discussions that are formed by the issue publics. This type of the discussion raises, arguably, the biggest amount of doubt 'whether Habermas is on Twitter' [18], as, from case to case, current research demonstrates varying results on echo chamber formation [5, 21, 42, 45], cross-group discussion potential [3, 4], and the roles of influencers detected by various means [6, 12, 25].

The reason for low comparability, as scholars argue, lies in the very nature of the discussions, as they are all *ad hoc* – that is, case-specific and random in formation. Such discussions have the outburst nature, may have varying patterns of dissolution, involve case-specific actors, and based on affect [40] – that is, are shaped by emotions rather than by rational argumentation, which may add to the non-comparability of the discussions. Thus, the major research question is – are *ad hoc* discussions comparable in topical and sociological terms also comparable in the discussion structures? Or, in other words, are there structural features characteristic of such discussions, that would, ideally, distinguish them from other discussion types on social networks and on the Web on the whole? And what could be the markers for such *ad hoc* discussion outbursts?

Today, in Twitter studies, there is scarce but growing evidence that *ad hoc* discussions differ in their nature from other discussion types. Thus, several papers underline the differences between calculated and *ad hoc* publics [17, 35], but no network parameters were used to prove the differences between these discussion types. Also, these studies were based on single cases, and we still lack the knowledge whether structural patterns for *ad hoc* discussions vary or repeat across cases, cultures, and types of vocabularies used for data collection. This creates a focus for our enquiry.

Degree centralities as a proxy for discussion comparability and a potential discussion quality metric. Degree centralities (in-degree, out-degree, and degree accumulating both of these) have been long ago recognized as the key structural metrics of user relations and random graph assessment [1, 10, 37]; degree distributions are recognized as a key variable describing users' interest toward each other [26]. At the same time, from the normative viewpoint relevant in social sciences, degree centralities are an important metric of user in-network influence [13], as well as the metric important in deliberative terms: the more users are reached by the same user (or reach a given user, which is the same for non-directed graphs but not for directed graphs), the more probable is the chance for cross-opinion discussion. More importantly, it can also characterize general user involvement into the discussion in comparison with other discussions; in terms of user involvement, degree centralities are, arguably, the most telling. A range of more topic-specific research papers have also argued that degree-based metrics are useful for network-based studies of social conflict and dangerous networks. Of those, one work [33] has shown the importance of relational measures (as the authors noted, 'who is related to whom') in detecting the key attackers in the terrorist network.

There are, of course, several criticisms about degree distributions, as they generalize the network structure without taking into account the role of influencers (for the reviews on detecting and comparing influencer structures, see [11–13]), or 'hub users'. Their roles, in various works, are assessed in different, if not directly opposite, ways. Thus, one stream of works insists on their disproportionately big role in information distribution (see [6, 25], and many others), while another line shows that such hubs may be inefficient due to their overload and incapability of transmitting information due to that [30]. But our goal is to test whether the discussions on the whole may be distinguished from Twitter on the whole.

As major research works in the area note [2, 14, 31], due to preferential connectivity in real-world networks, a certain power law (just as its absence) in degree distribution may become the characteristic that describes specific network structures. Thus, networks expressing power-law-like degree distributions are known as scale-free [1, 26], and 'scale-free tweet mention graphs would imply that a few Twitter users and 'mentioners' are responsible for a disproportionately high fraction of a community's discourse' [26: 587]. Thus, degree distribution may serve as a proxy for, e.g., core vs. periphery assessment in terms of aggregate influencer power: it can tell whether the network is dominated by a small number of users [46]. Thus, for Twitter, it would allow for: (1) differentiating one network type from another; (2) proving that the discussions are similar in their core vs. periphery relations, which suits our research goals. Moreover, our previous studies [8] have shown that power law exponents, indeed, vary for different types of web structures; e.g. for university websites, the average exponent is 1.8 instead of the expected 2.1.

Early works on World Wide Web topology mentioned above, as well as other research papers (see, e.g., [27]), all stated that power law distributions are characteristic for the Web on the whole. But another, later line of research papers has argued that the Web, by just evolving in time, has blurred the initial power-law-like degree distribution [20, 36]. Smaller-scale research, though, tells that degree distributions are still valid for smaller real-world samples.

Today, more and more criticism is raised about the explanatory potential and the very existence of scale-free networks – that is, of the power law in degree distributions – in large human-based datasets [15]. We, thus, want to test whether the certain power laws show up in *ad hoc* discussions, as distinguished from Twitter on the whole, cultivated long-term discussions, and random talk.

Power laws on Twitter: lack of comparative studies on the nature of the discussions.
So far, degree distributions have been studied for either the whole Twittersphere or its geographical segments; and, in these studies, the evidence on whether all the discussions on Twitter manifest power-law degree distributions is mixed.

Thus, cumulative degree distributions [39] for Twitter on the whole studied a decade ago [32] showed that the slopes γin and γout [were] both approximately −2.4. The authors have interpreted these figures as similar to those for the Web of those times (−2.1 for in-degree, cf. [24]) and blogosphere (−2,38, as based on Blogpulse conference dataset of 2003). With the latter claim, one could agree, but with the former one we would not, as our research shows that differences of 0.3 in exponent values may be characteristic of graph origin and/or discussion type [8]. Another study [44] based on the data dated mostly from April 2008 to April 2009 and featuring the most followed Twitter users from Singapore has also demonstrated power-law-like degree distributions of the user networks. A Twitter-large study [43] has shown a very different exponent value of −1.6. But authors [34] have crawled the whole Twittersphere and have not discovered any power law in link distribution on the global level, as other authors underline [29].

Several studies have focused on country-based segments of Twitter. Thus, one study [38] has dealt with the Twitter follow graphs. The researchers have shown that in-degree and degree on Twitter are best fit by power law, while out-degree is best fit by log-normal distribution. Besides analyzing the entire Twitter, it also did country-based segment studies for Brazil, Japan, and the United States; very little variance was found between the countries in terms of in-degree, out-degree, and degree distribution [38: 494]. Another study has dealt with the follow graphs of 10 country-bound Twitter segments around the world [41]. It has also demonstrated power-law in-degree and out-degree distributions, with power law coefficient ranging 5.91 to 9.51 and 8.12 to 13.62, respectively. But one also needs to note that user influencer status is linked much more to the number and activity levels of active followers (who retweet and/or mention a given user) than to the number of followers [19]; thus, one needs to look at the actual discussion graphs rather than at the graphs of following (much less linked to the real-world issue-based discussions) to more precisely determine the influential users, as well as to define the discussion type.

Even a fewer number of works have examined the degree distributions in hashtagged discussions. We can name one work [47] on Iranian elections; the discussion there also showed power law distributions, with in-degree being –2.85 and out-degree being –2.42, while the retweet-based network had in-degree of –1.94. Another group of authors [22] have also observed that retweet- and mention-based networks virtually did not differ in their scale-free topology.

From the review above, one can conclude that scholarly evidence for power laws in degree distributions is greater than that of the opposite. But, despite their potential, degree distributions have not been tested for *ad hoc* discussion outbursts in comparative perspective.

Our idea is to look how degree distributions work if we step-by-step eliminate the users with low degree index, starting from isolate users (D = 0), and look at the cumulative degree distributions for each case.

3   The Research Methodology

The research questions. Most of the available research proves that power laws are characteristic for Twitter 'calm' discussions, be it the whole Twitter or its parts, either hashtag-based or limited by region. But the exponent values of degree distribution vary highly, not allowing for any particular expectation. Thus, we ask: Will degree distribution of all the *ad hoc* discussion be fit by a power law? Will the exponent values be similar, thus indicating that the discussions are similar? Will the exponent values differ from |2.1|? Will the exponent values be similar enough across world regions, hashtag-only / keyword conglomerates, and neutral / affective hashtags?

The research hypotheses that emerge of these research questions look as follows:

H1. All the discussions under scrutiny will be characterized by power law in degree distribution.

H2. All the discussions will diverge from the |2.1| exponent value to the same direction and on comparable percentage. This includes hashtag vs. keyword conglomerates (**H3**).

H4. Neutral and affective (expressing emotions of either sympathy and compassion or negation and hatred) hashtags will be comparable in their power law exponents.

The cases under scrutiny and their substantial comparability. The cases of our attention all have the same set of features that is to ensure that the cases are comparable in sociological terms. Thus, all the cases have a violent trigger (a killing or rape/harassment) of inter-ethnic nature; the discussions are outburst – that is, the number of users involved has one sharp peak and then slows down gradually; media report the communities to split into the minority, pro-minority majority, and anti-minority majority groups; there is peaceful protest or mass commemoration of the victims in the aftermath of the conflict; there is direct involvement of authorities of several levels into conflict resolution; and all the conflicts provoke a discussion on Twitter that gets to national (sometimes also to global) Twitter trending topics.

We are looking at the following cases (in chronological order):

- A killing of a Russian Muscovite by an Uzbek immigrant and the subsequent anti-immigrant clashes in the Moscow district of Biryuliovo, Russia (2013);
- A killing of an African American teenager by a white police officer and the subsequent city riots in Ferguson, the USA (2014);
- The attack to the editorial office of *Charlie Hebdo* and the subsequent peaceful demonstrations in Paris, France (2015);
- The mass harassment of German females by male re-settlers from Middle East and North Africa in the New Year Eve in Cologne, Germany (2015–2016), and the subsequent protest meetings by PEGIDA and the 'Alternative for Germany' party;
- A bus attack at one of the Christmas markets in Berlin (2016).

In each case, the discussion data were collected and the respective web graph reconstructed; in the case of *Charlie Hebdo*, two graphs (one for a neutral hashtag and one for a compassion hashtag) were reconstructed.

Data collection. To collect the discussion bulk, we have created a specialized web crawler with adjustable modules [9]. It was done especially to overcome the well-known Twitter API limitations, like the ones on the number of requests to server and on the number of tweets available for download. It also bypasses the popularity algorithm and allows for human-like backfolding.

For bigger-scale discussions (Ferguson and *Charlie Hebdo*), one hashtag per graph was used. For smaller-scale discussions, snowballing reading of 1.000+ random tweets containing the primary hashtag/keyword was performed, thus bringing on keyword collections.

Table 1. The datasets collected.

The case	The number of nodes	The number of edges
Biryuliovo, 2013	11429	20106
Ferguson, 2014	169677	334050
Charlie Hebdo, 2015 (neutral)	952615	1782863
Charlie Hebdo, 2015 (affective)	719503	981131
Cologne, 2015–2016	40117	98508
Berlin, 2016	194937	298562

The initial datasets collected are represented in Table 1. Altogether, tweets by over 2 mln users were included into the research. Edges between the users were formed if any type of substantial interaction emerged between them (we counted retweets, mentions, and comments as such).

Web graph reconstruction and analytics. We have reconstructed the graphs for each case. The graphs were non-directed, as we were interested in the aforementioned deliberative aspects of the discussions. We have used the OpenOrd algorithm for the graph reconstruction. This methods is based on calculating the maximum distance (in %) between the two nodes and on a particular number of iterations (in all our cases, the algorithm converged the proposed 800 iterations to 750). This algorithm was chosen, as it brings to the discussion core (that is, to the center of the graph) the influential users defined by a wide set of variables, including absolute-figure ones (the number of tweets, likes, retweets, mentions, and comments) and centrality metrics (including degree, betweenness, parerank, and eigenvector centralities).

Then, the degree distribution exponents were calculated for each case, using eleven steps of isolate user elimination (for users with D = 0 to 10 in the initial graph).

The results and answers to our hypotheses are presented below.

4   The Research Methodology

As stated above, we have calculated degree distribution exponents for each case. In order to clearly highlight the differences with the expected exponent value of |2.1| based on previous Web studies, we have also calculated the exponents for the elimi-nated users. The exponent values for both active users and eliminated users, as well as their absolute deviations from |2.1|, are presented in Table 2. The degree distribution graphs are presented in Figs. 1(a, b), 2(a, b), 3(a, b), 4(a, b), 5(a, b) and 6(a, b) for the respective cases, as ordered in Table 2.

H1. All the cased that we have studied do, indeed, demonstrate power laws in degree distribution, and this is true for both active users (see Figs. 1(a), 2(a), 3(a), 4(a), 5(a) and 6a) and eliminated users (see Figs. 1(b), 2(b), 3(b), 4(b), 5(b) and 6(b). Thus, power law is indicative for *ad hoc* discussion outbursts across cultures and cases; H1 is supported.

Table 2. Exponent values for the cases and their divergence from the expects exponent value

The case	Exponent value, active users	Absolute deviation, active users	Exponent value, eliminated users	Absolute deviation, eliminated users
Biryuliovo, 2013	|1.56|	0,54	|2.04|	0,06
Ferguson, 2014	|1.39|	0,71	|2.16|	–0,06
Charlie Hebdo, 2015 (neutral)	|1.59|	0,51	|2.01|	0,09
Charlie Hebdo, 2015 (affective)	|1.80|	0,3	|2.45|	–0,35
Cologne, 2015– 2016	|1.28|	0,82	|1.91|	0,19
Berlin, 2016	|1.63|	0,47	|2.34|	–0,24

a)

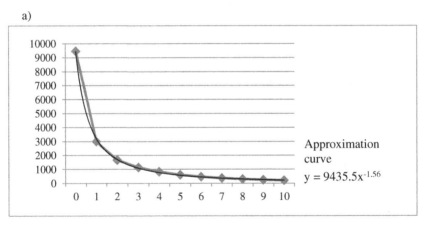

b)

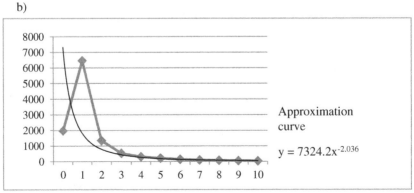

Fig. 1. Degree distributions for the Biryuliovo case: (a) active users; (b) eliminated users. *Source:* authors

a)

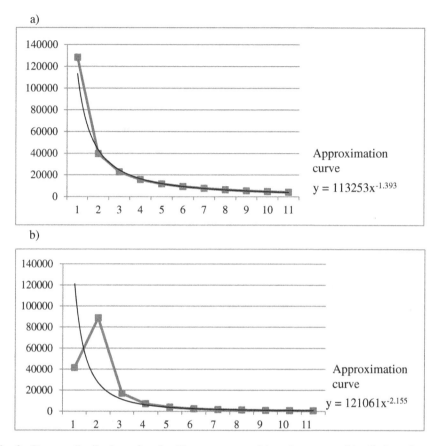

Approximation curve

$y = 113253x^{-1.393}$

b)

Approximation curve

$y = 121061x^{-2.155}$

Fig. 2. Degree distributions for the Ferguson case: (a) active users; (b) eliminated users. *Source*: authors

H2. All the cases do, indeed, diverge from the |2.1| expected exponent value to the same direction (see Table 2, column 2, exponent values for active users): the received exponent values are all equal or below 1,8, which indicates the lower power distance between the users. At the same time, the exponent values for eliminated users fluctuate around |2.1| to |2.4|, the figures indicative for the Web degree distributions in various studies mentioned above. Thus, power laws with exponent values definitely below those discovered for the World Wide Web are indicative for *ad hoc* discussions, which makes them, at least in terms of core vs. periphery relations, similar and comparable. The second part of the hypothesis is, though, not clearly supported: deviation from |2.1| substantially varies in percentage (from almost 40% for the Cologne case to 14,3% for #jesuischarlie), and thus we cannot indicate any figure more precise than the fluctuation

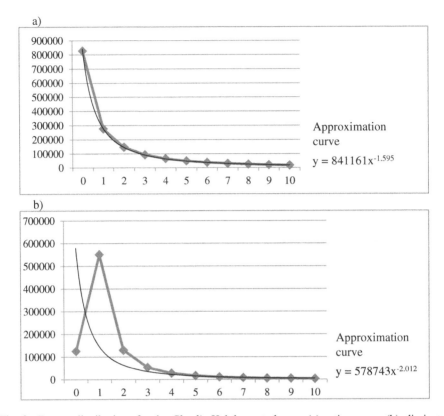

Fig. 3. Degree distributions for the *Charlie Hebdo* neutral case: (a) active users; (b) eliminated users. *Source*: authors

between |1.28| and |1.80| for *ad hoc* discussions; due to this, H2 is only partly proven. But we also need to state that, for three of six discussions, the exponent values were between |1.56| and |1.63|, which corresponds to earlier results in [43].

H3. For three of the six discussions (the Ferguson case and both Charlie Hebdo cases) the data were collected based on single hashtags (#ferguson, #charliehebdo, and #je-suischarlie, respectively), while other cases were collected by keyword conglomerates ranging from 6 keywords (for the Biryuliovo case) to over a dozen keywords (for both German cases). As we see from Table 2 and Figs. 1(a), 2(a), 3(a), 4(a), 5(a) and 6(a), there is no difference between single-hashtag and keyword-conglomerate data collection. H3 is supported, but this, paradoxically, might add not only to the evidence that *ad hoc* discussions have similar patterns of degree distribution but also to the evidence that Twitter as a platform fosters the power law degree distributions in any type of discussion. To answer this, more research is needed.

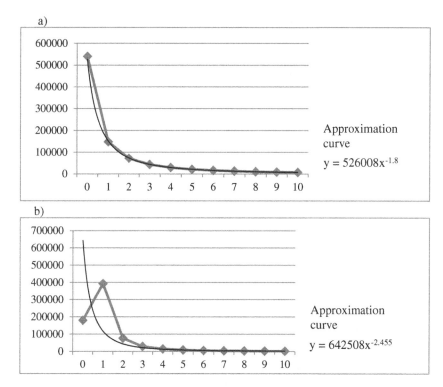

Fig. 4. Degree distributions for the *Charlie Hebdo* emotional case: (a) active users; (b) eliminated users. *Source*: authors

H4. As seen from Table 2 and Figs. 1(a), 2(a), 3(a), 4(a), 5(a) and 6(a), both neutral and affective hashtags are subjected to power law for both active and eliminated users, and exponent values deviate from |2.1| to the same direction. H4 is supported. We just need to mention that the compassion hashtag #jesuischarlie has shown the highest exponent values, thus demonstrating bigger gaps between the influential and 'peripheral' users, and this might create room for further comparative investigations.

Discussion. In search for proof of comparability of *ad hoc* online discussions, we have applied our idea of evaluating degree distributions to six datasets of five comparable Twitter discussions on inter-ethnic conflicts in four countries that happened in the 2010 s. What we have discovered is the following.

First, we have shown that all the discussions we have observed are subjected to power law in degree distributions. Second, we have shown that exponent values for degree distributions in the discussion graphs diverge from the figures indicated for the Web on the whole in previous research. Moreover, they diverge in the same direction and to varying but, to a certain degree, also comparable percentage. Taken together,

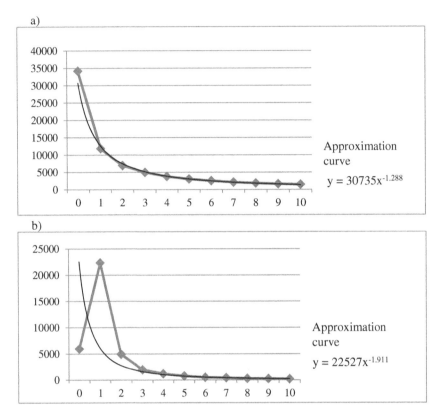

Fig. 5. Degree distributions for the Cologne case: (a) active users; (b) eliminated users. *Source*: authors

these findings indicate that, at least in terms of influence, interest, and/or power distribution in such discussions, they are comparable across countries and years, as well as across vocabulary types and neural/affective hashtags. Thus, exponent values may serve as indicators for the type of an online discussion. We have also shown that the peripheral part of the *ad hoc* discussions was always closer to |2.1| to |2.4| exponent values discovered earlier for the Web and some social networks, which may be a sign that the *ad hoc* discussions do differ from the 'average' network structure of the Web.

But we have also discovered that the exponent values for the discussions were fluctuating around |1.6| indicated earlier for Twitter on the whole; also, variance in value divergence from the expected figures was too big to state that a particular array of meanings may be indicative for ad hoc discussions and could become their structural marker. This needs further investigation, which might imply experimental design.

Last but not least, we have seen that *ad hoc* discussions, if judged by the exponent values, show the patterns of lower power distance between core and periphery. This may be due exactly to their spontaneous nature, as institutional actors who shape and frame the offline discussions compete with ordinary users, crisis witnesses, and

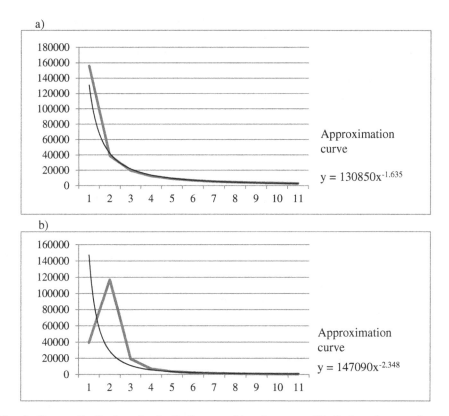

Fig. 6. Degree distributions for the Berlin case: (a) active users; (b) eliminated users. *Source*: authors

grassroots leaders. This may broaden our views upon discussion outbursts on social media.

Acknowledgements. This work was supported in full by Russian Science Foundation, grant 16-18-10125.

References

1. Albert, R., Barabási, A.L.: Statistical mechanics of complex networks. Rev. Mod. Phys. **74** (1), 1–54 (2002). https://doi.org/10.1103/RevModPhys.74.47
2. Barabási, A.L., Albert, R.: Emergence of scaling in random networks (1999). arxiv.org/abs/cond-mat/9910332v1
3. Barberá, P.: How Social Media Reduces Mass Political Polarization: Evidence from Germany, Spain, and the US, pp. 1–46. Job Market Paper, New York University (2014)
4. Barberá, P., Jost, J.T., Nagler, J., Tucker, J.A., Bonneau, R.: Tweeting from left to right: Is online political communication more than an echo chamber? Psychol. Sci. **26**(10), 1531–1542 (2015). https://doi.org/10.1177/0956797615594620

5. Bastos, M.T., Mercea, D., Baronchelli, A.: The Spatial Dimension of Online Echo Chambers. arXiv preprint arXiv:1709.05233 (2017)
6. Bastos, M.T., Raimundo, R.L.G., Travitzki, R.: Gatekeeping Twitter: message diffusion in political hashtags. Media Cult. Soc. **35**(2), 260–270 (2013). https://doi.org/10.1177/0163443712467594
7. Bennett, W.L., Segerberg, A.: The logic of connective action: Digital media and the personalization of contentious politics. Inf. Commun. Soc. **15**(5), 739–768 (2012). https://doi.org/10.1080/1369118X.2012.670661
8. Blekanov, I., Sergeev, S., Maksimov, A., Moskalets, R.: Research of university sites internal links distribution. In: Proceedings of the 3rd International Conference on Science and Computer Technology (ICST 2017), #8011875, pp. 182–185 (2017). https://doi.org/10.1109/icstc.2017.8011875
9. Blekanov, I.S., Sergeev, S.L., Martynenko, I.A.: Constructing topic-oriented web crawlers with generalized core. Sci. Res. Bull. St.Petersburg State Polytech. Univ. **5**(157), 9–15 (2012)
10. Boccaletti, S., Latora, V., Moreno, Y., Chavez, M., Hwang, D.U.: Complex networks: Structure and dynamics. Phys. Rep. **424**(4), 175–308 (2006). https://doi.org/10.1016/j.physrep.2005.10.009
11. Bodrunova, S.S., Blekanov, I.S., Maksimov, A.: Measuring influencers in Twitter ad-hoc discussions: active users vs. internal networks in the discourse on Biryuliovo bashings in 2013. In: Proceedings of the AINL FRUCT 2016 Conference, #7891853 (2017)
12. Bodrunova, Svetlana S., Litvinenko, Anna A., Blekanov, Ivan S.: Comparing influencers: activity vs. connectivity measures in defining key actors in Twitter *Ad Hoc* discussions on migrants in Germany and Russia. In: Ciampaglia, G.L., Mashhadi, A., Yasseri, T. (eds.) SocInfo 2017. LNCS, vol. 10539, pp. 360–376. Springer, Cham (2017). https://doi.org/10.1007/978-3-319-67217-5_22
13. Bodrunova, S.S., Litvinenko, A.A., Blekanov, I.S.: Influencers on the Russian Twitter: institutions vs. people in the discussion on migrants. In: ACM International Conference Proceeding Series, pp. 212–222. ACM (2016). https://doi.org/10.1145/3014087.3014106
14. Broder, A., et al.: Graph structure in the web. Comput. Netw. **33**(1), 309–320 (2000). https://doi.org/10.1016/S1389-1286(00)00083-9
15. Broido, A.D., Clauset, A.: Scale-free networks are rare. arXiv:1801.03400 (2018)
16. Bruns, A., Burgess, J.E.: The use of Twitter hashtags in the formation of ad hoc publics. In: Proceedings of the 6th European Consortium for Political Research (ECPR) General Conference 2011 (2011). http://eprints.qut.edu.au/46515/1/The_Use_of_Twitter_Hashtags_in_the_Formation_of_Ad_Hoc_Publics_(final).pdf. Accessed 20 Feb 2018
17. Bruns, A., Burgess, J.: Twitter hashtags from ad hoc to calculated publics. In: Hashtag Publics: The Power and Politics of Discursive Networks, pp. 13–28 (2015)
18. Bruns, A., Highfield, T.: Is Habermas on Twitter? Social media and the public sphere. In: Bruns, A., Enli, G., Scogerbo, E., Larsson, A.O., Christensen, C. (eds.)The Routledge companion to social media and politics, pp. 56–73. Routledge, New York (2016)
19. Cha, M., Haddadi, H., Benevenuto, F., Gummadi, P.K.: Measuring user influence in Twitter: the million follower fallacy. In: ICWSM Proceedings, vol.10, pp. 10–17 (2010)
20. Chen, Q., Chang, H., Govindan, R., Jamin, S.P.: The origin of power laws in Internet topologies revisited. In: 21st Annual Joint Conference of the IEEE Computer and Communications Societies Proceedings, vol. 2, pp. 608–617. IEEE, Piscataway (2002)
21. Colleoni, E., Rozza, A., Arvidsson, A.: Echo chamber or public sphere? Predicting political orientation and measuring political homophily in Twitter using big data. J. Commun. **64**(2), 317–332 (2014). https://doi.org/10.1111/jcom.12084

22. Conover, M., Ratkiewicz, J., Francisco, M.R., Gonçalves, B., Menczer, F., Flammini, A.: Political polarization on Twitter. In: ICWSM Proceedings, vol. 133, pp. 89–96 (2011)
23. Dahlgren, P.: Media and Political Engagement. Cambridge University Press, Cambridge (2009)
24. Donato, D., Laura, L., Leonardi, S., Millozzi, S.: Large-scale properties of the web graph. Eur. Phys. J. Condens. Matter Complex Syst. **38**(2), 239–243 (2004). https://doi.org/10.1140/epjb/e2004-00056-6
25. Dubois, E., Gaffney, D.: The multiple facets of influence: identifying political influentials and opinion leaders on Twitter. Am. Behav. Sci. **58**(10), 1260–1277 (2014). https://doi.org/10.1177/0002764214527088
26. Ediger, D., Jiang, K., Riedy, J., Bader, D. A., Corley, C.: Massive social network analysis: mining Twitter for social good. In: 39th International Conference on Parallel Processing (ICPP), pp. 583–593. IEEE (2010). https://doi.org/10.1109/icpp.2010.66
27. Faloutsos, M., Faloutsos, P., Faloutsos, C.: On power-law relationships of the Internet topology. ACM SIGCOMM Comput. Commun. Rev. **29**(4), 251–262 (1999)
28. Habermas, J.: Political communication in media society: Does democracy still enjoy an epistemic dimension? The impact of normative theory on empirical research. Commun. Theory **16**(4), 411–426 (2006). https://doi.org/10.1111/j.1468-2885.2006.00280.x
29. Hansen, L.K., Arvidsson, A., Nielsen, F.Å., Colleoni, E., Etter, M.: Good friends, bad news-affect and virality in Twitter. Futur. Inf. Technol. **185**, 34–43 (2011)
30. Harrigan, N., Achananuparp, P., Lim, E.P.: Influentials, novelty, and social contagion: The viral power of average friends, close communities, and old news. Soc. Netw. **34**(4), 470–480 (2012). https://doi.org/10.1016/j.socnet.2012.02.005
31. Huberman, B.A., Adamic, L.A.: Internet: growth dynamics of the world-wide web. Nature **401**(6749), 131 (1999). https://doi.org/10.1038/43604
32. Java, A., Song, X., Finin, T., Tseng, B.: Why we Twitter: understanding microblogging usage and communities. In: Proceedings of the 9th WebKDD and 1st SNA-KDD 2007 Workshop on Web Mining and Social Network Analysis, pp. 56–65. ACM (2007). https://doi.org/10.1145/1348549.1348556
33. Karthika, S., Geetha, R., Bose, S.: Whom to remove? Breaking the covert network. In: Fifth International Conference on Advanced Computing (ICoAC), pp. 348–354. IEEE (2013). https://doi.org/10.1109/icoac.2013.6921975
34. Kwak, H., Lee, C., Park, H., Moon, S.: What is Twitter, a social network or a news media? In: Proceedings of the 19th International Conference on World Wide Web, pp. 591–600. ACM (2010). https://doi.org/10.1145/1772690.1772751
35. Lynn, T., Rosati, P., Nair, B.: Calculated vs. Ad Hoc Publics in the #Brexit Discourse on Twitter. http://www.researchgate.net/profile/Pierangelo_Rosati/publication/319528530_Calculated_vs_Ad_Hoc_Pulics_in_the_Brexit_Discourse_on_Twitter/links/59b12cebaca2728472d0fbe4/Calculated-vs-Ad-Hoc-Publics-in-the-Brexit-Discourse-on-Twitter.pdf
36. Meusel, R., Vigna, S., Lehmberg, O., Bizer, C.: Graph structure in the web - revisited: a trick of the heavy tail. In: Proceedings of the 23rd International Conference on World Wide Web, pp. 427–432. ACM (2014). https://doi.org/10.1145/2567948.2576928
37. Mislove, A., Marcon, M., Gummadi, K. P., Druschel, P., Bhattacharjee, B.: Measurement and analysis of online social networks. In: Proceedings of the 7th ACM SIGCOMM Conference on Internet measurement, pp. 29–42. ACM (2007). https://doi.org/10.1145/1298306.1298311
38. Myers, S.A., Sharma, A., Gupta, P., Lin, J.: Information network or social network? The structure of the Twitter follow graph. In: Proceedings of the 23rd International Conference on World Wide Web, pp. 493–498. ACM (2014)

39. Newman, M.E.J.: Power laws, Pareto distributions and Zipf's law. Contemp. Phys. **46**(5), 323–351 (2005). https://doi.org/10.1080/00107510500052444

40. Papacharissi, Z.: Affective Publics: Sentiment, Technology, and Politics. Oxford University Press, Oxford (2015)

41. Poblete, B., Garcia, R., Mendoza, M., Jaimes, A.: Do all birds tweet the same?: characterizing Twitter around the world. In: Proceedings of the 20th ACM International Conference on Information and Knowledge Management, pp. 1025–1030. ACM (2011). https://doi.org/10.1145/2063576.2063724

42. Sunstein, C.R.: The law of group polarization. J. Polit. Philos. **10**(2), 175–195 (2002). https://doi.org/10.1111/1467-9760.00148

43. Welch, M.J., Schonfeld, U., He, D., Cho, J.: Topical semantics of Twitter links. In: Proceedings of the 4th ACM International Conference on Web Search and Data Mining, pp. 327–336. ACM (2011). https://doi.org/10.1145/1935826.1935882

44. Weng, J., Lim, E.P., Jiang, J., He, Q.: Twitterrank: finding topic-sensitive influential twitterers. In: Proceedings of the third ACM International Conference on Web Search and Data Mining, pp. 261–270. ACM (2010). https://doi.org/10.1145/1718487.1718520

45. Yardi, S., Boyd, D.: Dynamic debates: an analysis of group polarization over time on Twitter. Bull. Sci. Technol. Soc. **30**(5), 316–327 (2010). https://doi.org/10.1177/0270467610380011

46. Ye, S., Wu, S.Felix: Measuring message propagation and social influence on Twitter.com. In: Bolc, L., Makowski, M., Wierzbicki, A. (eds.) SocInfo 2010. LNCS, vol. 6430, pp. 216–231. Springer, Heidelberg (2010). https://doi.org/10.1007/978-3-642-16567-2_16

47. Zhou, Z., Bandari, R., Kong, J., Qian, H., Roychowdhury, V.: Information resonance on Twitter: watching Iran. In: Proceedings of the First Workshop on Social Media Analytics, pp. 123–131. ACM (2010). https://doi.org/10.1145/1964858.1964875

Topics of Ethnic Discussions in Russian Social Media

Oleg Nagornyy[(✉)] iD

National Research University Higher School of Economics, Saint Petersburg, Russia
nagornyy.o@gmail.com

Abstract. The paper reveals the topic structure of ethnic discussions in the Russian-speaking social media and explores how these topics are related to the post-Soviet ethnic groups. Analyzed more than 2.6 million texts from Russian-speaking social media published for two-year period from 2014 to 2015 and contained at least one of the post-Soviet ethnonyms, we conclude that ethnic discussions in these media are full of socially significant and potentially problematic topics (15 topics out of 97 can be regarded as problematic comparing to the 4 out of 150 topics on random sample from VK.com). The most salient topics are the topics about Ukraine-Russia relations over the recent conflict between two countries. We also found the racial bias in criminal topic towards peoples of the North Caucasus which are often mentioned in the context of crimes and terrorism.

Keywords: Ethnicity · Topic modelling · Social media · Russia

1 Introduction

The topic of ethnicity has been interested social scientists for a long time. The reason is that ethnicity traditionally refers to a very small number of prescribed social statuses that are acquired from birth, regardless of person's will and desire and cannot be changed. Such statuses are often perceived as an integral and unchanging part of the human personality hence they experienced by people more intimately and have a higher potential to emerge social conflicts. These circumstances make it important to study ethnic issues, especially in such polyethnic country like Russia.

Public opinion polls show inequality in attitudes towards ethnic minorities in Russia. According to the recent studies [4] Caucasian[1] and Central Asians arouse the most negative attitudes. What is more important, these attitudes may easily stream from the "offline" to the "online" space, influencing interethnic conflicts [11] and hate crimes in a real life [6]. Therefore, it is important to monitor ethnicity-related online content and to develop instruments for such monitoring. In this study, we seek to reveal topic structure of ethnicity-related discussion in the Russian language social media and compare public attention to different ethnic groups to shed light on the ethnic processes that take place offline.

---

[1] In Russian language the term "Caucasians" denotes the peoples of Caucasus region.

© Springer Nature Switzerland AG 2018
D. A. Alexandrov et al. (Eds.): DTGS 2018, CCIS 859, pp. 83–94, 2018.
https://doi.org/10.1007/978-3-030-02846-6_7

2   Related Work

With the explosive development of the Internet and the accompanying phenomena, many works devoted to the study of ethnic discursions in the global network have appeared. Based on content analysis of 83 Facebook profiles of African Americans, Latino, Indian and Vietnamese ancestry students, supplemented by 63 in-person interviews Grasmuck et al. [12] have found that ethnicity is very noticeable among Facebook users. African Americans, Latinos and Indians (Native Americans) create more elaborate personal pages that signal their ethnicity much more clearly than White or Vietnamese. Some users emphasize their ethnicity, opposing the tendency to conceal ethnic differences.

Ethnic studies on the Internet are not limited to just describing how ethnicity manifests itself in online space. Researchers from the United States have shown that this relation operates in both directions [6]. They found that Internet access is positively related to the number of crimes motivated by racial hatred: more people have Internet access have in certain area — higher level of racism it has.

It is also worth to mention a large amount of literature describing methods for identifying hate speech in social media [7, 24 and other works], but these works usually do not provide any substantive results, focusing on methodological part of the research.

There are also articles on ethnicity in Russian media. Most of them studying ethnic identity on largest national social networking site VKontakte. Dilyara Suleymanova, in her article "Tatar Groups in Vkontakte: The Interplay of Between Ethnic and Virtual Identities on Social Networking Sites" [21] studied communities formed around Tatars-related issues. Tatars is the second largest ethnic group in Russia. The author claims that Vkontakte is a powerful tool for constructing ethnic networks, connecting Tatars from all over Russia, and also functioning as a tool for building ethnic identity. The Tatar groups reproduce and mobilize traditional ideas about what it means to be a Tatar using topics of Tatar language and Islam. Moreover, some Tatar groups construct alternative versions of the Tatar identity.

The authors of the next work — researchers from the Higher School of Economics, Daniil Alexandrov, Aleksey Gorgadze and Ilya Musabirov — analyzed Caucasian groups in the same social network site [2]. With social network analysis they built a network of groups based on common membership. The authors identified different clusters of ethnic groups and determined the proximity of these clusters. Thus, it was revealed that clusters containing Armenian and Azerbaijani groups do not have common groups, apparently due to the Nagorno-Karabakh conflict between Armenia and Azerbaijan. Also, the authors used topic modeling to reveal the nature of relationships between groups in clusters. They found that religion, being not so salient topic, facilitates the establishment of links between clusters.

Researchers are also examined how various ethnic groups are presented in mass media. For example, according to some researchers, the image of Chechens (one of the ethnic groups from the North Caucasus) "has been subjected to such demonization in the Russian media that virtually any publication relating to Chechens — even if it has nothing to do with the armed conflict in Chechnya - turns out to be imbued with the language of hostility' [1]. This opinion was supported by Bodrunova et al. [5] in the research of Russian-language blogs. Authors tried to identify attitudes toward labor immigrants and have found that people from the North Caucasus are portrayed in the blogosphere as aggressive and hostile subjects. A deeper, qualitative analysis has shown that Daghestanians are portrayed as mighty barbarians, and Chechens as terrorists. Previous studies on this topic also show that peoples of North Caucasus cause the most negative attitude in Russians online media [4, 10].

One of the most relevant studies on the role of ethnicity in the mass media was carried out by Fadeev [9]. The hypothesis of this study was that more integrated ethnic groups (Ukrainians and Jews) are more positively represented in the local media of St. Petersburg compared to the relatively new ones (Tajiks and Chechens). This hypothesis was partially confirmed. Just as in our study the author used commercial media aggregator to collect texts with ethnonyms from the social. They also tried to link the use of ethnonyms with the general context of the message in social media, as we do. As the main method of analysis author chose to manually label attitudes to the ethnic groups mentioned in texts with the subsequent analysis of the percentage distributions of these labels for each ethnicity.

The aforesaid review on the topic of ethnicity in the Internet shows that the authors working in this topic mainly consider separate diasporas (groups in social networking sites, ethnic communities in cities and so on). Our work compares favorably with the fact that it is not limited to the study of one separate ethnic diaspora but aims to describe ethnic discussions on the macro level.

3   Data

The dataset for the research consist of texts on ethnicity from Russian-speaking social media. The data we needed must satisfy two conditions: (1) they should be relevant to ethnic discussions; (2) they should represent all Runet texts on this topic.

To achieve the first of these conditions, we compile a comprehensive list of ethnonyms used to search for texts related to ethnic discursions. This list included not only post-Soviet ethnonyms, but also other words related to them in one way or another. These words were (1) ethnonyms that denote ethnicity not belonging to the group of post-Soviet ethnicities but taking an active part in the life of the country (Jews, Gypsies), (2) racial slur, (3) words denoting a geographic location (Highlanders, Europeans, Asians), (5) obsolete words with ethnic meaning (Rusich[2], basurmanin[3]). For each word we generated male and female forms and built a list of bigrams (more than 30 bigrams for each ethnic group). Some words were removed from the list to prevent homonymy (homonyms are the words which sound or spelled in a similar way but have different meanings). Thus, together with relevant bigrams the final list of the words comprises over 4,000 units describing 97 ethnic categories.

To ensure representativeness we collect all texts related to ethnic discussions from the period of time from January 2014 to December 2015. The text is considered as relevant to ethnic discussions in case it contained at least one word or bigram from the

---

[2] An early medieval group, who lived in a large area of what is now Russia, Ukraine, Belarus.
[3] Hostile designation of a person non-Christian faith.

generated list. The texts were gathered using the social media monitoring service IQBuzz. IQBuzz monitor pages from thousands of websites looking for predefined words. Typically, such services are used by commercial companies for marketing purposes like to track effect of advertising campaigns or changes in business reputation, but we reoriented IQBuzz to solve research problems. The disadvantage of this solution is the uncertainty about how exactly the data was collected and how complete this sample. IQBuzz declares that it able to track "all the mentions in the Internet" but we cannot check this allegation. What about the advantages of data obtained with this service, we would like to note their saturation with additional information. Whenever possible IQBuzz provides the data about the author of the text, such as location, age and gender.

4   Distribution of Ethnic Texts by Regions and Domains

The 10 most representative regions can be seen in Table 1. It can be seen that 40% of texts are on ethnicity were generated by the users from two major Russian cities — Moscow and Saint Petersburg. Compared with them, each of the remaining Russian regions takes a much smaller share. This proportion reflects the fact that (1) these cities are the most populated subjects of Russia (the correlation between the share of the population in the region and the proportion of downloaded texts equals 0.8 which means more people live in the region more texts from that region we have), (2) the level of Internet penetration in these cities is higher, than in other regions [26]. However, even taking into account bigger number of Internet users in these cities, the share of these cities cannot be explained by the above factors only. We suggest, that a large proportion of reports on ethnic issues from Moscow and Saint Petersburg may indicate an intensification of ethnic processes in these densely populated cities, which is consistent with the theory of urbanists, who pointed to population density as a source of intense interactions between people in cities and greater complexity of these interactions [23].

Table 1. Distribution of texts by region (ten most significant regions)

Region	% of texts from that region in the dataset	% of population from that region in Russian population
Moscow	26,60%	8,43%
St. Petersburg	13,40%	3,60%
Perm Region	4,30%	1,79%
Krasnodar region	3,20%	3,79%
Sverdlovsk region	3,00%	2,95%
Rostov region	2,60%	2,88%
Samara Region	2,50%	2,18%
Tatarstan	2,30%	2,65%
Novosibirsk region	2,10%	1,89%
Chelyabinsk region	2,00%	2,39%

As for the texts distribution by source, it is not surprising that vast majority of the collected texts (82%) are produced by the users of Russia's largest social network site VKontakte.

While interpreting the obtained results, however, it is worth to remember that the origin of the revealed differences may be a limitation of the data collection tool, whose authors, while promising the widest possible coverage of sources, cannot guarantee this.

5   Topics of Ethnical Conversations Online and Their Metrics

We analyzed the topic profile of ethnic discussions, obtained with LDA. Topic modeling has already been successfully used to reveal the characteristics of ethnic discussions [2, 3]. Moreover, researchers from the Laboratory for Internet Studies created a modification of this algorithm called ISLDA, aimed, among other things, to facilitate extraction of ethnically relevant topics from texts [17].

As mentioned above, we built LDA topic model with 97 topics by the number of selected ethnic categories. To understand the meaning of the topics, we manually labeled each of them. The labeling included reading words and texts, related to the given topic with highest probability. Some topics could not be easily interpreted, the interpretation of 22 topic raised difficulties so they were labeled as "uninterpretable". The presence of such topics is a usual phenomenon in topic modeling. Moreover, as social media texts have short length and multimodal structure, topic models are often produce uninterpreted topics with this kind of data. The number of uninterpretable topics we obtained can be considered as a satisfactory result.

Topic labeling results showed that some topics are about ethnicity while the others do not touch ethnic issues at all — there are topics about crime, politics, work, cinema, family, economy, army, housing and so on. Comparing this set of topics with topics from random sample of Vkontakte texts, which represent a kind of "natural" topic structure, we see that our topic set contains a greater number of socially significant and potentially problematic topics (by that topics we assume the topics that affect whole society and are able to raise politic, social, ethnic and other kinds of conflicts) rather than topic about everyday issues like, for instance, games, music, cooking, health and beauty [18]. Users use the social network Vkontakte more like a place where they can save interesting culinary recipe, browse for new films and music and read gardening recommendations. This huge part of the posts is created with the instrumental aim to provide a quick access to potentially useful information. Other texts are automatically generated by numerous applications, most often for advertising purposes. Together with uninterpreted, these topics make up the bulk of the VKontakte. Although a small number of sociopolitical topics were identified in that study ("Christianity", "Islam," "Ukraine-Russia relations" and "City events"), their number is significantly less compared to the at least 15 potentially problematic topics found in our dataset. Most of that 15 topics are the topics about politics (Table 2).

Table 2. Most salient topics of ethnic discussions (25 most salient topics)

ID	Topic	Salience	Sentiment
97	Relations of Russians with Other Nations	162170	–0,08
32	Ukrainian-Russian relations	130185	–0,28
44	Russian Society, Russia as a National State	83245	–0,23
27	The revolution in Ukraine	77735	–0,66
48	Conflict relations between Russia and the West	75529	–0,61
15	Children and family	72469	0,17
12	Porn	70656	–1,55
28	Uninterpreted topic	69375	0,88
67	Uninterpreted topic	65616	–0,04
33	The activities of the Russian authorities	62222	0,61
39	Peoples of the North Caucasus, Islam, terrorism	56334	0,23
23	Uninterpreted topic	53387	–0,59
57	Poetry	50895	0,15
6	Economy	48274	0,86
79	Psychology and children	47089	0,14
14	Crimes and murders	46394	–0,22
16	Movies, festivals, performances	43962	–0,13
19	WWII	43618	–0,5
82	Ancient Slavs	41839	–1,15
73	Tourism and rest	41558	0,36
10	Christianity and Orthodoxy	40175	0,81
4	Uninterpreted topic	40032	–0,3
49	Tatars and other Turkic peoples	35559	0,07
63	US-Russian relations, US condemnation	34971	–1,38
37	The Internet	34688	–0,29

With respect to a large number of topics it is clear that, despite their connection with ethnonyms, they cannot be considered as ethnic topics. As an example, there are topics about sport (#26 Football, #41 Boxing), politics (#54 and others) and history (#69 and others). What about pure ethnic topics the most vivid examples are topics #29 (Armenian-Turkish relations, genocide of Armenians), #49 (Tatars and other Turkic peoples of Russia), #50 (Uzbeks), #71 (Jews and Judaism), #97 (Relations of Russians with other peoples) and #39 (Peoples of the North Caucasus, Islam, terrorism).

In addition to topics labeling we calculate two topic's metrics: salience and sentiment. The first one shows how widely this topic is covered by users, the second indicates most probable topics for texts with positive/negative sentiment scores. To obtain an index of the topic's salience we sum probabilities of all texts in a given topic. It turned out that the most vivid topics are topics #97 (Relationships of Russians with Ukrainians and Caucasians), #32 (Ukrainian-Russian relations), #44 (Russian Society, Russia as a national state), #27 (Ukrainian Revolution 2013-2014) and #48 (Conflict between Russia and generalized West). From this list it can be seen that the users of Russian-speaking social media discuss events in the east of Ukraine most of all. Since Russia's conflictual relations with the West are often discussed in the context of sanctions and the Ukrainian conflict in general, it can be considered that four of the five most notable topics affect this issue. As for the least represented topics, most of them are difficult to interpret.

The sentiment index for one topic was calculated in following way:

$$topic_t = \frac{\sum_{d=1}^{D}(propability_{td} \times sentiment_d)}{\sum_{d=1}^{D} propability_{td}},$$

where D — total number of documents, $propability_{td}$ — probability of document d in topic t, $sentiment_d$ — sentiment score of text d.

To get sentiment score for each text we employed SentiStrength software with LINIS Crowd sentiment lexicon [16]. SentiStrength ascribes two scores to each text — on negative and positive scales. This approach shows optimal results for short texts like tweets [22], that are close in length to posts from social media we seek to analyze. The overall sentiment score of each comment was calculated as a sum of the negative and the positive scores. Since the distribution of the sentiment index is close to Gaussian, we calculated z-scores to determine significance of the difference in sentiment between the topics. The analysis shows three significantly negative topics, two of them turned out to be about ethnicity (#29 and #59) and the last one is uninterpretable. The first of these two was formed around texts on Uzbek topic. The central ethnicities of the second topic are Armenians, Turks, Azerbaijanis, it highlights the conflicts (1) between Turks and Armenians in the context of the event known as Armenian Genocide[4] and (2) between Armenians and Azerbaijanis over the territories of Nagorno-Karabakh.

The next step was to identify ethnic groups most often found in topics about ethnicity. In order to know how widely ethnicity e is presented in topic t, we summarized the probabilities of the topic t in all documents in which ethnicity e occurs and divided the obtained number by the number of documents in which this ethnicity occurs and by the sum of the probabilities of topic t in all documents. The next step was to scale the values obtained on the z-scale which allows us to make well-founded conclusions about the significance of the expression of ethnic category e in the topic t, which, in turn, can be used to test a wide variety of hypotheses.

According to many works [8, 13], the media are a powerful source of racist discourse. In Russian media, peoples of the North Caucasus those who are often influenced by that kind of discourse [1, 14, 15, 19]. Given the topic of crime (#14) among the topics received, we can determine which ethnicities are more often mentioned in the context of communications on this topic, thus quantitatively testing the hypothesis of bias in the media, particularly to the peoples of the North Caucasus.

The most probable words for this topic are: "murder", "time", "business", "group", "detain", "crime", "police", "find", "killed", "employee", "day", "house", "court", "perish", "district", "place", "name", "get", "happen", "city", "dead", "death", "police", "kill", "February", "camp", "prison", "chief", "criminal". It can be concluded that this topic clearly refers to crimes and their consequences. Among these words, however, we do not meet the one that belongs to any ethnic category, which so far does not allow us

---

[4] Turkey denies the word "genocide" is an accurate term for these events.

to conclude that there is bias in the media. Let's take the next step and use the metric we calculated to see which ethnic categories are significantly more characteristic of the topic than the others (Table 3).

Table 3. Ethnicities in the topic "Crimes and Murders"

Ethnicity	Z-score
Ingush	2,0683
Mansi	1,9923
Kavkazec (Peoples of the Caucasus)	1,8843

Z-value equal to 1.8843 corresponds to a significance level of 0.06. So, we can argue, the topic "Crimes and Murders" is not an ethnically neutral topic. It is significantly more probable for texts in which three terms are mentioned: "Ingush", "Mansi" and "Caucasian", two of them — "Ingush" and "Caucasian" — belong to the peoples of the North Caucasus region. These results confirm the conclusions made in the above-mentioned works on the racial bias in the coverage of crimes, and are consistent with the results of the polls, which showed that most people, who believe into connection between crimes and ethnicity, consider peoples of North Caucasus to be responsible for these crimes [27].

Confirming the existence of a topic that speaks of ethnicities in the context of committed crimes, we trace the dynamics of its representation in the texts of our corps. So, we can identify significant outbursts of that topic in social media, which can show what kind of events excited the audience. To do this, we divide all the texts into intervals of a long week and calculate the average probability of the topic in the texts at each of the intervals. Since the values obtained are normally distributed, it is possible to scale them according to the Z-scale.

On the resulting graph (Fig. 1), no trend can be traced, fluctuations are seen throughout the entire time interval, the interpretation is difficult. There are two intervals when salience of the topic #14 significantly increases. The first is at the beginning of 2014, and the second one is on the 20th of July 2014. Monitoring media during this

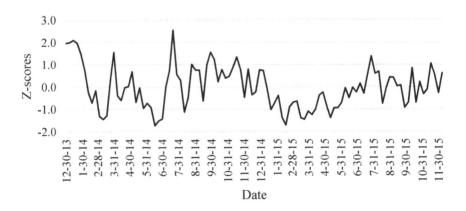

Fig. 1. Topic "Crimes and Murders"

period of time we have found several events that occurred in the second half of July and could cause a considerable discussion of the topic of crimes. The first event is a catastrophe in the Moscow metropolitan on July 15, the terrorist nature of which, although not confirmed, was actively discussed in social media; the second is the airliner catastrophe over the Donetsk region, also discussed in terms of crime, especially in Ukrainian media [25].

The topic "Peoples of the North Caucasus, Islam, terrorism" leads us to similar conclusions about the bias in social media towards the peoples of the North Caucasus. Let's look at the most likely words for her: "Chechnya", "Chechen", "Dagestan", "Muslim", "Kadyrov", "terrorist", "militant", "Caucasian", "Islam", "threatening", "Islamic", "republic", "Allah", "Muslim", "Ramzan", "Mansur", "Eivaz", "mosque", "brother", "war", "Ali", "imam", "IS", "Arab", "Russia", "Makhachkala".

There are three different topics actually mixed in one. The first one tells about the peoples and republics of the North Caucasus, the second is about terrorism and militants, and the third is about Islam. This neighborhood testifies that in Russian-language social media these three phenomena are closely related: both people from the Caucasus and the Islamic religion are perceived in the context of terrorism. In the eyes of the social media audience, terrorism acquires, therefore, ethnic and religious traits. The list of ethnic categories characteristic of the topic (Lezgins, Dargins, Avars, Kumyk, Chechens, Vainakhs, Dagestanians) only confirms these conclusions.

An analysis of topic salience dynamic shows one period, which falls to mid-November 2015, when the salience of the terrorist topic has increased significantly. This splash in the discussion on terrorism is easily explained, although the explanation is in no way connected with the peoples of the North Caucasus: on November 13, 2015, a series of major terrorist attacks took place in Paris, in which more than a hundred people were killed. Such a resonant event caused a noticeable discussion in the media, which was reflected in the graph (Fig. 2).

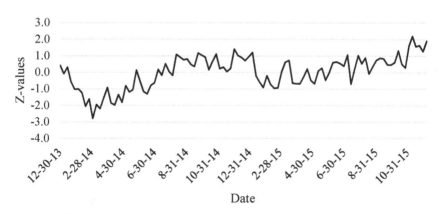

Fig. 2. Topic #39 "Peoples of the North Caucasus, Islam, terrorism"

6   Conclusion

In this work we explore topic structure of ethnic discussions in Russian-language social media. Despite the considerable number of works on ethnicity, since most of them focus their analysis on separate ethnic diasporas it is difficult to find large-scale studies devoted to ethnic discourse in a certain language and establishing relations between all major ethnic groups living on the territory inhabited by native speakers of the given language. Due to the rich data set of 2,659,849 texts from major Russian social media sites containing at least one of the 97 post-Soviet ethnic categories, this research claims to give comprehensive description of the ethnic discourse that was taking place in Russian-speaking social media from 2014 to 2015.

We have found that the topic structure of the ethnic discussions differs markedly from the topic structure of a typical social media site, which is the social network site Vkontakte. Texts in which ethnonyms are present contain more topics related to actual social and political issues, while topics of everyday life activities predominate in the "natural" topic profile built on randomly selected texts from Vkontakte social networking site. The revealed difference indicates that ethnic discussions is a problematic field, on which there is acute and often conflictual communication. It has the potential to grow into activities for constructing social problems as Spector and Kitsuse said [20].

In the topic structure of social media texts, we also identified the most salient and most negative topics. It is noteworthy that the largest share in the ethnic discussions is occupied by topics about Ukraine-Russia relations over the recent conflict between two countries. In this case, it is difficult to separate ethnic topics from political ones since the conflict between states is projected into the sphere of inter-ethnic relations. As for the significantly negative topics, we managed to find two negative and interpreted topics, the first is formed around Uzbek ethnicity and the second is about the Turkish-Armenian relations in the contexts of Armenian Genocide.

Acknowledgements. This article is an output of a research project implemented as part of the Basic Research Program at the National Research University Higher School of Economics (HSE) in 2018.

References

1. Ahmetieva, V.: Lyudi s pyos'imi golovami: obraz chechencev v rossijskih SMI [People with their heads in their heads: the image of Chechens in the Russian media]. In: Verkhovitsky, A. (ed.) The Language of Enmity Against Society [The language of enmity against society], p. 259. Sova, Moscow (2007). ISBN 5-98418-008-1 (in Russian)
2. Alexandrov, D., et al.: Virtual caucasus on VK social networking site. In: Proceedings of the 8th ACM Conference on Web Science, pp. 215–217. ACM, New York (2016). https://doi.org/10.1145/2908131.2908205
3. Apishev, M., et al.: Mining ethnic content online with additively regularized topic models. Comput. Sist. **20**(3), 387–403 (2016). https://doi.org/10.13053/CyS-20-3-2473

4. Bessudnov, A.: Ethnic hierarchy and public attitudes towards immigrants in Russia. Eur. Sociol. Rev. jcw002 (2016). https://doi.org/10.1093/esr/jcw002
5. Bodrunova, S., et al.: Are "Migrants" all the same? Mapping attitudes to the resettlers from Post-Soviet South in the Russian Blogosphere. Eur. Consort. Polit. Res. (2016)
6. Chan, J., et al.: The internet and racial hate crime: offline spillovers from online access. MIS Q. Manag. Inf. Syst. **40**(2), 381–403 (2016). https://doi.org/10.2139/ssrn.2335637
7. Davidson, T., et al.: Automated Hate Speech Detection and the Problem of Offensive Language. arXiv:1703.04009 Cs (2017)
8. van Dijk, T.A.: Prejudice in Discourse: An Analysis of Ethnic Prejudice in Cognition and Conversation. John Benjamins Publishing, New York (1984). ISBN 978-90-272-8003-9
9. Fadeev, P.: Etnicheskie gruppy Sankt-Peterburga v predstavlenii SMI [Ethnic groups of St. Petersburg in the media]. Mir Ross. World Russ. 1 (2017)
10. Foxall, A.: Ethnic Relations in Post-Soviet Russia: Russians and Non-Russians in the North Caucasus. Routledge, New York (2014). ISBN 978-0415833691
11. Gibson, S., Lando, A.L.: Impact of Communication and the Media on Ethnic Conflict. IGI Global (2001). ISBN 978-1-4666-9728-7
12. Grasmuck, S., et al.: Ethno-racial identity displays on Facebook. J. Comput.-Mediat. Commun. **15**(1), 158–188 (2009). https://doi.org/10.1111/j.1083-6101.2009.01498.x
13. Hartmann, P., Husband, C.: Racism and the Mass Media (1974). ISBN 978-0706700930
14. Ivanovich, A., Kovaleva, A.: Rasistskij diskurs ehlektronnyh SMI kak faktor konstruirovaniya mezhehtnicheskih konfliktov [Racist discourse of electronic media as a factor in constructing interethnic conflicts]. Upr. Konsult. Manag. Consult. **1**(73) (2015). ISSN: 1726-1139 (in Russian)
15. Karpenko, O.: YAzykovye igry s "gostyami s yuga": "kavkazcy" v rossijskoj demokraticheskoj presse 1997-1999 godov [Language games with "guests from the south": "Caucasians" in the Russian democratic press of 1997-1999]. In: Mul'tikul'turalizm i transformaciya postsovetskih obshchestv [Multiculturalism and Transformation of Post-Soviet Societies], pp. 183–188 ИЭА РАН, Москва (2002). (in Russian)
16. Koltcov, S.N., et al.: An opinion word Lexicon and a training dataset for Russian sentiment analysis of social media. In: Computational Linguistics and Intellectual Technologies, Moscow, pp. 277–287 (2016)
17. Nikolenko, S.I., et al.: Topic modelling for qualitative studies. J. Inf. Sci. **43**(1), 88–102 (2017). https://doi.org/10.1177/0165551515617393
18. Rykov, Y., Nagornyy, O., Koltsova, O.: Digital inequality in Russia through the use of a social network site: a cross-regional comparison. In: Alexandrov, Daniel A., Boukhanovsky, A.V., Chugunov, A.V., Kabanov, Y., Koltsova, O. (eds.) DTGS 2017. CCIS, vol. 745, pp. 70–83. Springer, Cham (2017). https://doi.org/10.1007/978-3-319-69784-0_6
19. Shnirel'man, V.: SMI, « ehtnicheskaya prestupnost' » i migrantofobiya [Media, "ethnic crime" and migrantophobia]. In: Verhovickij, A. (ed.) YAzyk vrazhdy protiv obshchestva [The language of enmity against society], p. 259. Sova, Moskow (2007). ISBN: 5-98418-008-1 (in Russian)
20. Spector, M., Kitsuse, J.: Constructing Social Problems. Cummings, Menlo Park (1977). ISBN-13: 978-0765807168
21. Suleymanova, D.: Tatar Groups in Vkontakte: the interplay between ethnic and virtual identities on social networking sites. Digit. Icons Stud. Russ. Eurasian Cent. Eur. New Media. **1**(2), 37–55 (2009). ISSN: 2040-462X
22. Thelwall, M., et al.: Sentiment in short strength detection informal text. J. Am. Soc. Inf. Sci. Technol. **61**(12), 2544–2558 (2010). https://doi.org/10.1002/asi.21416
23. Wirth, L.: Urbanism as a way of life. Am. J. Sociol. **44**(1), 1–24 (1938)

24. Xiang, G., et al.: Detecting offensive tweets via topical feature discovery over a large scale twitter corpus. In: Proceedings of the 21st ACM International Conference on Information and Knowledge Management, pp. 1980–1984. ACM, New York (2012). https://doi.org/10.1145/2396761.2398556
25. Bandits from DNR shot down the Boeing 777 in the Donetsk region! (conversation) (in Russian). https://www.youtube.com/watch?v=-LyD6FgHE8I
26. Razvitie interneta v regionakh Rossii [The development of the Internet in the regions of Russia] (2015) (in Russian). https://yandex.ru/company/researches/2015/ya_internet_regions_2015
27. Ukazyvat li natsionalnost prestupnikov v SMI [Should I specify the nationality of criminals in the media?] (in Russian), http://fom.ru/Bezopasnost-i-pravo/10586

Emotional Geography of St. Petersburg: Detecting Emotional Perception of the City Space

Aleksandra Nenko[✉] and Marina Petrova

Institute for Design and Urban Studies ITMO University, Saint Petersburg, Russia
al.nenko@gmail.com

Abstract. Emotional perception of the city space has a great share in subjective well-being and is one of the core subjective indicators of the quality of urban environment. Studies of emotional response towards the city space have recently gained popularity within digital humanities. In the paper we present a new system which allows collecting data on urban emotions - an interactive platform called *Imprecity*, which has been recently developed at ITMO University and integrated into a wider framework of *Smart Saint-Petersburg* project supported by city administration of Russian city Saint-Petersburg. When authorized through social networks *Imprecity* user receives a possibility to place emoji on St.Petersburg map as well as write comments on each emotion. Emotions are divided into 5 groups based on the typology of basic emotions defined by Paul Ekman - joy, sadness, anger, disgust, and fear. *Imprecity* functions as a mobile and desktop version of a website and will be further developed as a mobile app. The emotions and comments collected from users are processed to form recommendations for placemaking, moreover, active users of *Imprecity* have a possibility to unite together and propose projects for renovation of specific urban places with the help of experts. We consider methodological difference between studying emotional perception by processing spontaneous data generated by users online and study of emotionally loaded data created by users deliberately via *Imprecity*. We show visual analytical tools to process a test sample of data collected via Imprecity, such as emotional heatmaps, emotional ratings and word clouds. Analysis of data collected with *Imprecity* shows that users tend to express more joy than negative emotions; positive emotions tend to cluster close to the main points of attraction and major touristic routes. All types of emotions tend to cluster along the major mobility routes, in the city centre as well as in the sleeping quarters.

Keywords: Emotional perception · Social media · Urban environment
Interactive digital interface

1 Emotional Geography: Localization of Emotions

Louis Wirth has noted that urbanism is a way of life and that the city is an environment where a human being is immersed into a continuity of intensive and diverse encounters [1]. Since Wirth and his companions at Chicago School have elaborated an emphasis on human ecology, the main object of the urban studies is the human-environment interaction and emotions are the most visible marker of this relationship.

© Springer Nature Switzerland AG 2018
D. A. Alexandrov et al. (Eds.): DTGS 2018, CCIS 859, pp. 95–110, 2018.
https://doi.org/10.1007/978-3-030-02846-6_8

It has been long renowned that impressiveness of material forms, lines and planes, space and volume provoke emotional response in a human being. While the city while is a constellation of visual signs, sounds, scents, and tactile senses it is a context of rich phenomenological experience. Maurice Merleau-Ponty has proved the impact of architecture and material design on human body and soul [2]. In Lynch's *Image of the City* emotions are demonstrated as an invaluable constituent of mental maps of the city which become the actual basis for urban identity of its citizens [3]. Emotions are provoked not only by architecture and material forms, but also by almost irrational substance - atmosphere of the place. This atmosphere is often suppressed by violent powers of modernized cities which pervert natural feelings of people. As G. Simmel showed in "The Metropolis and Mental Life", written in 1903 and still relevant for our times, dynamics of urban life, never-ending interactions with human and non-human actors in the city provoke abundance of feeling [4]. Coping with this abundance lead to alienation of the 'urbanized' body from surrounding urban processes, as was depicted vividly in Simmel's term of 'alienated gaze'. At the same time environmental psychologists of today consider emotions and moods felt every day in urban space as prerequisites of the quality of life in the city. Normalizing urban-human interaction and returning human scale back into urban space has become the biggest endeavour of the new urbanism school, which fought loss of intimacy in human life in the world of skyscrapers and highways [5]. 'Urban' emotions burst out in the intensive dynamics of the contemporary city - informational turnover, transport and pedestrian movements, circulation of goods and services, communication processes and technological progress forward challenges for human perception. Intensity of the city provokes emotional intensity in the human being. After John Urry, who has proclaimed mobility as a paradigm of the contemporary city [6], we argue that the paradigm of the modern citizen is feeling.

The city is not a unity of one emotion or feeling only. Its different units have its own atmospheres, which was remarkably proved by situationist movement in Paris in 1960s. Pioneers of psychogeography Guy Debord and his fellows explored urban environments through a playful *drifting (dérive)*. Psychogeography was defined in 1955 by Debord as "the study of the precise laws and specific effects of the geographical environment, consciously organized or not, on the emotions and behavior of individuals" [7], and as "a whole toy box full of playful, inventive strategies for exploring cities… just about anything that takes pedestrians off their predictable paths and jolts them into a new awareness of the urban landscape [8]. In Debord's *Psychogeographic Guide of Paris* (1957) the city is presented through a new partition formed out of spontaneity of the dérive [9]. The map of Paris is cut up in different areas that are experienced by some people as distinct unities (neighbourhoods). By wandering, letting oneself float or drift each person can discover his or her own ambient unities of a specific city. The mentally felt distance between these areas are visualized by spreading out the pieces of the cut up map.

Sustainable human-urban development requires from researchers to investigate the urban emotions in more detail. The frontier for this research nowadays is urban data analysis: each day users of social media, information sharing platforms, micro-blogging create enormous volume of geolocated data loaded with their evaluations of and emotions about the city space. Another line of cutting edge research and development is online services and mobile applications for emotional analysis and sharing.

Recent research is trying to connect analysis of emotions with rich data coming from social networks and generated online by the users. Researchers have shown that the mobile data can used to extract emotional state of the user [10] and mobile phone usage patterns can help to predict negative emotion [11]. Researchers are often creating their own systems for extracting emotions and visualizing them on maps based on geotagged social media data, such as Twitter posts [12]. Some well-known works include those by L. Manovich and his team in frames of *Selfiecity* project, analyzing 3,200 selfies shared via Instagram from five global cities: Bangkok, Berlin, Moscow, New York, and Sao Paulo. According to results Manovich and Tifentale (2014) prove that Bangkok and Sao Paulo smile more - in Bangkok 0.68 average smile score and Sao Paulo - 0.64, while people taking selfies in Moscow smile the least - only 0.53 on the smile score scale [13]. Quercia et al. (2014) consider emotions which emerge from visual features of London neighborhoods. In their crowdsourced project they have collected votes from 3.3 K individuals on photographs representing different street landscapes, which let them define the features of the streets that are perceived as beautiful, quiet, and happy. Thus the amount of greenery is associated with each of three qualities, while broad streets, fortress-like buildings, and council houses tend to be associated with ugliness, noise, and unhappiness [14].

2   Emotional Ambiguity of Social Media Data

Social media universe is already enormous and is constantly growing and evolving taking on different shapes of human behaviour. Globally, more than 2.8 billion people or 37% of the world's population use social media on everyday life basis [15]. Personal profile in social media is becoming a digital avatar of a person, and her movements through the city space tracked through check-ins and geolocated posts become her digital footprint. According to Brand Analytics, on May 2017 the number of social media users was 38 million person, together have generated 670 million messages. In May 7 143 thousand new users have registered in Instagram, 1 171 thousand new authors have published 78 372 thousand tweets in Twitter, 1 953 thousand new authors have sent 53 413 thousand public messages in Facebook. For the same time period 25 722 thousand users of the most popular Russian social network VKontakte have sent more than 310 795 thousand messages [16]. The growing trend is "mobilization" of the social media, i.e. active users prefer to use applications for smartphones not their desktop versions.

Spontaneous user-generated data provides a number of advantageous features for the study of emotions in the city space:

1. *Coverage and volume.* Social media data is fastly growing while more and more users become active with their personal computers and smartphones, and more and more people share their emotions online [17].
2. *Details.* User-generated data provides an account on urban behaviour and its emotional aspects on individual level. For example, urban data allows to track routes of individual mobility in the city, use of different urban facilities, attitudes and

preferences towards city space as well as social and demographic details of a certain user.

3. *Expressivity.* Almost every behaviour in social media is emotionally loaded while likes, check-ins, shares, recommendations, comments, emoticons reflect certain emotions and moods expressed towards urban space as such or specific places and venues.

4. *Richness.* Social media provides extensive information on different aspects of urban behaviours: mobility, space use (types of on-site activity), attitudes towards space. The range of different spaces is not limited and covers formal and informal environments, venues for work and leisure, third places and transit spaces, etc.

5. *Availability.* Social media data is stored on servers and in online archives and is often available for researchers [17].

Thanks to these advantages, emotional analysis of big urban data is rapidly developing along several methodological lines:

1. Analysis of semantic data
   a. Analysis of texts of commentaries. Tonal and sentimental analysis of text is used to define emotional load, which combine tasks of identifying sentiment expressions and determining the polarity or *valence* of the expressed sentiment [18]. One of the main drawbacks of sentiment analysis is low sensitivity to detecting different emotions, in the majority of cases analysis results in defining positive, negative or neutral emotional load of the text. Some of the authors identify a broader range of sentiment classes expressing various emotions such as happiness, sadness, boredom, fear, in addition to positive or negative evaluations [19]. Works on sentiment analysis focus both on full text as well as phrasal and sentence level [20, 21]. Analysis of semantic data can be upgraded with the help of machine learning, which can create a more finely tuned tool to recognize emotions in a specific dataset. Another way to elaborate textual analysis is multilayer analysis of the texts: quantitative analysis through sentimental analysis and qualitative analysis through discourse analysis of emotional expressions.
   b. Analysis of hashtags. Hashtags as highly suggestive and expressive bits of information can be parsed and considered separately along with other informational layers of the posts. Hashtags created by users themselves strongly vary, so it is recommended to use only frequent hashtags which appear in a significant dataset of different posts. Sentimental load for hashtags and smileys can be manually annotated by human judges, in particular, by Amazon Mechanical Turk service subjects. Analysis of hashtags is quite "noisy" and should be compared with other items, for example, Davidov et al. (2010) present multi-class analysis of hashtags together with analysis of smileys retrieved from Twitter [22].
2. Analysis of visual data
   a. Analysis of facial expressions. Widescale analysis of selfies taken with Instagram was conducted in frames of the *Selfiecity* project which resulted in development of an interactive interface for navigating in New York [23]. While moving through the streets the user can see a set of photographs of the local popular and unique places taken by other users. Analysis of facial expressions is helpful for

preventing violence and aggression at certain urban settings, such as stadiums or during certain events, such as Olympic games [24].

b. Analysis of smileys, emoticons and emoji. Another set of data is provided by icons expressing emotions in the text. These icons are much used in social media to eliminate the gap between online and offline communication and enhance understanding and interpretation of the text by the communication counterparts. Smileys might provide better results than semantic analysis or analysis of hash-tags while they are originally limited and assigned particular meanings before-hand [22].

Though social media have become a powerful data generator for studying subjective perception of the city, analysis of urban emotions based on this data has certain draw-backs:

1. *Loss of depth due to online anonymity.* Online communication is different from face-to-face since non-verbal communication is absent. Emotions are a major part of nonverbal communication and are most effectively translated through its mecha-nisms, such as the vocal tonality, intensity and volume, facial expressions, tactile feelings, etc. While emotions are expressed online in different ways - semantically through specific words or hashtags and visually through emoticons and photographs, detection of emotions through procession of the texts, photos or emoticons does not give the precise picture of face-to-face interviews, observations or other experiential methods.

2. *Loss of truth due to conspicuous behaviour.* Alike T. Weblen's famous notion on "conspicuous consumption" [25] behaviours online could be named conspicuous or demonstrative, oriented towards visibility of the social status, political or cultural attitudes, fascination with fashion. The perfect framing of the social media of all types leads to concentrated visibility and specific rules of online conspicuous behav-iour (considerate, socially conformal, goal and success-oriented).

3. *Loss of representation due to lower scope.* The major age of the Internet audience as for now is below 50 years, the average age has not yet reached 45 years. Different social media are specific for different age groups, for example, Facebook is more popular within older people than other social media, age group of 45–55 years old makes up 15% of its users, 55 and above –9% [16].

4. *Loss of interpretative power due to fuzziness of subjective indicators.* Concentration of the emotions of a specific kind in a certain location might not have a direct connection to the place and its environmental features due to a multitude of other reasons, such as personal psychological state of a user, so they can't be regarded as firm explanatory variables or predictors of the emotions expressed in social media.

These considerations lead to an understanding that though social media is a fine resource for emotional analysis of the city, there are considerable drawbacks. The latter might be reduced if analysis of spontaneous social media data is conducted alongside with analysis of deliberately expressed emotions. Luckily emotions are the hot topic for many users and researchers, hence emotional sharing has become the core of many specialized online platforms and mobile-based applications. Emotional data generated in them is framed for emotional mapping, emotional sharing and even creating connections with people experiencing similar emotions. These data can be used together with emotional analysis of social media data for comparative analysis, which gives more reliable results.

3   Digital Solutions for Urban Emotions

Regarding viability of emotions many online services and apps focus on creating opportunity for people to share emotions which emerge in the course of their everyday life, in particular, while interacting with the city space. Such services, as foursquare[1], flamp[2], 2GIS[3], Local experts by Google maps[4] give possibility to check-in, leave impressions and rate venues and places. Online tools and apps create different engaging mechanisms to keep users connected with the app, for example, customized suggestions on the places to visit formed by built-in algorithms along with the user estimated preferences in price, quality and type of products and services. The multitude of online platforms and mobile-based apps prioritizing emotional perception of the city constitute more or less definitive groups.

Services with diagnostics of personal emotions. Great share of existing applications are focused on measurement of emotional state of human body based on physical parameters and opt for elimination of risk of psychological disease, in particular, during the process of movement through the city. One of the examples is Emotionsense is a platform created by Cambridge University scholars to trace the emotional state of the person by processing audio data collected by personal smartphone, in particular, phone calls and audio chats [26]. To detect emotions - happiness, sadness, fear, anger or neutral reaction - the application analyzes laughter, pauses and voice timbre of the speech. Spatial emotional awareness[5] (SEA) provides a portable electronic wearable which allows a person to move throughout space based on locational proximity of people who can elicit good feelings [27]. As users navigates through the city space, portable device shows them the warmth coming from friends or supportive people and cold coming from people better to avoid.

Services creating emotional maps. Some of the services are more oriented towards mapping emotions in connection to certain locations and routes a person chooses in the city. These services are valuable for place-making, geo-marketing and tourism spheres because they help attracting people to certain places in the city. One of the examples, *Emotion Map* is a multifunctional application which gives users an opportunity to record and monitor emotions in relation to their activities and locations. In the app users place marks at locations they visit, choose their emotional label and label of their activities in that places. Marks can be shared with friends or with all app users. While travelling through the city users can check the location for interesting emotional marks created by their friends, all app users or themselves. Users can chat with each other and also receive

---

[1]   https://foursquare.com/.
[2]   https://spb.flamp.ru/.
[3]   https://2gis.ru/.
[4]   https://maps.google.com/localguides.
[5]   http://emotionsense.org/.

daily reports showing personal emotional statistics [28]. *Bio-mapping* project by artist Christopher Nold (also see his book *Emotional Cartography*), though not an app, is a catchy example of emotional maps creation with the help of volunteers wired up with GPS and polygraph technology who wander around a neighbourhood area, noting feelings and reactions to their surroundings [29, 30]. Querias et al. have created a crowdsourcing platform to prove that there is more variability in ways which people prefer to take than just the shortest one, such as beautiful, silent and happy ones.

Services creating emotional communities. *WiMo* is a mobile app that allows people to share their emotions about specific places and store their emotional feelings about places. *WiMo* also creates a social network based on common interests and enables users to share opinions, experiences and passions about urban places [31]. *Tillbaka* app is a location-based storytelling and story experiencing system for web-enabled mobile phones. The system is based on a novel concept of pervasive play where stories emerge and develop on several dimensions, in particular, geographic one. The app was created to empower youth in a largely immigrant and lower-income neighborhood to share memories, feelings, and attitudes and engage in civic discussions about the neighborhood.

4   Imprecity: Interactive Platform for Mapping Emotions

Imprecity interactive platform was developed for analysis of subjective perception of the urban environment (the name combines words "impression" and "city")[6]. Imprecity is designed to combine focuses and advantages of platforms and apps described above. On Imprecity map its authorized users - citizens or city guests - can leave emoji which symbolize 5 basic emotions about urban places and comments explaining these feelings. 5 basic emotions are defined along with Paul Ekman's theory and are namely joy, sadness, anger, disgust, and fear [32]. Collected 'soft' data is integrated into a coherent algorithm of expert-citizen interaction. Imprecity is one of the projects in frames of St. Petersburg Smart City programme supported by the city administration. One of its focuses is elaboration of the 'smart citizen' concept and creating various tools to receive more feedback from the city dwellers and enhance their participation in urban development processes. Imprecity is tested and oriented towards St. Petersburg users, but can be scaled for other cities.

Imprecity is developed to engage multiple stakeholders: citizens can share their impressions and discuss urban places, they can also form groups of action aimed at upgrading places they love or vice versa places they find the worst and which hamper their emotional state. Experts in urban studies and urban development receive a unique subjectively loaded data which help them analyze citizen's perception of the city space. For the city administration emotional data is valuable to get an overview of citizen's subjective well-being and to define lists and ratings of places which need to be recovered in frames of city-led projects of urban regeneration, or saved as they are from developmental pursuits.

---

[6] Imprecity website - www.imprecity.ru.

Engaging with Imprecity has three stages: (1) Mapping emotions (Fig. 1), (2) Forming community and (3) Taking action. At the 1st stage user can authorize through social networks (facebook, Vkontakte and Google+), put emoji and comments on the map, see the heatmaps for each emotion and a general one (Fig. 2). Mapping emotions stage is inheriting services and apps which are creating emotional maps. In nearest future Imprecity will provide its users with illustrated analytics about emotions of St.Petersburg, as well as comments and recommendations from experts about certain places, suggested places and routes. It is also planned to integrated personal statistics toolkit for the user, to take advantage of personal diagnostics of user emotions. Stages 2 and 3 are recently under development. It is planned that users which will be similar in emotional attitudes will be highlighted and suggested to each other. Authorized users will have a possibility to see their friends emotions. Imprecity will provide a chat function. Thus it is planned to create framework for forming of groups of people sharing common emotional experience. A group of people will be provided a possibility to suggest an idea for improvement of a certain place and submit it through Smart City St. Petersburg portal, where it will be considered by Program Office consisting of experts and city officials. The 'project group' will be provided with methodological tips, expert consultation and also information about different possibilities for citizen participation in St.Petersburg.

a.   b.

Fig. 1. Imprecity mobile version. User Interface. (a) Starting page (b) Emotional heatmap for sadness (turquoise).

a.   b.

Fig. 2. Imprecity mobile version. Authorized user. (a-b) Choosing emoji from the list and placing it on the map. The map is filled with personal emoji: joy (yellow), anger (red), sadness (turquoise), fear (blue) and disgust (violet). (Color figure online)

Such an algorithm differs Imprecity from other apps alike: those, which allow to share emotions about the city space, but do not presuppose further socially meaningful actions (like Foursquare, Ushahidi[7]); those, which support community formation, but do not link them with the experts community and city administration (like WiMo and Tillbaka discussed above, as well as Russian-based apps Locolo[8] and Kto Vokrug[9] (Who's around)); those providing opportunity to report the problems, but not create ideas for urban renewal (like St.Petersburg-based platforms citizen-led one - Krasiviy Peterburg[10] (Beautiful Petersburg) and governmentally supported one - Nash Peterburg[11] (Our Petersburg)); those which allow to vote for urban problems to be solved, but do not foresee community formation (like Moscow-based platforms supported by government Aktivniy Gorozhanin[12] (Active Citizen) and My Street[13] (Moya Ulitsa)).

5 Impressions of St.Petersburg: Analytical Tools and Test Results

Based on data generated through Imprecity we can use several analytical tools to visualize emotions of the city. Here we describe the tools and provide illustrative results on a pilot sample of data (400 emoji and comments from 15 users collected in September

---

[7] https://www.ushahidi.com/.

[8] https://locolo.me/.

[9] https://play.google.com/store/apps/details?id=com.whoisaround.android&hl=ru.

[10] красивыйпетербург.рф/.

[11] https://gorod.gov.spb.ru/.

[12] ag.u-ude.ru/.

[13] https://www.mos.ru/city/projects/mystreet/.

2018). The test sample was created predominantly by MA students of the Institute for Design and Urban Studies at ITMO University, St. Petersburg.

Emotional Heatmaps. Emotional diversity and richness is an indicator of the usability of the space, of its ability to generate impressions at all. Urban places shouldn't be sterile or only filled with emotions of joy - it would signify artificial situation when other 'negative' emotions are controlled or suppressed. Interpretation of the emotional maps can be given along with specific environmental features of the given places which were emotionally valued. Imprecity test heatmaps show that users share joy much more than any other emotion, as well as it is also the most disperse and covers the whole of the city, even the Kronstadt island, however the biggest hotbed of joy is in the central city areas and locates close to the central streets with beautiful architecture, cafes and bars, and vibrant public life (Fig. 3a). Concentration of other emotions is also biggest in the central city area, showing that along with good feelings space properties here elicit bad feelings. There are also some regularities, for example, sadness map shows a clot of sadness collocated with the former industrial area of the city, which is now mostly in ruined and abandoned state (Fig. 3c).

Emotional Ratings of Urban Place. Emotional ratings can be used as informative base for multiple tasks of urban planning and development, as well as different cultural and economic spheres. For venues and touristic sphere emotional ratings suggest choice of the best venues for creating touristic routes. For the urban planners and decision makers at various levels ratings hint the need to optimize conditions at specific venues. Ratings induce competitive thinking and thus stimulate development, though they should be always backed up with on-site analysis to receive deeper interpretation of the situation. Emotional ratings can be created not only for discrete points, but also for areas, for example, administrative city districts or neighborhoods.

The emotional ratings created for the test sample of Imprecity show that public spaces, such as squares, parks, embankments, are often rated both positively and negatively (Fig. 4). The most rated in joy is Aleksander Nevsky Square, a place favourite with citizens for walks in big historical cemeteries situated here (Fig. 4a). The most rated with anger is Zenit Arena Stadium, which was built for the Football Match 2018 and has become a site for massive corruption by the city administration and construction companies with many negative feedback many from the citizens (Fig. 4b).

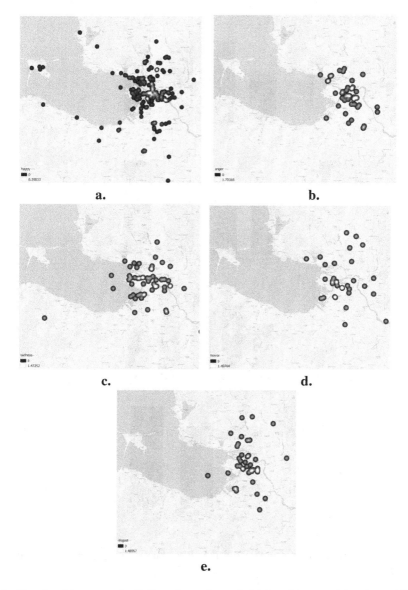

Fig. 3. Emotional heatmaps for St.Petersburg city scale (a) joy, (b) anger, (c) sadness, (d) fear, (e) disgust.

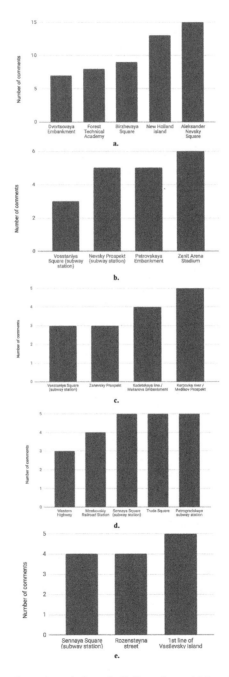

Fig. 4. Emotional ratings for top-rated places in St.Petersburg: (a) joy, (b) anger, (c) sadness, (d) fear, (e) disgust

Semantic Analysis of Comments. Along with emoji Imprecity generates its own semantic data through comments left by users when they place an emoji on a map. Semantic data can be subjected to tonal and sentimental analysis, which can be processed in different versions, for example, applying existing expert vocabularies of emotionally loaded semantics to user-generated comments or creating a subset of comments evaluated by experts and then using it for machine learning and consequent tonal analysis of the whole dataset. Comments given together with emoji serve as a source of verification for the conclusions made according to emotional heatmaps and ratings, as well as of unique stories which make understanding and interpretation of emotional values much more vivid. In particular, user stories hint the reasons why the emotion is formed or why it occurs in this place now. Here we show an example of a word cloud based on textual data from comments which are associated with the emoji of joy; the words are sized in accordance with the frequency of usage of the word. The most frequent words used are features of the places: adjectives and pronouns 'favourite', 'beautiful', 'good', 'cosy', 'cool', 'pleasant', 'public', 'free'; nouns signifying urban places: namely 'place', 'park', 'home', 'quiet', 'area', 'university'; nouns signifying properties of the place: 'panorama', 'musicians', architecture', 'greenery', 'landscape', 'coffee', 'cinema'; verbs signifying practices in the place: 'live', 'walk', 'skiing', 'biking'; nouns signifying diminutive for the names of the districts of the city, namely Petrogradskaya and Vasilevsky island (Vasechka and Petrogradochka) (Fig. 5).

Fig. 5. Word cloud with comments for happy marks

6   Conclusion

Analysis of emotional perception of the city space has long roots in different fields of research, and nowadays is a trendy area within data analysts as well as application developers. Emotional analysis is a way to improve quality of urban life and urban environment. Results coming from emotional analysis can be employed by urban planners, designers, as well as researchers to diagnose the 'subjective' problems of the city. However retrieval of subjective perceptions and emotional evaluations based on data coming from social media and other spontaneous data is still yet underexplored, and there are quite many drawbacks caused by the nature of data itself, as well as many methodological gaps, in particular, in analysis of particular emotions in connection to the city space, such as joy, anger, fear, etc. We suggest that analysis of social media data should be conducted together with data received deliberately from users with the help of services and apps, such as Imprecity.

Imprecity enables collecting a unique set of data to illustrate direct emotional attitude of people towards the urban environment. In this way Imprecity allows analyzing perceptions and impressions on particular urban places and support decision making with targeted interventions to improve the latter for the sake of human well-being. The data can be processed in a number of ways - emotional maps, emotional ratings of places and word clouds with verbalized features of places triggering particular emotions.

7   Discussion

We are designing Imprecity as an interactive platform which combines advantages of other services and apps, as well as delivers an algorithm for connecting major stake-holders of the city - citizens, experts, and city officials. However, Imprecity is a work in progress and there are some risks it can face. First risk is deficit of usability if the motivational system is not created correctly. Many of the applications alike do not become popular. We are planning to overcome this barrier through creating a system of engaging infographics and analytics on the city level, which will be publicized by our mass media partners, and on the personal level, by developing tools for personal emotional diagnostics for Imprecity users. Second risk is that users won't actively engage into creating content - sharing emotions and ideas - and will rather consume information. We are going to overcome this by creating a system of tokens or bonuses, such as discounts for cultural facilities and other venues, to praise active users. Third risk is failure of the coherent and sound relation between active citizens and city officials when it comes to the realization of project ideas created by initiative groups. We are planning to eliminate this risk by introducing a C2G mediator namely experts who consult and share a wide spectre of information on how citizens can participate in urban planning processes.

Acknowledgement. This work is financially supported by Ministry of Education and Science of the Russian Federation, Agreement #14.575.21.0165 (26/09/2017). Unique Identification RFMEFI57517X0165.

References

1. Wirth, L.: Urbanism as a way of life. Am. J. Sociol. **44**(1), 1–24 (1938)
2. Merleau-Ponty, M.: Phenomenology of Perception. Humanities Press, New York (1962). Routledge & Kegan Paul, London
3. Lynch, K.: The Image of the City. The MIT Press, New York (1960)
4. Simmel, G.: The metropolis and mental life. In: The Sociology of Georg Simmel. Free Press, New York (1976). (1903)
5. Gehl, J.: City for People. Island Press, Washington (2010)
6. Urry, J.: Mobilities. Polity Press, Cambridge (2007)
7. Debord, G.: Introduction to a Critique of Urban Geography. Les Lèvres Nues, Paris (1955)

8. Hart, J.: A New Way of Walking. Utne Reader, New York (2004)
9. Debord, G.: Psychogeographic Guide of Paris. Permild and Rosengreen, Denmark (1957)
10. Nielek, R., Wierzbicki, A.: Emotion aware mobile application. In: Pan, J.-S., Chen, S.-M., Nguyen, N.T. (eds.) ICCCI 2010. LNCS (LNAI), vol. 6422, pp. 122–131. Springer, Heidelberg (2010). https://doi.org/10.1007/978-3-642-16732-4_14
11. Hung, G.C.-L., Yang, P.-C., Chang, C.-C., Chiang, J.-H., Chen, Y.-Y.: Predicting negative emotions based on mobile phone usage patterns: an exploratory study. JMIR Res. Protoc. **5**, 3 (2016)
12. Guthier, B., Alharthi, R. Abaalkhail, R., El Saddik, A.: Detection and visualization of emotions in an affect-aware city. In: Proceedings of the 1st International Workshop on Emerging Multimedia Applications and Services for Smart Cities (EMASC 2014), pp. 23–28. ACM, New York (2014)
13. Quercia, D., O'Hare, N., Cramer, H.: Aesthetic capital: what makes london look beautiful, quiet, and happy? In: Proceedings of the 17th ACM Conference on Computer Supported Cooperative Work & Social Computing (CSCW 2014), pp. 945–955. ACM, New York (2014)
14. Tifentale, A., Manovich, L.: Selfiecity: exploring photography and self-fashioning in social media. In: Postdigital Aesthetics: Art, Computation and Design, pp. 109–122. Palgrave Macmillan, London (2015)
15. Gallagher, K.: The Social Media Demographics Report: Differences in age, gender, and income at the top platforms. http://www.businessinsider.com/the-social-media-demographics-report-2017-8. Accessed 20 Feb 2018
16. Brand Analytics: Social Networks in Russia, summer 2017: Numbers and Trends. http://blog.br-analytics.ru/sotsialnye-seti-v-rossii-leto-2017-tsifry-i-trendy/. Accessed 20 Feb 2018
17. Zhan, X., Ukkusuri, S.V., Zhu, F.: Inferring urban land use using large-scale social media check-in data. Netw. Spat. Econ. **14**(3–4), 647–667 (2014)
18. Riloff, E.: Learning extraction patterns for subjective expressions. In: Proceedings of the Conference on Empirical Methods in Natural Language Processing. Sapporo (2003)
19. Mihalcea, R., Liu, H.: A corpus-based approach to finding happiness. In: AAAI Symposium on Computational Approaches to Analysing Weblogs. AAAI Press, Palo Alto (2006)
20. Wilson, T., Wiebe, J., Hoffmann, P.: Recognizing contextual polarity in phrase-level sentiment analysis. In: Proceedings of the Human Language Technology Conference and Conference on Empirical Methods in Natural Language Processing. Vancouver (2005)
21. Titov, I., McDonald, R.: A joint model of text and aspect ratings for sentiment summarization. In: Proceedings of the Association for Computational Linguistics and Human Language Technology Conference. Columbus (2008)
22. Davidov, D., Tsur, O., Rappoport, A.: Enhanced sentiment learning using twitter hashtags and smileys In: Coling. Poster volume, pp. 241–249 (2010)
23. Tifentale, A., Manovich, L.: Selfiecity: exploring photography and self-fashioning in social media. In: Berry, D.M., Dieter, M. (eds.) Postdigital Aesthetics: Art, Computation and Design, pp. 109–122. Palgrave Macmillan, New York (2015)
24. Boychuk, V., Sukharev, K., Voloshin, D., Karbovskii, V.: An exploratory sentiment and facial expressions analysis of data from photo-sharing on social media: the case of football violence. Procedia Comput. Sci. **80**, 398–406 (2016)
25. Veblen, T.: The Theory of the Leisure Class: An Economic Study of Institutions. Penguin Books, New York (1994). [1899]

26. Rachuri, K.K., Musolesi, M., Mascolo, C., Rentfrow, P.J., Longworth, C., Aucinas, A.: Emotion sense: a mobile phones based adaptive platform for experimental social psychology research. In: Proceedings 12th International Conference on Ubiquitous Computing, pp. 281–290. Association of Computing Machinery (2010)

27. Kapelonis, C., Cassiano, L., Fuste, A., Ansu, L., Singh, N.: SEA: Spatial Emotional Awareness for Relationship Nurturing. http://augmentation.media.mit.edu/2017/05/19/sea-spatial-emotional-awareness-for-relationship-nurturing/. Accessed 20 Feb 2018

28. Emotion Map. https://apkpocket.pw/emotion-map/edu.syr.ischool.orange.emotionmap.apk. Accessed 20 Feb 2018

29. Bio Mapping - Emotion Mapping. http://biomapping.net. Accessed 20 Feb 2018

30. Emotional Cartography - Technologies of the Self. In: Nold, C. (ed.) (2009). www.emotionalcartography.net. Accessed 20 Feb 2018

31. Mody, R.N., Willis, K.S., Kerstein, R.: WiMo: location-based emotion tagging. In: Proceedings of the 8th International Conference on Mobile and Ubiquitous Multimedia. Cambridge (2009)

32. Ekman, P.: Basic Emotions. In: Dalgleish, T., Power, M. (eds.) Handbook of Cognition and Emotion, pp. 45–60. Wiley, Sussex (1999)

E-Humanities: Arts & Culture

Art Critics and Art Producers: Interaction Through the Text

Anastasiia Menshikova[1], Daria Maglevanaya[1], Margarita Kuleva[1（✉）],
Sofia Bogdanova[1], and Anton Alekseev[2]

[1] National Research University Higher School of Economics, St. Petersburg, Russia
{aamenshikova,dvmaglevanaya,sabogdanova}@edu.hse.ru, mkuleva@hse.ru
[2] St. Petersburg Department of V. A. Steklov Mathematical Institute,
St. Petersburg, Russia
anton.m.alexeyev@gmail.com

Abstract. As well as the most areas of social life, the field of art is now extended to the cyberspace. In this study, we analyze online reviews of Russian art critics with two objectives. On the one hand, we investigate the patterns of the interactions between critics and artists (both contemporary and recognized ones) in the Russian Art. Since the Russian school of art critique is still in the process of formation, an analysis of web data we offer a significant contribution to the scope of Russian Art studies. On the other hand, we use social network analysis and text mining tools in order to gain more insights from the data and affirm the applicability of the modern tools to the classic research tasks. In this study we analyze data from the 5 Russian art magazines, in particular articles, authors and named entities from this texts. As a result, we explored different patterns of the critics production that could divide this area of web interaction both by geographical and textual characteristics of agents and articles.

Keywords: Cultural production · Art critics
Social network analysis · Sentiment analysis · Text mining

1 Introduction

The field of art, including both those who create pieces of the art and those who have a need to consume it, has a way of constant changes, according to the transformations in social and political context over time. However, the changes are conditioned not only by an external context but equally by an interaction of the actors who play in the field of art in order to achieve the most prestigious and influential positions. Nowadays, with the spread of the Internet, this interplay, in terms of Bourdieu [1], partly has moved to the virtual space. There are both plethora of art-projects, performed on the web, and the participation of the various online media in the critique of works of art.

A. Alekseev—List of authors is in reverse alphabetical order.

D. A. Alexandrov et al. (Eds.): DTGS 2018, CCIS 859, pp. 113–124, 2018.
https://doi.org/10.1007/978-3-030-02846-6_9

In this paper, we investigate the reviews of art critics from St. Petersburg and Moscow, who discussed the events in the art world, such as public and private art exhibitions, art performances etc. The aim of our research is twofold. Firstly, we try to implement classical methods of text mining, including sentiment and semantic analysis, as a tool for the investigation of the specific texts. Obviously, critical reviews appear to be an uncommon text material, that is why we test the effectiveness of the method for web reviews to evaluate the sentiment in comparing with the manual one. We also check if the traditional predictors of sentiment are relevant for the assessment of specific for the web space textual data. Secondly, we study the relationship between artists and critics with the help of social network analysis. We are interested in the critics from the same city or publishing house tend to discuss the same artists. Additionally, we map the result of the sentiment analysis as the node attributes in order to evaluate the positive or negative evaluation and the position in the network.

2   Theoretical Background

2.1   Cultural Production Paradigm

In the middle of the 20th century, thanks to the scientific contribution of Richard Peterson, researchers were able to talk about the field of art as a sphere of cultural production, which showed its autonomy [12]. Peterson offers to study the processes in the art market, in which the following agents are involved: artists, critics, collectors, connoisseurs, spectators etc., who carry out their functions in the production of culture. Therefore, there is a need to analyze art as a system of agents interactions, for example, the artists as a derivate part. Following the logic of Peterson's theory, we believe that any participant of the cultural market perform a certain product: artists produce paintings, consumers produce opinions. However, the result of the activity of art critics (censors) - critical articles, texts published at different times arising on a par with the work of art is equally significant and partly determine the further success of the work. Critics were included in the art system much later than artists, together with art dealers [16], they attracted the interest of researchers later. The relevance of the work presented is mainly based on the fact that studies focusing on the role of the critic in cultural industries are not so widespread. However, those studies that designate criticism as the main object of interest are mostly localized in other industries, such as the film production [5], music [14], literature [11] and others [15]. In many areas critics perform certain functions, depending on the sphere, they could acting judges, experts, evaluators, analysts and professional reviewers [5]. It is useful to look at the assessment devices, understand, which mechanisms precede the result of criticism, identify the constructions, which describe the artistic product. Critics establish the quality of the product with the help of professional judgments, which is not a subjective assessment, but the statement based on the previous legitimations [11]. Contemporary criticism is divided into genres: news art journalism, scientific research (for example, a thesis by one artist), criticism as a dictionary for the layman - a critical article with an educational function, depending on the genre of the critic, an independent evaluation, assumes different roles: to compare, to open, to teach and to recommend [13].

2.2   Hypothesis

We put forward several hypotheses inspired by previous work on the work of art critics. Bourdieu suggested that all products created by artists base on the established norms and conventions, which implies an orientation toward the classical tradition of performance and the use of references to the classics [2]. Hence, we expect that art critics mention the authors of the classic works more frequently than contemporary artists, that is why recognized masters will connect the rest discussed artists in the network of art producers.

The second hypothesis is related to the evaluation of pieces of art: describing the mechanisms for recognizing the art product by critics, Kattani et al. emphasize that (1) actors on the periphery of the field of art production would rather be recognized by the critics community, while the actors in the core will receive recognition by their colleagues, and that (2) those critics that do not accept a new trend in art, risk losing their reputation if this new direction becomes popular [3]. Consequently, we expect to see the positive sentiment more frequently related to the contemporary artists.

Thirdly, we suppose that the manual marking is the more efficient tool than the common sentiment lexicon for the sentiment assessment. Additionally, we will try to assess what features of text are the good predictors of the phenomenon when the critical review has value judgements in it. Do "classical" predictors work here: punctuation marks, elongated consonants, pronouns, etc., or does professional lexics remove these differences? Does the publishing house, as the critics affiliation, influence the level of sentiment and personal judgment in his/her texts? Are some authors more prone to writing in a more positive/negative manner than others?

3   Methodology and Data

This study uses a networked semantic analysis of texts, which assumes that the language and knowledge can be modeled as a network of words, connected with certain relationships - ties. Since in the analysis it is important for the analysis not to focus on one author or criticism, but take into account all the judgments of critics (for the selected period) that assess all authors [10], 5 web resources about the art were chosen: "ArtGuide" "Colta", "Aroundart", "TheArtNewspaperRussia" and "Kommersant". The data was collected for a period of 6 months (2016–2017). Publications were selected by the following criteria: (1) the publication relates to the event of the Russian art field; (2) articles are published regularly (several times a month). All five editions' staff consists of the art critics both from St. Petersburg and Moscow. If the edition is not entirely dedicated to the art production, we extracted articles from the archive using only "art" section. What we also took into account: the events or objects relate directly or indirectly to the fine/applied/performance art were mentioned in the article. This implies that the dataset did not include reviews of theatrical productions or concerts, film and music festivals, and so on. We selected authors that write reviews regularly in the editions listed: the total number of characters in reviews per last six months for each of them was greater or equal to 30,000. Thus, the final data included 13 authors from 5 editions, which are listed in Table 1.

Table 1. Sample of art critics

Critics		
Media	Author	City
Artguide	Anna Matveeva	Saint Petersburg
Artguide	Ekaterina Allenova	Moscow
Artguide	Valentin Dyakonov	Moscow
Colta	Anastasia Semenovich	Saint Petersburg
Colta	Natalia Serkova	Moscow
Aroundart	Valeriy Ledenev	Moscow
Kommersant	Anna Tolstova	Saint Petersburg
Kommersant	Kira Dolinina	Saint Petersburg
Kommersant	Alexey Mokrousov	Moscow
Kommersant	Sergey Hodnev	Moscow
ArtNewspaper	Anastasia Petrakova	Moscow
ArtNewspaper	Vadim Mikhailov	Saint Petersburg
ArtNewspaper	Pavel Gerasimenko	Saint Petersburg

3.1   Named Entities Extraction and Their Context Evaluation

Lemmatization was the first stage of the work, with the use of Pymorphy Python tool [7] and the extraction of named entities mentioned by critics in their works. It was implemented using the Natasha Python library [8], where the extraction of personalities (Person and ProbabilisticPerson) was enabled with additional manual filtering and extra manual corrections. Since we were intended to see the context (positive or negative), where critics speak about the persons from the art field, the next task was to extract the sentiment of these contexts. To do this, it was necessary to put those entities we didn't have positions in the text (using a heuristic based on Levenshtein distance [9]) and then obtain the textual symmetric word span (window, context) around person's mention. Then the sizes of spans with various widths (2, 5, 10 and 20 words to the left and to the right of the word). After that LINIS sentiment dictionary [6] was applied to calculate the number of positive, neutral and negative words in contexts set by selected spans. The distributions of positive (coral) and negative (lilac) words are displayed on figures below. We stopped at 10 as the suitable word span size, because unlike the contexts with fewer words, with spans of this size the results were interpretable and there was more data. Spans of size 20 showed similar results (Fig. 1).

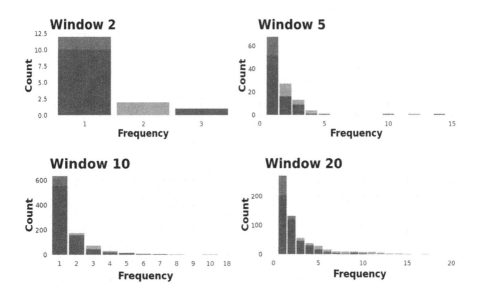

Fig. 1. Text selection

3.2   Networks and Sentiment

Next, as we were interested to examine the similarity of critics based on the artists mentioned in their articles, a bimodal network was built where the critic was linked with the entity mentioned in the texts. For this network, as well as for both unimodal projections, we decided to map the positive and negative sentiments on the graph according to the sentiment of words included in the window near the entity. Therefore, using the sentiment dictionary developed by the LINIS laboratory [6], we counted the number of positive and negative words caught by the 10+10-words-"window", which surrounded the name of the artist. In this way, we considered the relationships of artists and critics, as well as the relationships between artists and critics separately, meaning the sentiment.

3.3   Value Judgment Prediction

The next goal was to understand what features of the text hint that it possibly contains value judgments. For this purpose, the following series of procedures were carried out. In the texts, parts of speech were counted, the frequency of each of which was a separate feature. As a dependent variable, we took the sentiment of the whole text (initially on the scale -1 to 1, then joined into just three categories of critical texts: negative, neutral and positive). Also as variables were coded: media, critic's city and critic's gender. In addition, it was also important to try to find out which factors affect the overall emotional response. To estimate the sentiment, in this case, Chetviorkin lexicon [4] was applied.

4   Results

4.1   Networks and Sentiment

Before proceeding to the semantic analysis of the articles, we would like to dwell on an analysis of the Russian art criticism field on the basis of visualized networks with the type of critic-artist relationship. The two-mode network, shown above, based on mentions of various authors of artists, historical personalities, works of art, etc. In the network (Fig. 2) squares displayed art critics, and rounds depict artists, mentioned in the articles. The visualization represents the connections between the critic and the named entity extracted from his text are colored in accordance with the sentiment of the words that have fallen into the "context window" (the number of negative words has been subtracted from the number of positive words). The red color indicates the presence of positive words in the context, blue color indicates the presence of negative words. Those works or artists, described more neutrally, connect them and the named entity with gray edges.

Critics are designated as nodes of larger size, where the circle depicts St. Petersburg professional, and the square depicts Moscow. It is interesting that the critics from St. Petersburg were in the center of the network, while Moscow is closer to the periphery. The set of names surrounding the critic is associated with exhibitions that took place in St. Petersburg or Moscow during the analyzed period. It is possible that Moscow reviewers were on the periphery due to the fact that the variety of exhibitions and performances in Moscow is greater. We expected to see more positive relations in the center of the network, but positive and negative sentiment distributed almost equally. However, those critics, who have the smaller number of shared entities and actually are on the periphery of the network, have more negative terms in their reviews. It also was interesting to find out to what extent the critics were emotional and inclined to include the elements of personal judgment in the reviews. There are two networks below, where one can see a network of critics (nodes) based on the common mentions of the same entities: the weight of the edge is the number of entities mentioned by the two authors (Fig. 3). Averaged numbers of negative words in reviews by an author are drawn with a gradient fill (from white/neutral to blue/negative) on the left, similarly for positive words (from white/neutral to red/positive) – on the right. It can be concluded that the texts of different authors differ by either general emotionality or by general neutrality. Interestingly, the most emotional authors (Kira Dolinina, Valentin Dyakonov,

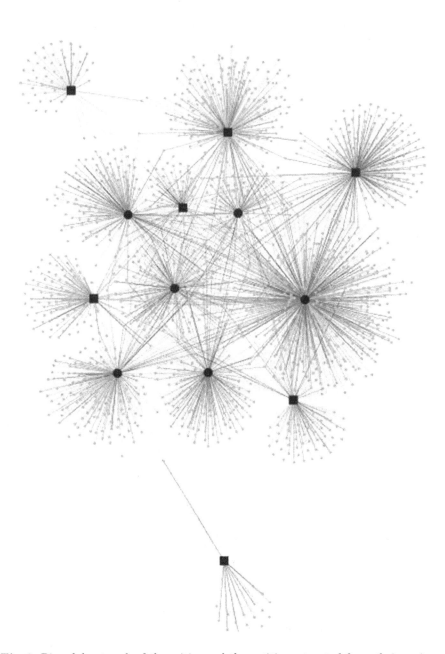

Fig. 2. Bimodal network of the critics and the entities extracted from their reviews

Anna Tolstova) are employees of Kommersant, which does not specialize in art, unlike all other magazines. Natalia Serkova, who is on the periphery of our network, apparently, describes specific exhibitions or does not refer to the classics in large volumes, however, she also works in a journal not specializing on art.

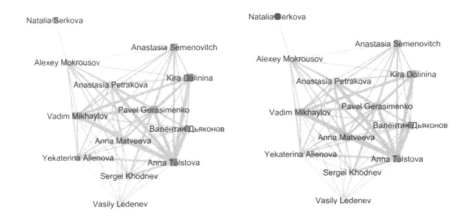

Fig. 3. Unimodal projection: network of the critics. Positive sentiment at the left, negative sentiment at the right

Lastly, we consider unimodal projections, the network of entities which were mentioned by the same authors in the same review. The center is, mostly, consists of classics (Leonardo, Pushkin, Raphael, Picasso, Rembrandt etc). Positive tone prevails over the negative one, and they are almost evenly distributed throughout the network, with a slight shift towards the periphery. The hypothesis on the recognition of creative persons by the periphery of the network is thus confirmed (Fig. 4).

If we look at the presence of negative words, critics here neutrally speak about classics authors, who are mostly placed at the core of the network and use negative sentiment words for those who are closer to the periphery on the network of entities. It becomes rather controversial, referring to our hypothesis about the recognition and positive evaluation of the periphery actors in the field of cultural production (Fig. 5).

4.2   Value Judgment Prediction

We run two regression models with the aim of predicting the level of evaluation judgment detection (Tables 2 and 3). As we hypothesized, the model of the critic's review where vocabulary were scribed by assessors according to specific lexicon of the art field explains the presence of positive or negative assessment performs significantly better and describes about 32% of observations (Table 3) than the Chetviorkin lexicon of emotional response with the prediction power of only 5% (Table 2) ($R^2 = 0.319$ and $R^2 = 0.055$).

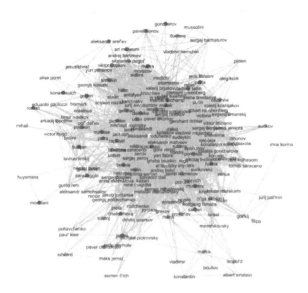

Fig. 4. Unimodal projection: named entities network. Coloured according to the positive sentiment presence (Color figure online)

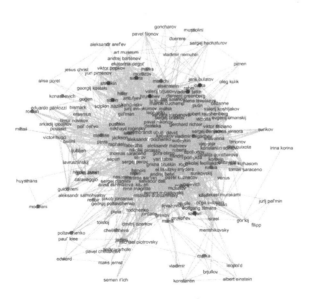

Fig. 5. Unimodal projection: named entities network. Colored according to the negative sentiment presence (Color figure online)

Table 2. Regression model 1, predicting the emotional polarity of critic judgment

	Dependent variable:
	Sentiment by lexicon for the web reviews
City Saint Petersburg	−0.061 (−0.125, 0.004)
Magazine Colta	−0.266*** (−0.404, −0.128)
Magazine Artguide	−0.011 (−0.098, 0.076)
Magazine ArtNewspaper	−0.078 (−0.121, 0.013)
Magazine Kommersant	0.152*** (0.063, 0.241)
Male	0.043 (−0.031, 0.117)
Adjectiveness	0.011*** (0.007, 0.015)
Verbness	−0.011** (−0.019, −0.004)
Constant	2.139*** (2.047, 2.231)
Observations	1,093
R^2	0.061
Adjusted R^2	0.055
Residual Std. Error	0.560 (df = 1001)
F Statistic	9.317*** (df = 8; 1001)

Note: *p < 0.1; **p < 0.05; ***p < 0.01

Table 3. Regression model 2, predicting the sentiment polarity of critic judgment

	Dependent variable:
	Sentiment by the art field vocabulary markup
City Saint Petersburg	0.003 (−0.029, 0.034)
Magazine Colta	−0.215*** (−0.273, −0.156)
Magazine Artguide	−0.086*** (−0.125, −0.047)
Magazine ArtNewspaper	−0.093*** (−0.210, −0.072)
Magazine Kommersant	−0.075*** (−0.119, −0.031)
Male	−0.044** (−0.079, −0.009)
Adjectiveness	0.286*** (0.228, 0.344)
Verbness	0.226*** (0.201, 0.250)
Constant	0.078*** (0.039, 0.117)
Observations	1,093
R^2	0.323
Adjusted R^2	0.319
Residual Std. Error	0.284 (df = 1084)
F Statistic	74.031*** (df = 8; 1084)

Note: *p < 0.1; **p < 0.05; ***p < 0.01

The best sentiment analysis model (Table 3) demonstrated that in determining whether the review is neutral, positive or negative the proportion of adjectives or prepositions in a review play a significant role and increase emotional coloring in a positive way. In addition, artists are not advisable to ask the critics from the magazine ArtGuide for a review, as this can lower the positive sentiment in reviews mentioning them. Also, it is important to admit the fact that if the article was written by male it decreases positive background of the review.

5 Discussion

In the sphere of cultural production, it is necessary to study not only all agents participating in the field of art but also the products of all agents participating in the field. With the help of a deep analysis of critical articles over a long period, one can try to explain the success of an artistic figure, looking at the state of the art field, trying to predict its further development. That is why, it is necessary to look at the categories of current development of the field, which the field gatekeepers converge in opinion, and what is the basis for their disagreement. In this paper, we made an attempt to describe the field of art within the framework of two types of field agents: artists and critics. Critics also tend to have an exclusive sympathy for artists who do not intersect with anyone. We have discovered that critics are more likely to agree with each other on the legitimate artists they mention, whose works have become classics, however, young artists also appear in these lists of recognized geniuses. Also, it is clear that critics have exclusive/special sympathies to those artists that never appear in the articles of other critics. In addition, the city as a location of critic's work affects one's choice of the object or the object for criticism, because one can see from the graphs that critics from one city have a greater cardinality of the intersection on artists mentioned by them. This study is far from being comprehensive in the description of the field of production of criticism, however, the authors have tried to set a start to a new study by testing a possible methodology for studying the discussed field with specific vocabulary. In the future, for a more accurate study, we plan to construct up a custom dictionary of sentiment words, which should include art terms, since the context span in our experiment contained too few 'sentimental words', and include in the study of living artists, who, for example, most often will meet in networks to see the problem from all sides.

References

1. Bourdieu, P.: Theory of Symbolic Power. Culture/Power/History: A Reader in Contemporary Social Theory, vol. 155 (1994)
2. Bourdieu, P.: Outline of a sociological theory of art perception. Int. Soc. Sci. J. **20**(4), 589–612 (1968)
3. Cattani, G., Ferriani, S., Allison, P.D.: Insiders, outsiders, and the struggle for consecration in cultural fields: a core-periphery perspective. Am. Sociol. Rev. **79**(2), 258–281 (2014). https://doi.org/10.1177/0003122414520960

4. Chetviorkin, I., Loukachevitch, N.: Extraction of Russian sentiment lexicon for product meta-domain. In: Proceedings of COLING 2012, pp. 593–610 (2012)
5. Eliashberg, J., Shugan, S.M.: Film Critics: Influencers or Predictors? J. Mark. **61**(2), 68–78 (1997). https://doi.org/10.2307/1251831
6. Koltsova O., Alexeeva S., Kolcov S.: An opinion word lexicon and a training dataset for Russian sentiment analysis of social media. In: International Conference Dialogue 2016. Computational Linguistics and Intellectual Technologies, pp. 277–287 (2016)
7. Korobov, M.: Morphological analyzer and generator for Russian and Ukrainian languages. In: Khachay, M.Y., Konstantinova, N., Panchenko, A., Ignatov, D.I., Labunets, V.G. (eds.) AIST 2015. CCIS, vol. 542, pp. 320–332. Springer, Cham (2015). https://doi.org/10.1007/978-3-319-26123-2_31
8. Kukushkin, A.: Natasha/named entity recognition library. GitHub. https://github.com/natasha/natasha. Accessed 7 Mar 2018
9. Levenshtein, V.I.: Binary codes capable of correcting deletions, insertions, and reversals. Sov. Phys. Doklady **10**(8), 707–710 (1966)
10. de Nooy, W.: Social networks and classification in literature. Poetics **20**(5), 507–537 (1991). https://doi.org/10.1016/0304-422X(91)90023-I
11. de Nooy, W.: A literary playground: literary criticism and balance theory. Poetics **26**(5–6), 385–404 (1999)
12. Peterson, R.: The Production of Culture A Prolegomenon. Am. Behav. Sci. **19**(6), 669–684 (1976). https://doi.org/10.1177/000276427601900601
13. Shrum, W.: Fringe and Fortune: The Role of Critics in High and Popular Art. Princeton University Press, Princeton (1996)
14. Swanwick, K.: Musical criticism and musical development. Br. J. Music Educ. **8**(2), 139–148 (1991). https://doi.org/10.1017/S026505170000824X
15. Uzzi, B., Spiro, J.: Collaboration and creativity: the small world problem. Am. J. Sociol. **111**, 447–504 (2005). https://doi.org/10.1086/432782
16. White, H.C., White, C.A.: Canvases and Careers: Institutional Change in the French Painting World, vol. 1. University of Chicago Press, Chicago (1993)

Digitalization as a Sociotechnical Process: Some Insights from STS

Liliia V. Zemnukhova[1,2(✉)]

[1] Sociological Institute of the FCTAS RAS, 7-ya Krasnoarmeyskaya 25/14, St. Petersburg, Russia
l.zemnukhova@gmail.com
[2] European University at St. Petersburg, Gagarinskaya str. 6A/1, St. Petersburg, Russia

Abstract. The production and implementation of digital technologies face multiple restrictions, limitations, obstacles, and barriers. The more ubiquitous they become, the more social situations and interactions they take part in. One of the possible ways to understand more about digitalization is to deconstruct the process of dissemination of technologies and innovations as well as intangible knowledge. The paper represents a mixture review of methodological perspectives, which help to grasp the complexity of sociotechnical relations. It involves studies of the knowledge production, artifacts spreading, and innovation diffusion in order to approach digitalization as sociotechnical phenomena.

Keywords: Digitalization · Sociotechnical process · Users · STS

1 Introduction

Social studies of technological development and dissemination are rooted in two major fields – innovation studies (IS) and science and technology studies (STS). The former comes closer to economic modeling and investigates larger systems of production and spreading of innovations, the latter takes a more microscopic view of knowledge and artifacts transfer. In order to approach digital technology, it might be helpful to trace the process of knowledge, artifacts, and technology investigations.

I begin with some encyclopedic examples to show the mechanics of the deconstruction work and then turn to more recent findings, which provide a variety of optics towards digital technology and their barriers. The important precondition of the article is that digital technologies are to be treated as both knowledge and artifacts in a sense that it is intangible and science-intensive. I will look up to several traditional approaches in STS, such as Laboratory Studies (LS), Studies of Scientific Knowledge (SSK), Actor-Network Theory (ANT), and Social Construction of Technology (SCOT) to provide a basis for the investigation of digital technologies. Later, I will introduce some concepts from IS and the intersection of IS and STS.

2 The Knowledge Production

The end of the 70s is marked by the classical studies of laboratories, such as Lynch's examination of neuroscience laboratory [1], Knorr-Cetina's study of a food science

D. A. Alexandrov et al. (Eds.): DTGS 2018, CCIS 859, pp. 125–133, 2018.
https://doi.org/10.1007/978-3-030-02846-6_10

laboratory [2], and Latour & Woolgar's "Laboratory life" in La Jolla [3]. These studies were conducted in California independently of each other, but with a shared aspiration to understand how knowledge is produced and goes out of the laboratory. They provided certain background and rules for the following studies of knowledge and technology transfer.

Laboratory studies were trying to overcome difficulties and traditions of science study prevailed at that period. Rationalistic philosophy of science put in the corner demarcation between scientific and other types of knowledge and treated the scientific process as linear and formalized with criteria, principles, and rules [4]. The institutional theory of science was interested not in the knowledge itself, but rather in the processes around, such as norms, systems of benefits, communications and collaborations between scientists [5]. History of science provided narratives on scientific development as changing conceptualizations and paradigms [6, 7]. Sociology of knowledge insisted on social determinism of knowledge, and that natural or logical knowledge is impossible to explain sociologically, as it has own rules [8]. All of these fields of study left the knowledge itself, its content, out of the focus of research.

Sociology of scientific knowledge (SSK) was different and the closest to LS: it suggested to go inside of the knowledge and to reconstruct how different operations happen and change knowledge [9]. Owing to SSK, science studies got an opportunity to examine the content of knowledge. The strong program of SSK has its roots in Edinburg School (David Bloor) and Bath school (Harry Collins). While Bath School suggested going into the quarrel between different positions of knowledge claims [10], Edinburgh School provided the principles of SSK, which were adopted by LS and actor-network theory (ANT). These principles became crucial for the further theorizing: causality examines the reasons and context of knowledge production, symmetry makes equal explanations for successful and unsuccessful knowledge claims; reflexivity makes sociological knowledge applicable to sociology itself; impartiality denies espousing successful knowledge [9]. It is necessary to keep in mind these principles when considering the path of knowledge from the academy to the world.

The first LS aimed at describing of scientific knowledge in its completeness. It was a brand-new idea to "go and look" what really happens in the production of knowledge as opposed to previous views on science studies. It became possible to examine scientific knowledge without boundaries between science and non-science. Moreover, scientific knowledge greatly depends on the context of discovery, where the context of justification is not fully rational. Finally, scientific knowledge is constructible, and not only by a researcher.

Initially, the book "Laboratory Life: the construction of scientific fact" [3] was oriented to sociologists of science and held an idea of social construction of scientific knowledge. Latour being an anthropologist and sociologist working with African tribes, got opportunity to spend 2 years in the laboratory of endocrinology at Salk Institute. The scientists and technicians at laboratories were treated as a tribe and all the everyday practices were observed through the lens of anthropological strangeness (p. 40).

The problem statement was the following: while anthropologists thoroughly study exotic tribes with their cultures and myths, civic and highly valued tribes of scientists are out of research attention (p. 17). By the end of 70s, an interest to the significance of science and its influence on contemporary life was already formed, but the focus of the research was the context and external effects of scientific activities, such as financing or politics of distribution. Latour and Woolgar aimed at a reflexive description of scientists' everyday life, and thus demagnification it. For this reason, the authors use two main principles for the study: symmetry and reflexivity taken from SSK but with some modifications. The former is a similar approach to explaining true and false claims with the same relation to the context and to the content of science. The latter principle originally means that explanatory model created by sociologist must be applicable to sociology itself, but Latour and Woolgar put it in the following way: "We attempted to address the issue of reflexivity by placing the burden of observational experience on the shoulders of a mythical 'observer'" (p. 283). By the end of the book, Latour follows his second principle and applies the approach to himself by describing the process of building his own order. He concludes that science is never wittingly special practice.

While anthropology of science refined thoughts about the connection to primitive and pre-scientific, sociology of science concentrated on talks, gossips, and scandals rather than content. Latour insists that all the dichotomies produced by scientists have to be reconsidered from a practical point of view – who does what. It is the form of agnosticism – not to take any definition or argument of trust. The purpose is to construct scientific knowledge and the process of regulation and making order out of chaos (p. 34). The main common problem of scientists is a mess of alternative descriptions. All of them have equal chances to become reliable up to a certain point. There are no true or false descriptions a fortiori, so scientific progress is in the elimination of alternative descriptions.

Latour goes through the process of knowledge production and studies in details the construction – that is everyday practices of recordings, descriptions, classifications, tables, and graphs, and then articles (p. 153). A recording is what laboratory produces. Scientists produce unfounded claims (artifacts), which compete with other arguments and claims (p. 123). The purpose is to turn an argument into the fact, which reproduces without any context. Artefacts (before facts) are taking part in competitions, scientific search, virtual objects.

The fact is something fabricated, and as soon as fact was constructed, the process of its construction erased. Reification or materialization of an argument is the result of its stabilization (p. 238). To contest the fact resulted from apparatus or machine means to contest the whole story behind this machine. Black boxing means that the fact is unproblematic and is used for other purposes.

All the claims and arguments are produced in some circumstances, that is modality. The aim is to provide a claim without modality. For this reason, scientists try to convince each other and reproduce the same experiments. Science is not the description of reality, but the way of convincing colleagues. Contention (as an agonistic field – in terms of Latour and Woolgar (p. 238)) unites politics and epistemology contexts. At the same time, the volume of costs for alternative descriptions production depends on credibility. It unites economics and epistemology. There is no big distinction between internal properties of reputation and trust and external aspects of credits and funding.

Science is made of circumstances because the local constellation of practices and networks with technologies and tools allows showing what has to be shown. If there is no optics for seeing it, it does not exist there. The science is very local and needs machinery for that. In "Pasteurization of France", Latour shows that social circumstances do not explain how science is made [10]. Agnosticism in telling who does what makes Latour use means of semiotics (actant), without imagining, but observing. Science and politics are the same for methods.

Scientists have to detach signals from the noise against a background of equally possible events – the ratio of signal to noise (p. 240). The construction of fact is the process of transformation of the set equal claims into the set of unequal probabilities. Inequality depends on the cost of production of the claim. The blackboxing is possible with credibility. Some of the claims are becoming facts, but others remain artifacts. The transition between in- and equality is the creation of order (p. 241). A laboratory is a way of turning or putting facts into a black box. Writing is not a translation of information, but an operation of ordering. To reach a signal there is a need for order to avoid noise inside and outside of the laboratory.

Latour continued exploring the issue of knowledge production and its social effect. The context for the science is another science, in the case of Pasteur – hygienists, who become the first translator of the problem [11]. Every side has its own interest, but one group make others follow their interests. It is a way of how small laboratory raised the world. In a laboratory scale, Pasteur created the environment and proved that illness is the result of infection and environment, he constructed the illness by paralleling of real and laboratory life. The term of translation is the key (and it was not used in "Laboratory Life": it is the first translation – to rule microbes because of knowing how it works (mediator). The second translation – is virulence change – Pasteur provided hygienists by arguments for the further funding (status rose). "The Pasteurization of France" by Latour is used here to show how differently (hi)stories of science and technology might be written. He explores controversy as a matter of concern of different networks.

3   Moving Knowledge and Artifacts

The research by Steven Shapin and Simon Schaffer [12] is conducted in the framework of SSK, and deals with historical epistemology. They purpose to reconstruct the story of Boyle's air-pump experiment and how it became a fact (in Latour's terms). They position their book as an "exercise in the sociology of scientific knowledge" (p. 15) in Edinburgh School tradition. The analysis called "experimental polity" was connected with interest groups at that period, such as the Royal Society and problematic of Restoration polity. At the same time, the emphasis was put "the origins of a relationship between our knowledge and our polity" (p. 343), on the nature of knowledge and validation of experimental philosophy. Though the issues of science are connected to the history of politics, the focus was put on a problem of knowledge as the problem of social order.

They follow Latour, using his vocabulary, and analyze the nature of credibility from stranger's perception. By accepting the principle of symmetry, they deconstruct the debated between Hobbes and Boyle. In their book, it is not just internal argue, but it mostly works as political debate. Unlike Latour in *Pasteurization of France*, concentrating on translations and examining actants, Shapin and Schaffer try to understand external context as a struggle of philosophies – naturalistic and experimental, meaning different epistemological positions of Hobbes and Boyle, respectively.

Boyle's position was that natural philosophy foundations should be based on experimental matters of fact. His work on sequential and detailed reproducing of experiment and constructing by that matters of fact was to put up a building of defense. The strength of Boyle was in inviting other natural philosophers to support "defense of experiment and of the engine that was its powerful and emblematic device" (p. 207). It was just one of many parts of the complete ideological assault, as he involved not only air-pump as a material technology, but also transferable factual data and his networks with other scientists. This was a political game with groups of interest and institutions, where "he who has the most, and the most powerful, allies wins" (p. 342). Later, Shapin will show how Boyle worked on reputation and struggled for truth in authorship [13].

The network or community included those who share the idea, trust, and conventions of the experiment. They accept matters of fact by working on the same technology and reproduce the same social order: "the effective solution to the problem of knowledge was predicated upon a solution to the problem of social order" (p. 282). By these means, practices became institutionalized and experiments continued to produce scientific knowledge. Shapin and Schaffer relate Restoration polity and experimental science as being "a form of life" (p. 342). Attention to internal issues of arguing and external cultural context makes this study valuable, as it crossbreeds material culture and presents the story of technology.

Robert Kohler demonstrates more social story around the laboratory object. By animating the fly and giving it social dimension, he traces social relations around it. Kohler calls up Shapin and Schaffer in his attention to experimenting, however, he stops on the working process without the meaning of results. The study of Drosophila by Kohler brings together nature, science, and society [14]. Focusing on "living instrument" and "those who share a particular organism, rather than… a theory, problem, or discipline" (p. 14), Kohler separates his work from other lab studies. Though he still acts as constructivist.

His research was intended to provide genetic mapping, and instead discovered the experimental life of a fly group. However, it is not about experimenting, but more about technosocial relations and communal "moral economy" as a part of laboratory culture meaning experimental systems, focusing on daily craft in the laboratory. Kohler uses Thompson's term "moral economy" for a description of implicit or tacit rules, norms and customs of work, exchange, and communication within laboratories and between communities of experimenters-drosophilists, for managing the visible material process as well as growing social status leading to the success of this fly movement. Moral economy is about "how unstated moral rules define the mutual expectations and obligations of the various participants in the production process" (p. 12).

Moral economy worked in the laboratory as well as outside of it. Inside, there was an atmosphere supported by each other and lab's ethos. The father of modern Drosophila genetics, Thomas Hunt Morgan worked as a manager expanding his fly network (translated his interests) for exchange with distant colleagues, accepting and training students, earning authority and credibility. A cycle of credit is used by Kohler in order to show economic and symbolic meaning of scientific production. Morgan's group at some point established control of the field and scientific discoveries because other groups were not competitive enough.

The symmetry between people and things reproduces and transforms in interdependently. The fly, "fellow laborer" (p. 23), is endowed with the role in the process of knowledge construction. It is "a biological breeder reactor, creating more material for breeding experiments than was consumed in the process" (p. 47). As a result of domestication of the fly, it became impossible to make planned experiments on development and evolution. Before the second-generation drosophilists Theodosius Dobzhansky appeared, it looked like the symbiosis of people, flies, and laboratories, with a deadlock of "novel experimental systems opportunistically, whatever original intentions" they had (p. 211). After he came, the development of evolution study restarted, as crossed the boundary between the laboratory and the field.

The real aim of Lab studies, to believe Latour, was to watch and observe scientists as tribes and science as it is made in everyday life. In the same way, Lynch was closer to ethnomethodology, and Knorr-Cetina went through system theory. The object of study experiences the influence of researcher's background. The principle of reflexivity in this sense could be expanded towards meta-study – to study those who study labs. These pioneer works gave a sense to and a push of social sciences towards the epistemological position of the researcher. As we saw on the examples of Shapin and Schaffer, Kohler, these types of research are very different from classical lab studies.

New generations of lab studies rather lost the edge of pioneers and description language. Anthropologists [15] and historians [16] of science and technology easily adopted particular ideas but added new conceptualizations and social relations. Instead of going to observe, they fully armed with hypotheses and analytical apparatus. Critics of Laboratory studies [17] confirms that laboratory studies are transforming from case to case, from author to author, but still influenced in a great respect the STS itself. "The first thing any new lab study should do is go directly for what laboratory studies have missed—a particular fact—and wrestle with how its endurance obtains within the "in situ" world of practice" (p. 291). And particularly interesting for me is the question of contemporary laboratories with digitalized technologies – whether lab studies powerful to uncover the black box named hardware or even software.

4   The Use of Technology

Unlike early STS, innovation studies had another approach to technologies and the process of their production, development, and diffusion. The idea was to find out general explanation of how science works, where innovation goes, and who is in charge of technological development. Early IS scholars rather defined the direction of innovations, going either from academia or from market demand – these two approaches appeared to be known as the science-push/demand-pull debate [18]. However, later research showed that there is no sense to conceptualize technological development as a linear process. The turn to interactive signed the emergence of the next generation of innovation models [19]. Notably, the review of generations by Nicolov and Badulescu provides no room for users separate from market and consumption.

Users became inevitable parts of the innovation process since the late 1980s [20]. Their active role was not only in terms of feedback and market power, but also they are more and more often seen as innovators [21]. Even if users do not have enough expertise, they can arise the problem to be solved with a technology. Social construction of technology (SCOT) deals with these kinds of users and their groups showing the dynamics of their influence, participation, and reshaping of certain technologies, even after their stabilization.

SCOT addresses many different categories of users, such as relevant groups, active users, consumers, citizens, and even non-users [22]. Each of these categories puts a certain lens towards the development of technology, highlighting specific political, economic, cultural, social features. Redefining users through these contexts helps to map the needs, positions, skills, feedbacks of users and their practices. Crossbreeding users and producers is another step further towards an understanding of how things work in technological development. Mutual shaping of technology, the early participation of users, fast feedback and cooperation – these are the elements of sociotechnical ensembles [23]. It is about similar understanding, making sense, giving meaning, and coordinating efforts.

Some aspects of these continuous interactions between users and technologies were also elaborated within the framework of actor-network theory. Innovation as a sociotechnical system [24], which is highly heterogeneous and strives for setting relations between diverse elements. ANT suggests a concept of 'inscription' to define what is already embodied in the artifact to understand how to use it [24]. It is helpful to detect barriers in the early stages of technology production, and to analyze inscribed models, their possibilities and limitations. There are also specific ways how producers try to discipline future users [25]. They do not just impose the best-usage norms, but also design the artifact with material restrictions and put restrains from the wrong usage.

There are some more perspectives on the development of innovation, which more or less define roles and areas or logic of interaction. For example, the socio-politics of usage [26] combine the concept of a socio-technical configuration and the concept of user representation. The former includes technical and social logics, and the latter unites production and use logics [27]. Another is the idea of a socio-technical frame [28], where producers, users, and other groups build networks of interaction with stable patterns.

Of a special interest are the approaches, which appeared on the intersection of IS and STS, as they are trying to grasp and hold a complexity of technological development. One such example is the concept of sociotechnical configuration [29], which includes social relations as those connecting, using, and making sense for technological artifacts. This multilevel approach includes sociotechnical landscapes (macro), regimes (meso), and nishes (micro). Each of this has its roles, rules, practices, symbolism.

5   Conclusion

This massive review was conducted in order to show how hard it is to conceptualize, huge networks of technological dissemination. Digital technology being intangible, science-intensive, are better to treat as both knowledge and artifact. Laboratory studies showed the very detailed microscopic view on how knowledge is produced and disseminated, and how much it depends on micro practices of everyday life – there are competitions between claims (artifacts), recordings and orderings, classifications and articulations. The production process consists of these many iterations, each of which is able to redefine knowledge and to construct alternative universe of arguments and claims. And as soon as these claims are closing into facts, they start to travel out of the laboratory to accompany themselves with supporters and followers.

SCOT and ANT develop this logic further and include many more additional principles of building the networks of allies and groups of participants. The attempt to view a larger scale of the production leads to the idea that production does not finish with the black boxing and dissemination is another stage or step to the construction of knowledge. The complexity of innovations, and especially digital technology, turns the attention of innovators to look more carefully into the production process and the following development, taking into account many alternative perspectives and contexts.

For researchers and practitioners, it means that it is necessary to take into account difficulties and peculiarities of the previous studies with certain backgrounds, networks, groups, users, and acting entities. In each case, there would be working apparatus to find constraints and decisions to move digitalization further, and the previous research experience seems inevitable here.

Acknowledgement. The research is supported by the Russian Science Foundation grant (RSF No17-78-20164) "Sociotechnical barriers of the implementation and use of information technologies in Russia: sociological analysis."

References

1. Lynch, M.: Art and Artifact in Laboratory Science: A Study of Shop Work and Shop Talk in a Research Laboratory. Routledge and Kegan Paul, London (1985)
2. Knorr-Cetina, K.D.: The Manufacture of Knowledge: An Essay on the Constructivist and Contextual Nature of Science. Pergamon Press, Oxford (1981)
3. Latour, B., Woolgar, S.: Laboratory Life: The Construction of Scientific Facts. 2nd edn. Princeton University Press, Princeton (1986 [1979])
4. Popper, K.: Conjectures and Refutations: The Growth of Scientific Knowledge. Routledge, London (1963)
5. Merton, R.: The Sociology of Science: Theoretical and Empirical Investigations. University of Chicago Press, Chicago (1973)
6. Fleck, L.: Genesis and Development of a Scientific Fact. University of Chicago Press, Chicago (1979 [1935])
7. Kuhn, T.: The Structure of Scientific Revolutions. 3rd ред. University of Chicago Press, Chicago (1996 [1962])

8. Mannheim, K.: Ideology and Utopia: An Introduction to the Sociology of Knowledge. Harcourt Brace, New York (1985 [1936])
9. Bloor, D.: Knowledge and Social Imagery. 2nd edn. Routledge, London (1991 [1976])
10. Collins, H.M.: Changing Order: Replication and Induction in Scientific Practice. Sage, London (1985)
11. Latour, B.: The Pasteurization of France. Harvard University Press, Harvard (1988)
12. Shapin, S., Schaffer, S.: Leviathan and the Air-Pump: Hobbes, Boyle, and the Experimental Life. Princeton University Press, Princeton (1985)
13. Shapin, S.: A Social History of Truth: Civility and Science in Seventeenth-Century England. University of Chicago Press, Chicago, London (1995)
14. Kohler, R.: Lords of the Fly: Drosophila Genetics and the Experimental Life. University of Chicago Press, Chicago (1994)
15. Traweek, S.: Beamtimes and Lifetimes: The World of High Energy Physicists. Harvard University Press, Cambridge (1988)
16. Clarke, A., Fujimura, J.: The Right Tools for the Job: At Work in Twentieth-Century Life Sciences. Princeton University Press, Princeton (1992)
17. Doing, P.: Give me a laboratory and i will raise a discipline: the past, present and future of laboratory studies. In: Hacket, E. (ed.) The Handbook of Science and Technology Studies, pp. 279–296. MIT Press, Cambridge (2008)
18. Godin, B., Lane, J.P.: Pushes and pulls: Hi(S)tory of the demand pull model of innovation. Sci., Technol. Hum. Values **38**(5), 621–654 (2013)
19. Nicolov, M., Badulescu, A.D.: Different types of innovations modeling. In: Annals of DAAAM for 2012 & Proceedings of the 23rd International DAAAM Symposium, vol. 23, No. 1, pp. 1071–1074. DAAAM International, Vienna, Austria, EU (2012)
20. von Hippel, E.: The Sources of Innovation. Oxford University Press, New York (1988)
21. Bogers, M., Afuah, A., Bastian, B.: Users as innovators: a review, critique, and future research directions. J. Manag. **36**(4), 857–875 (2010)
22. Oudshoorn, N., Pinch, T. (eds.): How Users Matter: The Co-construction of Users and Technologies. MIT Press, Cambridge (2003)
23. Bijker, W.E.: Sociohistorical technology studies. In: Jasanoff, S., Markle, G.E., Petersen, J.C., Pinch, T.J. (eds.) Handbook of Science and Technology Studies, pp. 229–56. Thousand Oaks, London, and Sage, New Delhi (1995)
24. Akrich, M.: The description of technical objects. In: Bijker, W., Law, J. (eds.) Shaping Technology/Building Society: Studies in Sociotechnical Change, pp. 205–224. MIT Press, Cambridge (1992)
25. Thévenot, L.: Essai sur les objets usuels: Propriétés, fonctions, usages. Raisons Prat. **4**, 85–111 (1993)
26. Vitalis, A. (ed.): Médias et Nouvelles Technologies: Pour une Socio-Politique des Usages. Éditions Apogée, Rennes (1994)
27. Boudourides, M.A.: The politics of technological innovations: network approaches. In: International Summer Academy on Technological Studies: User Involvement in Technological Innovation, Deutschlandsberg, Austria, 8–14 July, pp. 31–41 (2001)
28. Flichy, P.: Understanding Technological Innovation. A Socio-Technical Approach. Edward Elgar Publishing, Northampton (2007
29. Rip, A., Kemp, R.P.M., Kemp, R.: Technological change. In: Rayner, S., Malone, E.L. (eds.) Human Choice and Climate Change. Vol. II, Resources and Technology, pp. 327–399 (1998)

Selection Methods for Quantitative Processing of Digital Data for Scientific Heritage Studies

Dmitry Prokudin[1,2]([✉]) [iD], Georgy Levit[2], and Uwe Hossfeld[3]

[1] St. Petersburg State University, Universitetskaya nab. 7/9, 199034 St. Petersburg, Russia
hogben.young@gmail.com
[2] ITMO University, Kronverksky Pr. 49, 197101 St. Petersburg, Russia
georgelevit@gmx.net
[3] Jena University, Am Steiger 3, 07743 Jena, Germany
uwe.hossfeld@uni-jena.de

Abstract. The methods of newly appeared field of Digital Humanities are getting more and more popular in the history of science. These methods influence the establishing of digital information resources accumulating and aggregating huge amount of metadata and full text publications. In a previous publication we used an example of a Russian evolutionary biologist and ecologist Georgy F. Gause to preliminary estimate the potential of digital resources for the science studies including history of science. We selected prioritized resources to be used in further research.

The present study explores the methods of selection, processing and quantitative analysis of data extracted from digital information resources. Our concentration is on the digital information resources offering structured metadata. We selected, processed and visualized extracted metadata. Based on the analysis of the achieved results we came to the conclusion on the potential of using digital information resources in the history of science. Besides, the possibility of extracting unstructured metadata has been explored.

Keywords: Scientific information · Digital information resources
Extraction of structured data · Methods for quantitative data processing
Digital scientific heritage · Georgy F. Gause

1 Introduction

The growth of humanities in the digital era led to their convergence with informational-communicational technologies. The use of digital technologies in the social sciences and humanities has given rise to digital humanities. At present there are four major directions within digital humanities:

1. Textological studies (mostly linguistic).
2. Factological studies of e-collections.
3. Multi-media objects research (including virtual reconstructions).
4. Impactanalysis of the digital environment on humanities in general [39].

© Springer Nature Switzerland AG 2018
D. A. Alexandrov et al. (Eds.): DTGS 2018, CCIS 859, pp. 134–145, 2018.
https://doi.org/10.1007/978-3-030-02846-6_11

Digital Humanities in Russia are on the increase and represented by the following scientific centers and societies:

- "History and Computer" association (AIK) (http://aik-sng.ru);
- Russian Association for Digital Humanities(http://dhrussia.ru);
- HSE Centre for Digital Humanities (https://hum.hse.ru);
- Historical Informatics Department of Moscow State University(http://www.hist.msu.ru/Labs/HisLab/);
- Department of information systems in Arts and Humanities of Saint-Petersburg State University (http://arts.spbu.ru/fakultet/kafedry/kafedra-informatsionnykh-sistem-v-iskusstve-i-gumanitarnykh-naukakh);
- Design and Multimedia Center of ITMO University (http://cdm.ifmo.ru);
- Center for Digital Humanities, Perm University (http://dh.psu.ru).

Digital Humanities have recently been actively discussed in Russia [8, 16, 32, 33, 44]. Several conferences on Digital Humanities were held [1, 10, 14, 20]. Approaches and methods of Digital Humanities have been actively used by Russian scholars in various fields of humanities. Results of some studies can be found in [4–7, 25].

Proceeding from the newly developed methods we have conducted a study of the scientific heritage of the outstanding Russian ecologist, evolutionary biologist and anti-biotics researcher Georgy F. Gause by appealing to digital information resources [34]. On the initial stage of the research we have selected digital resources offering instruments for extracting structured data to be further processed. In the present paper we offer methods of extracting and processing metadata applicable in the history of science and science studies.

2   The Use of Digital Data in Science Studies

The application of digital methods in historical studies resulted in the establishment of a new scientific trend « Digital History". It was connected with the growth of e-libraries and e-archives which became important sources of historical studies [16]. At present, popular Digital History directions are the following [2, 3]:

- virtual reconstructions of objects representing cultural heritage; spatial representation and establishment of GIS-apps;
- design of new internet resources (digital encyclopedias, dictionaries, atlases etc.);
- application of interactive hypermedia-technologies;
- joint development of internet resources by a professional group.

One of the major research trends of application of information technologies in the historical studies is the search within digital databases, accumulating huge amount of text information. The search within such resources implies specific methods and technologies, first of all, technologies of searching and extracting certain data. This kind of technology is known as "text mining" or "data mining" [12, 13, 19, 21, 36, 37, 40].

These methods are tightly connected to the technologies of textual analysis elaborated within linguistics [38]. Extracted information is also used to detect consistency determined by the processing of big data [22, 29, 30].

Visualization of information is of great importance for contemporary humanities as well. Visualizing bibliometrics and designing spatial representations of historical information on maps with the help of GIS-apps (geographic information systems) have become routine operations in digital humanities [11, 27, 28, 41, 42].

A more complex approach combining various methods and technologies is applied in the current studies. Such approach is especially effective in science studies and history of science when an impact of an outstanding scholar of the past should be estimated. Various instruments and technologies are used to produce similar methods [9, 17, 18, 24, 26, 31, 35].

3   Objectives

The major objective of the present study is to apply quantitative methods of metadata processing to scientific publications found in various digital information resources. The results of quantitative processing are used to estimate the abilities of using digital information resources to analyze the influence of certain personalities and their ideas on the development of history of science.

To achieve our objectives, we perform the following tasks:

1. Extraction of structured metadata of scientific publications from various digital information resources;
2. Quantitative processing of the extracted data in accordance with the objectives of the study;
3. Visualization of the results;
4. Interpretation of the results.

We have concentrated on methods based on relatively simple technologies, so that the majority of researchers in humanities could use them considering their costs and simplicity.

4   Methods and Approaches

Our approach includes development of methods and instruments available to an average researcher with the background in humanities. In the present paper we employ methods often used in Digital Humanities:

– data search in digital information resources;
– data extraction from digital information resources (data mining);
– Big Data processing.

The methods used can be subdivided into two groups:

1. Methods of data extraction from digital informational resources.

2. Methods of quantitative processing of the extracted data.

The first group (1) includes:

– Automatic extraction of structured data by means of digital information resources;
– Manual data extraction from digital information resources offering no means for data export.

The second group (2) includes:

– Selective minimization of metadata sets to achieve a research goal;
– Standardization of metadata sets extracted from various digital information resources;
– Consolidation of extracted metadata into the united resource.
– Quantitative processing of data;
– Interpretation of the results.

5   Source Databases

The first stage of our study researched the possibilities of applying digital technologies in the history of science. The major criteria were the representativeness of extracted data in accord with certain search queries [34].

We selected Georgy F. Gause, who was one of the most outstanding Russian/Soviet biologists and medical scientists, as an object of case study [15]. Gause was chosen because of several reasons. First of all, he was a prominent scientist, who lived in the Stalinist and post-Stalinist USSR and was neither associated with bizarre anti-scientific currents such as Lysenkoism [23] nor prosecuted. He was scientifically active through his whole life and enjoyed high credibility in both the USSR and abroad. Second, Gause published in English in the Western scientific media and was well known in the West. His major book *The Struggle for Existence* was initially published in English in the USA. This allows to compare an influence of a scientist living in an isolated totalitarian society on science development within and outside this society. Third, Gause was active in several research fields. He is best known for his contribution into evolutionary theory and ecology, as a first scientist who experimentally demonstrated the struggle for existence. Gause's law, known as a *competitive exclusion principle*, proves that two species competing for a limited resource cannot coexist at constant population values. It is less known that Gause was also a prominent anti-biotics researcher who developed the first soviet anti-biotic Gramicidin S in the early 1940s. Our historical objective was to compare Gause's influence in the USSR and abroad and to find consistency to be later explained by traditional means of history of science.

According to our objectives we have estimated various digital information resources in relation to their ability to offer structured data:

– SCOPUS (http://scopus.com) covering all scientific fields and offering over 22,748 peer-reviewed journals, of which more than 4,470 are full open access.

- Besides, over 558 book serials are covered in Scopus, accounting for 34,000 individual book volumes, 1.3 million items and more than 138,000 non-serial books (https://www.elsevier.com/solutions/scopus/content).
- Academic Search Complete (EBSCO, https://www.ebscohost.com/academic/academic-search-complete) is currently the most complete multidisciplinary database of scientific publications and includes full texts of more than 8500 journals of which 7300 are peer-reviewed.
- SpringerLink (http://link.springer.com) is a digital resource of Springer publishing house embracing more than 2700 scientific journals and more than 4000 publication series, as well as 100000 books starting from 1842.

The choice of these resources was determined by the following criteria: (1) instruments of structured metadata extraction; (2) number of records.

Polythematic indexing and abstracting data base Web of Science Core Collection (http://webofknowledge.com) was not included in our analysis, as it offered only 9 records. However, Google Scholar was included in terms of the possibility to use various filters, although it does not allow extracting structured metadata.

6   Extraction of Structured Metadata

To extract metadata, we have employed instruments for importing data from digital information resources. In all systems in question we have used the search query "Gause AND competitive exclusion principle". As a result, we received 575 records in SCOPUS. These records were saved as a CSV-text file (Comma-Separated Values).

After that we explored EBSCO and found 103 bibliographic records. These records were spread over three pages with 50 records per page. EBSCO does not allow to extract all results of the search. Due to this fact, we initially saved the results displayed on each page in a special file and only then imported metadata of publications and united them into a single CSV-file.

The search in SpringerLink resulted in 462 records. These records were saved in the csv format as well.

7   Quantitative Processing of the Extracted Data

To quantitatively process the extracted data, we have chosen Microsoft Excel spreadsheet. This instrument combines both powerful calculation means for data processing and possibilities of data visualization. This is necessary to present and interpret the results. We have imported results from digital electronic resources into separate worksheets. It turned out that all sets have different structure and the order of columns. In this connection we have determined a minimal set of data necessary for further data processing.

The following columns were included:

- Item Title
- Authors

- Publication Year
- Publication Title
- Journal Volume
- Journal Issue
- Content Type
- Item DOI

All three sets of data (EBSCO, Scopus, Springer) on separate worksheets were manually arranged in accord with the order of columns. Then, all the data was copied and pasted on a separate worksheet. The resulting set included 1140 records. Then we have sorted out the records in accord with the column "Item Title". It turned out that there were duplicated records in the resulting dataset. These duplicated sets were manually removed. The final data set embraced 996 records. For the analysis of the publication dynamics we have used a pivot table instrument. We have placed data from the column "Publication Year" in the pivot table. After that we have designed a graph summarizing rows with similar meaning (Fig. 1).

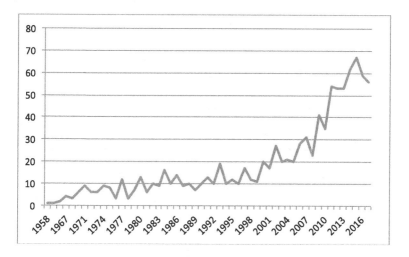

Fig. 1. Temporal distribution of publications

The pivot table also allowed to classify data in accord to the column "Publication Title". By doing this we have selected only journal articles ("content type" = "article") by including only journals with more than five publications. The results are presented as a histogram (Fig. 2).

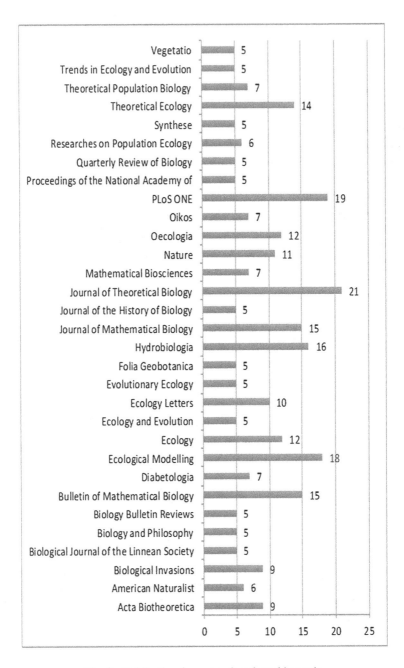

Fig. 2. Publication frequency in selected journals.

Finally, we have classified publications in accord with content type (column: Content Type) using the same method (Fig. 3).

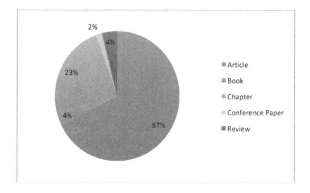

Fig. 3. Publication frequency by content type.

8   Information Resources Offering Unstructured Metadata

Information resources offering unstructured metadata have a great research potential as well. The initial search by using the search query "Gause AND competitive exclusion principle" resulted in the following records:

– OpenDOAR - 6 520 records;
– Google Scholar – 1 710 records excluding citations.

It turned out, that the use of OpenDOAR for further research is impossible because this information resource does not offer instruments for extended search and additional filtering of information, as well as a possibility to extract final results.

By contrast, Google Scholar offers a possibility to filter search results. In addition, this system presents search results by using references to full-text journal articles, technical reports, preprints, dissertations, books, and other documents, including selected web pages that are deemed to be 'scholarly' [43]. All the data are collected and indexed from all sources available.

We used a basic search in Google Scholar to get 1930 references to publications (with citations) beginning with 1960. Employing a temporal filter, we determined references for each year from 1960 till 2017. Altogether 58 search queries were used. The results were listed in Microsoft Excel spreadsheet and presented in Fig. 4A.

Since Google Scholar is indexing all available internet resources of scientific publications, we used this system to search within Russian-language publications. To do this we designed a search query in Russian: "* конкурентного исключения" Гаузе. The symbol "*" was used to replace any word, because in Russian the term "principle" and "law" can be used interchangeably. This search query resulted in 252 records beginning with 1984.

The filters allowed to calculate the temporal frequency of publications. The results are presented in Microsoft Excel spreadsheet and summarized in a graph (Fig. 4B).

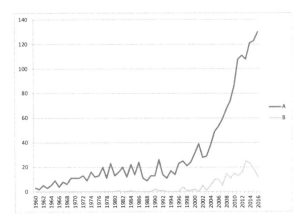

Fig. 4. Frequency of publications per year; data extracted from Google Scholar (A – search query "Gause AND competitive exclusion principle", B – search query "* конкурентного исключения" Гаузе).

9   Conclusions

Our case study based on selected digital information resources offers several conclusions.

1. To achieve big data, it is necessary to use as many resources as possible. For example, the use of only three resources in the present study led to only 13% duplicated results.
2. The achieved results of the quantitative processing of information demonstrated stable positive dynamics. At the same time, the temporal frequency of publications per year in three resources employed (EBSCO, Scopus, Springer) correlated with the data from Google Scholar.
3. The quantitative analysis of journal articles and their distribution in journals allows concluding that the highest frequency of publications correlated with the following disciplines: biology, ecology.
4. The quantitative analysis of publications in relation to the publication type allows concluding that scientific articles embrace 67% of all publications. In fact, the percentage of articles should be even higher, as many publications, classified as "chapter", are papers published in conference proceedings.
5. The comparative analysis of temporal frequency of publications extracted from Google Scholar demonstrated positive dynamic of Russian-language publications compared to the dynamic of English-language publications (Fig. 4). This can be explained by the fact that Russian digital informational resources have embraced materials beginning with the early 1980s (earliest publication date is 1981). Until that time Russian-language publications are very scarcely available. At the same time there are gaps between publications: 1982–83, 1985–89, 1993–96, 2001. Negative dynamics from 2014 can be explained by a delay in indexing of Russian-language publications.

Further development of the current study presupposes an investigation into the possibilities of the Russian digital databases. Quantitative processing of these data will allow to compare Russian and worldwide resources in relation to the history of science and science studies.

References

1. Antonjan, K.G., Ron, M.V.: Mezhdunarodnaja konferencija "Nauki o kul'ture v perspektive digital humanities": problemy gumanitarnyh nauk v jepohu cifrovyh tehnologij. Universum **3**, 141–142 (2013). (in Russian)
2. Arthur, P.: Exhibiting history: the digital future. Recollect. J. Natl. Museum Aust. **3**(1), 33–50 (2008). http://hdl.handle.net/20.500.11937/26207
3. Arthur, P.: Virtual strangers: e-Research and the humanities. ACH Int. J. Cult. Hist. Aust. **27**(1), 47–59 (2009)
4. Borisov, N.V., Nikitin, A.V., Smolin, A.A., Trushin, V.A., Chepurov, A.A., Chepurova, O.A.: Multimedia reconstruction of a stage event. "The Seagull" on Alexandrinsky Stage, First Night, 17 October 1896. Int. Cult. Technol. Stud. **1**(1), 15–23 (2016). (in Russian). http://cat.ifmo.ru/ru/2016/v1-i1/65
5. Borisov, N.V., Volkov, O.G., Nikitina, L.L., Nikolaev, A.O., Smolin, A.A., Stolyarov, D.A.: Application of video 360° technology for the presentation of the Solovetsky monastery cultural heritage. Int. Cult. Technol. Stud. **1**(1), 24–31 (2016). (in Russian). http://cat.ifmo.ru/en/2016/v1-i1/88
6. Borodkin, L.I.: Virtual reconstruction of Moscow monasteries: projects in the context of digital humanities. Perm Univ. Herald Hist. **3**(26), 107–112 (2014). (in Russian)
7. Borodkin, L.I.: Digital humanities and virtual reconstructions in the museum space. In: Role of Museums in Information Support of Historical Science, Eterna, Moscow, pp. 386–395 (2015). (in Russian)
8. Bryukhanova, E.A., Rygalova, M.V.: Historical GIS on-line: a review of foreign and Russian projects. Altai State Univ. J. **2**(90), 56–59 (2016). https://doi.org/10.14258/izvasu(2016)2-08. (in Russian)
9. Budantseva, N.V.: Local History Electronic Resources of Libraries: some Results of Bibliometric Analysis of Documentary Flow. Tambov Univ. Rev. Ser. Humanit. **10**(102), 187–192 (2011). (in Russian)
10. Chaminova, A.A.: The conference « Digital Humanities 2016 » as the reflection of the development of digital humanitaristics. Humanitarian Inform. **12**, 96–101 (2017). https://doi.org/10.17223/23046082/12/11. (in Russian)
11. Chen, C.: Science mapping: a systematic review of the literature. J. Data Inf. Sci. **2**(2), 1–40 (2017). https://doi.org/10.1515/jdis-2017-0006
12. Damerow, J., Peirson, B.R.E., Laubichler, M.D.: The giles ecosystem – storage, text extraction, and OCR of documents. J. Open Res. Softw. **5**(1), 26 (2017). https://doi.org/10.5334/jors.164
13. Donina, O.V.: The application of data mining methods in linguistics. In: Proceedings of Voronezh State University. Series: Systems Analysis and Information Technologies, vol. 1, pp. 154–160 (2017). (in Russian)
14. Gagarina, D.A.: Russian-French seminar « Textometry and Corpuses of Russian Texts». Perm Univ. Herald Hist. **4**(31), 222–226 (2015). (in Russian)
15. Gall, J.M.: Georgy Franzevich Gause. Nestor-Istorija, St. Petersburg (2012). (in Russian)

16. Garskova, I.M.: Information support of the humanities research in digital era: models of formation and development. Perm Univ. Herald Hist. **3**(26), 76–86 (2014). (in Russian)
17. Garskova, I.M.: Thematic analysis of historiography АИК. Inform. bjul. associacii "Istorijaikomp'juter", vol. 5, pp. 8–10 (2008). (in Russian)
18. Hinrichs, U., Alex, B., Clifford, J., Watson, A., Quigley, A., Klein, E., Coates, C.M.: Trading consequences: a case study of combining text mining and visualization to facilitate document exploration. Digit. Sch. Humanities **30**(1), 50–75 (2015). https://doi.org/10.1093/llc/fqv046
19. Huijnen, P., Laan, F., de Rijke, M., Pieters, T.: A digital humanities approach to the history of science. In: Nadamoto, A., Jatowt, A., Wierzbicki, A., Leidner, Jochen L. (eds.) SocInfo 2013. LNCS, vol. 8359, pp. 71–85. Springer, Heidelberg (2014). https://doi.org/10.1007/978-3-642-55285-4_6
20. International Scientific Conference « Digital Humanities: Resources, Methods, and Research » (in Russian). http://2017.dhconf.ru. Accessed 19 Feb 2018
21. Jockers, M.L., Underwood, T.: Text-mining the humanities. In: Schreibman, S., Siemens, R., Unsworth, J. (eds.) A New Companion to Digital Humanities, pp. 291–306. Wiley, Chichester (2015). https://doi.org/10.1002/9781118680605.ch20
22. Kaplan, F.: A map for big data research in digital humanities. Front. Digit. Humanit. **2**(1) (2015). https://doi.org/10.3389/fdigh.2015.00001
23. Kolchinsky, E.I., Hossfeld, U., Kutschera, U., Levit, G.S.: The revival of Lysenkoism in Russia and epigenetics. Curr. Biol. **27**(19), 1042–1047 (2017). https://doi.org/10.1016/j.cub.2017.07.045
24. Kornienko, S.I.: Izuchenie istorii gosudarstvennogo upravlenija I samoupravlenija v dorevoljucionnoj Rossii (na osnove sovremennyh informacionnyh tehnologij).Vlast 2011 (2009). (in Russian)
25. Kukovyakin, A.V., Lyapin, S.H.: Lomonosov's universum: the experience for functional integration of virtual atlas and full-text library. Int. Cult. Technol. Stud. **2**(1), 6–12 (2017). (in Russian). http://cat.ifmo.ru/ru/2017/v2-i1/98
26. Laubichler, M.D., Maienschein, J., Renn, J.: Computational perspectives in the history of science: to the memory of Peter Damerow. Isis **104**(1), 119–130 (2013). https://doi.org/10.1086/669891
27. Leydesdorff, L., Carley, S., Rafols, I.: Global maps of science based on the new web-of-science categories. Scientometrics **94**(2), 589–593 (2013). https://doi.org/10.1007/s11192-012-0784-8
28. Liao, S.: The comparative study on the scientific knowledge mapping tools: VOSviewer and Citespace. Sci-Tech Inf. Dev. Econ. (2011). http://en.cnki.com.cn/Article_en/CJFDTOTAL-KJQB201107061.htm
29. Manning, P.: Big Data in History. Palgrave (2013)
30. Mayer-Schönberger, V., Cukier, K.: Big Data. A Revolution That Will Transform. How We Live, Work, and Think. MIF, Moscow (2014). (in Russian)
31. Peirson, B.R.E., Bottino, E., Damerow, J.L., et al.: Quantitative perspectives on fifty years of the journal of the history of biology. J. Hist. Biol. **50**(4), 695–751 (2017). https://doi.org/10.1007/s10739-017-9499-2
32. Pogorskiy, E.K.: Features of digital humanities. Inf. Portal Humanities. Knowledge. Understanding. Ski **5**, 7 (2014). (in Russian)
33. Popova, S.M.: Analysis of foreign and Russian experience in the development of digital infrastructure of socio-humanitarian researches. Genes. Hist. Res. **1**, 208–251 (2015). (in Russian). https://doi.org/10.7256/2409-868x.2015.1.13820

34. Prokudin, D., Levit, G., Hossfeld, U.: Selection methods of digital information resources for scientific heritage studies: a case study of Georgy F. Gause. In: Bolgov, R.V., Borisov, N.V., Smorgunov, L.V., Tolstikova, I.I., Zakharov, V.P. (eds.) Internet and Modern Society: Proceedings of the International Conference IMS-2017, St. Petersburg; Russian Federation, 21–24 June 2017, ACM International Conference Proceeding Series, pp. 69–74. ACM Press, New York (2017). https://doi.org/10.1145/3143699.3143739

35. Pyankov, A.S.: History of Zemstvo Institutions: Digital Historiography Research. Perm Univ. Herald Hist. **3**(30), 52–60 (2015). (in Russian)

36. Sinn, D., Soares, N.: Historians' use of digital archival collections: the web, historical scholarship, and archival research. J. Assn. Inf. Sci. Tec. **65**, 1794–1809 (2014). https://doi.org/10.1002/asi.23091

37. Sinn, D.: Impact of digital archival collections on historical research. J. Am. Soc. Inf. Sci. **63**, 1521–1537 (2012). https://doi.org/10.1002/asi.22650

38. Sopina, A.L.: Text analysis in the field of digital humanities. Curr. Issues Linguist. **1**, 140–143 (2016). (in Russian)

39. Thaller, M.: Controversies around the digital humanities. Hist. Inf. **1**, 5–13 (2012). http://kleio.asu.ru/2012/1/hcsj-12012_5-13.pdf (in Russian)

40. Toon, E., Timmermann, C., Worboys, M.: Text-Mining and the history of medicine: big data, big questions? Med. Hist. **60**(2), 294–296 (2016). https://doi.org/10.1017/mdh.2016.18

41. van Eck, N.J., Waltman, L.: Citation-Based clustering of publications using CitNetExplorer and VOSviewer. Scientometrics **111**(2), 1053–1070 (2017). https://doi.org/10.1007/s11192-017-2300-7

42. van Eck, N.J., Waltman, L.: Visualizing freely available citation data using VOSviewer. https://www.cwts.nl/blog?article=n-r2r294

43. Vine, R.: Google Scholar. J. Med. Libr. Assoc. **94**(1), 97–99 (2006). PMC 1324783

44. Volodin, A Yu.: Digital Humanities in Search of Self-defining. Perm Univ. Herald Hist. **3**(26), 5–12 (2014). (in Russian)

The Use of Internet of Things Technologies Within the Frames of the Cultural Industry: Opportunities, Restrictions, Prospects

Ulyana V. Aristova[1](✉) ⓘ, Alexey Y. Rolich[2] ⓘ,
Alexandra D. Staruseva-Persheeva[1](✉) ⓘ, and Anastasia O. Zaitseva[1] ⓘ

[1] School of Design Faculty of Communications, Media and Design,
National Research University "Higher School of Economics", Moscow, Russia
{uaristova,apersheeva}@hse.ru, aozaitseva@gmail.com
[2] Department of Computer Engineering, Tikhonov Moscow Institute of Electronics
and Mathematics, National Research University "Higher School of Economics", Moscow, Russia
arolich@hse.ru

Abstract. The article presents an analysis of the possibilities and limitations of the use of information and communication technologies, in particular the Internet of things as an effective tool for artistic and sociocultural practices in the context of transformations of cultural industries. It is revealed that such radical transformations lead to a change in the formats of cultural objects, their content and form. The prospects of technological development are analyzed and the framework of interdisciplinary research is set.

Considering two main trends in the field of culture - the fusion of art with science and the high demand for viewers' participation in art-projects, we emphasize the role of technology in the development of media and focus on the prospects that can provide the Internet of things. In addition, analyzing the perspectives of contemporary technological tools as creative tools, we argue that the Internet of things and derivative technologies can have a strong influence on design, education and culture: today the society faces exponential innovative growth in all areas, but the most promising among them are those which provide the user with an active position, ability to provide feedback and an option to become co-author of the responsive, recipient-oriented projects that engage complex technical excellence in order to meet the expectations of a contemporary adaptive user, viewer or student.

Keywords: Internet of things · IoT · The culture industry
Cyber-physical systems · RFID · NFC · iBeacon · Communication society
Creative production · Design · Creative industry · Trends · Contemporary art
Theater · Interactivity · Participation · Immersion · Media art

1 Introduction

Intensive technological development of the society contributed to the activation of the processes of introducing digital technologies both in the sphere of production and of

© Springer Nature Switzerland AG 2018
D. A. Alexandrov et al. (Eds.): DTGS 2018, CCIS 859, pp. 146–161, 2018.
https://doi.org/10.1007/978-3-030-02846-6_12

cultural industry. In the cultural industry, this has led to a radical transformation at all levels: organizational, marketing, technological, substantive, personnel etc. The very structure and form of the cultural product, which today has a hybrid character and combines classical artistic mediums and the latest information and communication technologies, has changed [1].

In this regard, the problem of application of certain technologies in the cultural industry, taking into account their opportunities and limitations in use, and indicating the possible prospects for their development, becomes particularly urgent for a research. One of the key technologies today we consider an information-communication technology of the Internet of things.

In recent years, the flow of scientific publications on the issue of digital technologies in such areas as economics, medicine, education and other industries has been increasing [2–10], we also focus on their application in the field of cultural industry. In our article, we point a scientific niche in which we consider information and communication technologies – the Internet of things (IoT) in particular – as an effective tool of artistic and socio-cultural practices in the context of the development of cultural industries that require understanding of the possibilities of their practical use in a communication society.

Cultural industries today have become an important part of the national economy of many countries, creating an environment for the implementation of creative work, the introduction of information and communication technologies at all levels of human life [1]. The leading role in this process today is played by Internet technologies, which not only accelerate the production and distribution of goods and information and images, but also create a feedback effect which may not be compared to anything from the past experience of consumption.

Considering the impact of information and communication technologies in the cultural industry (including wireless sensor networks, the IoT technology and etc.) we deliberately avoid extrapolating the problems of their linear development. Our key idea is that the intense technological development at each phase involves a new plan of connecting technology with cultural queries of the industry and to rearrange elements between scientists, engineers, designers and artists. That, in turn, will lead to the emergence of brand new combined products.

Basing on this, we will focus on the current state of the Internet of things technology itself, its communicative and aesthetic opportunities in the field of culture, will determine its capabilities and limitations, and describe the range of tasks that need to be addressed in the nearest future.

2   Internet of Things in the Cultural Industry

The Internet of Things (IoT, Internet of Things) is a system of integrated computer networks and connected physical objects (things) with embedded sensors and software for data collection and reporting, with capability of remote monitoring and control in the automated mode without human intervention [10–15]. The technology of wireless sensor networks, consisting of wireless sensors and control devices is capable of self-organization with the help of intelligent algorithms, shows large-scale prospects in the cultural industries. We see them as leading technologies in this field.

In this paper, IoT is understood more specifically as a multilevel system that includes sensors and controllers associated with specific exhibits, works of art or elements of a theatre stage, means for transmitting collected data and their visualization, powerful analytical tools of interpretation of the received information with the capability of remote monitoring and control in the automated mode, without human intervention.

These technological features make it possible to use the Internet of things as an effective technology of communication with the consumer in a space of diverse cultural objects. Nowadays, the production of cultural industries is focused not only on making objects, but also on the events that enhance our vision, convey vivid impressions and memorable images, make viewers' understanding of the world more profound and form new perceptual habits. This is achieved primarily through multimedia – images, sounds, tactile sensations and the transmitted texts (narratives) are intertwined to create new hybrid forms of art.

Due to the unprecedented extension of discursive networks and the accessibility of content created by the cultural industry, as well as by amateurs, the central place in the system no longer belongs to an Author but is grasped by a Viewer who independently makes his "curatorial" choice in the vast field of texts, images and sounds, makes high demands for the content, and therefore, becomes a driver of creative production [16]. Today these expectations are primarily bound up with interactivity and the immersive effect of a project, the involvement and engagement of a viewer becomes a quality mark in the fields of education, art and entertainment.

In this work we distinguish the following forms of cultural objects, each of which synthesizes an artistic, media and technological component:

1. museums and exhibition spaces;
2. interactive installations (including total ones) in combination with interactive art objects;
3. theatres and show-programs;
4. interactive audio guides (except museums and exhibitions).

In order to enhance the participation and immersive effect of the communication process associated with the experience of the cultural object space and immersion in its content in museums, exhibition projects, theaters, show programs, the Internet of things technology gets associated with the management of light, sound, stage, impressions and experiences of the viewer.

Communication with the audience takes on a synthetic character which allows to broadcast the main content of the cultural product with the greatest accuracy and efficiency. A person receives information through five sensory organs: eyes (vision), ears (hearing), tongue (taste), nose (sense of smell), skin (touch, tactile sensations, etc.). Today it is partially possible to carry it out by the means of the Internet of things.

Light control takes place through lighting electrical systems, automated laser systems and led technologies with program control and wireless communication.

Sound management takes place through analog and digital amplifiers, sound recording and broadcasting, modern audio systems capable of synchronizing with on-premises or cloud services.

The stage in the theater is a set of controlled stage elements, such as curtains, scenery, created manually or with the help of digital technologies, mechanized stage space, etc.

Impressions and experiences of the viewer are associated with the ability to participate, change and even create the development of the plot of any play. Depending on the physical abilities of the viewer, theatre venues can radically change the classical idea of the theater itself.

During a visit to a museum or exhibition space the sequence the viewer's visual range, which includes works of classical and modern art (paintings, sculptures, art-objects, installations) is of particular importance. In some cases, visual impact can be complemented by tactile sensations, the impact on the olfaction, vestibular system, etc.

Thus, with the help of the listed systems and processes, it is possible to design a new reality and form a space with specified characteristics and properties. This will help to attract the audience, increase the number of visits to cultural sites and develop the cultural industry in the direction of the art of the latest media.

3 Opportunities and Limitations of Internet of Things Technology Application in Cultural Objects

Owing to this accelerating trend of interdisciplinarity, cultural industries attract an increasing number of diverse specialists, including experts in the field of high technology, scientists, engineers, the merging of cultural strategies and scientific methods is clearly illustrated in such a striking phenomenon of the present, as media art.

Every day we are being struck by streams of images, but if in the information society it was in the mode of one-way broadcast, in the modern communication society social patterns have developed in such a way that the individual requires reaction, feedback in various forms, from "likes" and votes to participate in serious discussions. In the 20th century, the cultural industry has taken such a form, that main objective was a creation of the immediate effect of ecstasy, but today the developers of product faces new challenges: to arouse the viewer to a dialogue. The combination of these two goals requires investments in high technology, with which it is possible to create impressive media projects in formats never seen before. It is noteworthy that for the first time these tasks were set by artists who in the 1960s focused their attention on creating interactive environments and communicative situations (actions and happenings), where it was important to bring the viewer out of the state of passive absorption of images, to invite them to become co-authors in the creative process. Media art and the art of the latest technologies in 1970–1990s also as one of the key artistic strategies placed the emphasis on interactivity [17]. Today, this method of communication with the audience integrated into projects in the field of mass culture, and it should be noted that the speed with which the industry of entertainment combines formal findings and methods of contemporary artistic practices is getting higher every day. The innovations presented at major international exhibitions such as Documenta or Arts Electronica, are almost instantly included in the toolkit of the cultural industry.

Within the diversity of institutions, museums and art galleries occupy a special place; today, they have a new, more flexible structure than before and their role goes far beyond the classical understanding of museum as a guardian of national heritage, nowadays a "new museology" is developed [18]. Participation sets one of the most important trends in the evolution of not only museums, but also cultural institutions in general: all institutions – from film clubs to libraries and archives – seek to engage the viewer in a dialogue and active exploration, to offer one the role of a partner and even co-author of cultural events [19].

Elements of the IoT technology are already quite actively used for the organization of exhibitions. In this vein, such technologies as iBeacon, RFID and NFC are the most widely used at the moment.

The technology of placing Bluetooth beacons (iBeacon etc.) around the perimeter of the exhibition space and close with the exhibits allows to organize visitors' local positioning as well as their indoor navigation. The tracking of users allows to not only effectively organize an exhibition space (identify the most popular routes around the exhibition space, total time of visit), but also to obtain data on the amount of time spent in front of certain exhibits, and accordingly, to track the level of interest in specific types of exhibits. The collection and analysis of the listed data may contribute to decision-making aimed at improving the efficiency of the exhibition space and in some cases increase incomes. A combination of technologies based on Bluetooth beacons and Wi-Fi positioning is used quite often to improve the efficiency of local positioning. This technology is widely used to increase the level of sales in shopping centers and malls.

In museums, such beacons automatically deliver the content depending on the zone the visitor is in. Sensors, placed in different parts of the exhibition space, interact with mobile devices (audio guides and smartphones with established museum applications) and activate specific content when approaching the desired location in space. The application of Bluetooth beacons can be found in the Metropolitan Museum [20], the Guggenheim [21] in New York, the Museum of Rubens [22] in Antwerp and in Russian Multimedia Art Museum Moscow (MAMM) where a successful implementation took place in 2015 [23].

RFID (Radio Frequency Identification) is a method of automatic identification of objects by radio signals which are read, recorded and stored in so-called transponders or RFID-tags. In fact, the emergence of RFID technology was the starting point for the development of the Internet of Things. In museums, visitors are given cards (similar to transport cards) or bracelets with RFID tags. They allow you to control the number of viewers, time of visit, to rank items on the basis of marks; it means that if the visitors like the object, it is possible to attach your bracelet to the scanner next to it, and your "voice" will be taken into account. This technology is used in the collection and analysis of users' behavior and adoption of appropriate solutions to improve the efficiency of the organization of the exhibition space. RFID is used at the Cleveland Museum of Art [24], Science Museum The Exploratorium (San Francisco) [25], and in Russia the technology in this role was implemented at the exhibition of high technologies SMIT [26] in Moscow.

Near field communication, NFC (Near-Field Contactless Communication) is a technology for wireless data transfer of small radius of action which enables the exchange of data between devices over a distance of about 10 cm. In Russian museums, the NFC technology is not as widespread as abroad. In the world, NFC is used along with QR codes, for example, in the Australian Museum, Sydney [27] or the Museum of London [28]. NFC tags are placed next to the objects and paintings, and a mobile device with a special chip (a smartphone or a dedicated audio guide), reads the label, gives users access to the content in the form of images, videos, texts and links.

The abundance of recipients and consumers of cultural industries makes special demands to the formation and design of such facilities. Working on a concept, designers and artists are now modelling a behavioral situation with a person in a particular semantic field. Everything that happens in «the project» one strongly feels, he lives it through plunging into the exhibition venue where the set affects all senses and enhances the immersion. It is worth considering that in addition to educational and informational influence, visiting a museum or exhibition intends to optimize the emotional state of a person. In fact, one's state at the beginning of the visit can be easily changed at the "exit" by means of the appropriately organized space [28]. The exhibition itself and exposed objects are the effective tools for such changing, tuning. The emerging recipient's condition is influenced by various factors: perception of works in classical forms of visual and plastic art, as well as visual content, time, the use of light (light installation), sound (sound installation), fragrances, touches and etc. The overlay of technology and blend of different mediums on the visual content in accordance with the given scenario forms a special atmosphere of immersion and participation at all levels of perception.

4 Participation and Immersion: The Communication Process in Various Forms of the Cultural Projects

In order to present the "landscape" of interactive cultural products with the greatest completeness, we will follow below as the task of enhancing immersion and participation was solved in the previous period of the cultural development (based on the material of contemporary art of the 1960s–2000s) and what role technologies played in this process, and then turn to the examples of recent years.

The problem of participation holds a central place in the discourse of contemporary art in the 1960s due to the development of Neo-Dadaism and Fluxus, the philosophy of which presupposes blurring of the boundaries between art and life. Artists criticized rigid forms of art establishment, creating works not for sale on the art market, but for immediate impact on a viewer that no longer was separated from the artist by insurmountable border. The work of art ceases to be autonomous and distracted, it goes "off the pedestal", becomes open to the viewer, and interactive [30, 31]. Minimalist, post-minimalist and neodada objects have become the first striking phenomenon of this kind, that influenced the viewer not like a traditional sculpture which has to be understood on the level of its subject (narrative), but as material bodies present in the same space as the body of the viewer and thereby forcing him to pay attention to his own presence and also to abolish habitual schemes of behavior, to deconstruct patterns of perception [32]. The degree of participation was drastically raised by Allan Kaprow in the invented format of happening, which was in fact not the inspection of art objects, but adventure of spectators in a specially arranged exhibition space (A. Kaprow, "18 happenings in 6 parts", 1959; "Yard", 1961).

Soon the interactive component of an art work became closely related to the use of modern technologies. In the early 1970s, video was the most up-to-date one and artists made use of its main innovation – the effect of feedback – to create installations, key elements of which were a camera-screen closed circuit on which the captured objects were projected. A viewer standing before the lens was literally facing himself in such an estranged reflection, seeing, however, not a mirrored, but a converted reflection, and it made him feel the aesthetic distance from his image and at the same time experience his own here-and-now presence. Such works were created by Peter Campus, Bruce Nauman, Dan Graham and other artists, they sought to maximize the use of a viewer's body and therefore organized extended space around the camera-screen system in which a viewer could move then receiving both visual and motor impressions. A contemporary art researcher François Parfait emphasizes that an interactive installation is always "a construction in space and of space" [33].

The next step in the development of interactive works has become the use of the computer and the Internet in such projects as "Zapping Zone" (1990–1994) and "Imme-mory" (1996) by Chris Marker, where in the gallery space the artist placed dozens of monitors, creating a polyphono-chaotic flow of information, where the viewer is already occupied not with contemplation, but with search-oriented activity, attempting to pave their way in this saturated audiovisual field. The metaphor of the way is made play by Jeffrey Shaw in his interactive installation "The Legible City" (1988–1991), where the viewer was asked to ride a real bike, the movements of which were recorded by special sensors and transformed into a journey through the virtual city, which was shown on the screen.

The 2000s are characterized by the creation of responsive interactive environments through the use of motion sensing technologies. Focusing on the behavior and bodily reactions of a viewer, the artists create interactive works the aesthetics of which is based on contact and figuratively given feedback ("Probe" by B. Debacker 2008; enactive cinema by P. Tikka 2005, virtual reality video).

The next step taken by artists and designers is creation of virtual environments where a viewer may experience a state of maximum immersion and influence the world designed by the author ("Bjork Digital" 2017) as well as create interactive environments in real exhibition space with the help of wireless technology. Single works, as well as environments are responsive to the presence of the viewer, consisting of a series of elements that interact by the principle of Internet of Things and will form a new stage of aesthetic development of the feedback phenomenon as a way to strengthen partici-pation, likewise, awareness of the viewers, their sense of themselves as elements in a living system of the world.

The development of information and communication technologies in this regard is the driver of the transition to participation projects and immersive artistic solutions, allowing to transfer the aesthetic experience to a new level. No doubt the cultural industry supplies products of different quality and purpose, and all of the above-described technology and artistic techniques are often used as a banal theme, however, we set ourselves the task to show that these same technologies and techniques in the hands of an artist can be a tool to create such "art", which will require the active intellectual and spiritual work of the viewer, his/her inner transformation. In particular the Internet of things, as will be shown further, is the technology allowing to solve one of the most actual problems of both modern art, and culture: the problem of art creation not "for all", but "for everyone", or, in P. Virno's terminology, for «masses» [34].

Modern theatrical performances and other show programs also abound in a variety of devices and sensors that can be organized online. The most striking example is light costumes, combining modern light technology, Internet of things technology and theater, a variety of optical effects. Modern virtuoso juggling and twisting of neon props, combined with bright space suits creates a futuristic atmosphere. The technology of "cold neon" is used to create such an effect: an electro-luminescent wire covered with a phosphor that shines in an electromagnetic field. Each suit is a complex multi-channel light system handmade. Costumes depending on the requirements can be controlled wirelessly transmit Bluetooth data [35], ZigBee [36], Wi-Fi [37] or even to organize a body network standard (Body Area Network, IEEE 802.15.6 [38]). The microcontroller costume program can be loaded both written for each dancer, implying the performance of certain synchronized with lighting effects costume dancer movements and dynamic software system that analyzes the position of the dancers relative to the scene and each other, as well as having the ability to dynamically adjust to the style of dance (for example, freestyle). In the first case the choreographer thinks over not only movements for each participant, but also sequence of when what detail of a suit has to be lit. A separate application on the PC synchronizes the operation of all controllers with music — all this is calculated to milliseconds. The second case is more suitable for improvised performances and dances, but has a high complexity of implementation [39, 40].

An example of the use of light costumes is the theater of light performance "Svetlitsa" [41], which arranges not just dance numbers, but dance and theatrical performances. Light performance theatre "Bright faces" uses its own technological developments, including programmable micro-controllers that allow you to control the led costumes with neon elements and inflatable constructions, creating a complex, self-organizing network with the ability to control different elements of the systems depending on the specific situation. In this case, specialists program lighting effects for each costume and art object individually. The program is transmitted to them in real time synchronously with music and video installations in accordance with the Director's idea.

Another example of the use of light costume is the Prague Black theatre or the Theatre of light and shadows [42] - a new unusual kind of theatre art, based on the play of light and shadows, with the addition of bright color effects. Black scene, actors in black costumes, "reviving" stage props. There are no words, there is only music, pantomime and acting plastic. This theater embodies the idea of harmonious synthesis of shadow theater, music, dance, pantomime and modern information technologies in order to give the audience aesthetic pleasure. Perhaps because there is no language barrier characteristic of drama or Opera, the IMAGE theatre has become surprisingly popular among tourists from all over the world, and tickets for its performances are booked in advance. More than 7000 performances played and almost one and a half million spectators within twenty years, this is not the only scale of the success of the theater.

Today, classical stage scenery is traditionally associated with theatrical, circus or cinema decorations that imitate reality or create a new one — depending on the artist's idea. But over the past couple of decades the technology of video removal penetrates in stage scenery more and more. Modern video technologies can create great entertainment, but at the same time they are on the way to simplify the technical component of the use of scenery in the theater or show. In this regard, the market of today's theater follows two main trends that directly affect the scenography of the play: the first – the use of multimedia technologies; the second – the use of complex mechanized/robotic structures together with lighting effects. In other words, many performances undergo fundamental changes aimed at the "wow-effect" on the part of the viewer, which is easier to achieve with the help of scenography than through an intricate plot. An example is such a technology of creating stage space as Black box theater [42], which is a mobile and flexible system not only for organizing stage and working space, but a rather complex mechanized system with a huge number of sensors and actuators connected to each other with the help of Internet of things technologies.

Internet of things technologies in combination with virtual and augmented reality technologies in the cultural industry solve the actual problem of finding and revealing new ways of communication between the audience and the theater.

Two years ago, the Parisian Comedy theatre (Théâtre Le Comedia [44]) held the opening of the next season under the slogan "Breaking language barriers!". A pair of augmented reality glasses, connected to the Internet, helps spectators who want to enjoy the French theater, but do not speak the language. Thanks to this device, everyone can see the English or native translation of the phrases spoken by the actors.

In addition to the augmented reality (AR) [45], the theatre uses virtual reality (VR). In the traditional sense it is, of course, rather a film, because there is no direct contact with the actors, but the new genre to the theater is very close. First of all, due to the effect of presence, when the viewer is inside the action. In addition, if in the cinema our mind is manipulated by the operator and the editor, while in the theater the viewers decide what to pay their attention to: to the actors, to the scenery or to the neighbor. The same freedom is given to the viewer in the VR-production: the viewer can at least ignore the whole performance of the characters and look at the sky above his head. In more complex productions, the audience is not even limited to a static position in the chair, but allows you to navigate the virtual world — which is already close to the now popular immersive performances. A serious problem for the development of VR theaters and productions is the imperfection of communication and data transmission technologies for VR helmets and virtual reality glasses. Due to the low speed of streaming data transmission at the network level, it becomes impossible to organize collective interaction in VR-theater space or staging, which from the point of view of the network architecture is the inability to create a wireless network within the concept of the Internet of things, the nodes of which are VR-helmets. These restrictions are currently being solved by many leading companies, such as TPCast [46].

The theatre is in itself an art, the art of the real and rich, and, therefore, the virtual reality of the sometimes to anything. However, thousands of people around the world have limited physical capabilities, which often prevents them from touching the beautiful. It is obvious that in this case, the development of technology works exclusively for the benefit of physically challenged people, helping everyone to hear and see the play as it is perceived by a healthy person. Today, the development of personal screens with subtitles, augmented reality glasses, personal audio headphones and even remote broadcasting in virtual reality serve people with disabilities, breaking the boundaries between art and disease of each of us.

The IoT is used not only for the organization of the exhibition space but also for improvement of the effect of immersion due to the total automation of the exhibits and staging during the whole visit to the exhibition space. The main task is to collect and analyze data coming from both the interactive exhibits at the exhibition, and visitors. The analysis of extensive data from exhibits is connected to the network and operates within the framework of the concept of the Internet of Things solving the following problems: the detection of faults or partial failure of equipment used at the exhibition; remote monitoring of equipment status in real time on a PC or mobile device; predicting and foretelling of a malfunction or partial failure of the equipment; automation of power systems, the procurement process, scheduling, logistics optimization, control and maintenance by predicting results. The solution of these tasks leads to a change in an employee (from an agent to analyst-controller intervening in the process in exceptional cases) and a significant reduction of economic losses due to untimely identified and/or predicted situations of failure or accident.

5 Prospects for the Development of New Technologies in the Cultural Industry

Having considered the application of the IoT technologies in the cultural industry, we finally arrived at the conclusion that the inferences we draw, lie in different planes of its functioning.

It is obvious that the Internet of Things is currently effectively used not only to improve the performance of the global digital economy and cultural industry as an independent region. It is an effective technology, the use of which results in increasing immersion of exhibition spaces and museums, as well as strengthening the communication "dialogue" component, which in turn affects the achievement of a qualitatively new level of development of commercialization models in the field of art that can be interpreted as an absolute trend. In this regard, we can identify a number of trends that will determine its development in the near future.

Owing to the growing trend of interdisciplinarity, cultural industries attract an increasing number of diverse specialists, including experts in the field of high technology, which can be seen on the example of such a bright phenomenon of our time as media art.

It opens up prospects and the fact that the artist, the designer and the curator know the possibilities and limitations in the use of information and communication technologies, including the Internet of things. It is well known that artists and designers were among the first to react to technological breakthroughs. Experiments with the digital medium were carried out by them for decades before the official digital revolution [17].

Obviously, the possibility of their use has great potential and exceeds existing limitations. At the same time, we take into account that in due course any technology shows its limitations in use when new challenges arise. The emergence of new tasks is associated with the emergence of new cultural practices, scenarios of interaction with the environment and communication with the audience.

We assume that possible applications of the IoT technology in the cultural industries lie in the way to use them as an effective means of communication with the audience. Here we distinguish two strategies:

(1) an institution collects data about the activity of visitors using the latest technologies, and this information is used to improve the efficiency of the project;
(2) with the help of the latest technology, the viewer gets the opportunity to get engaged in a dialogue with the authors of the project, to set up the exhibition space upon one's needs or share his opinion about the exhibition.

If the first strategy is an elaboration of the classic methods of cultural institutions working with the audience (mandatory option), then the second trend is part of a general movement towards democratization of culture. Exhibition space or a separate exhibit "responds" to the appearance of the viewer and through this engages him in a dialogue, in intellectual work, in a game, the result of which will be a valuable experience, unique to each viewer. In other words, the Internet of Things can be used as a monitoring system or as a tool to shape out single (depending on specific qualities of the viewer) "texts", addressed not to the crowd, but to everyone personally. It seems that in the framework of the development of communication society this type of cultural products will have great value.

As the communication society develops, the future acquires the characteristic of invariance, presenting to the person more changeable than in previous years [47]. And this, in turn, leads to a new understanding of the communication process, which should become plastic, that will contribute to the evolution of traditional processes of interaction with the audience. From the point of view of the art theory this example leads culture to function in "the death of the Author" mode: a monologue is replaced by a dialogue with the audience. It is through dialogue, through the recognition of the uniqueness of each personality that cultural industry can make a major and long-awaited turn from the formulaic mass production to producing multi-faceted art for the multitude [34].

The presence of a variety of branches in the development of both technologies and cultural practices enhances the trend of nonlinear development in these areas. High probability to face a situation, when tested starting from socio-cultural practices of the artist, designer and curator will order a technology before the current one will put a limit and it sets an independent vector of technological development.

At the same time, there are a number of problems at the level of collaboration in the field of technological development and production of cultural products, innovation management: the lack of a common established terminology understandable to representatives from different industries, the minimum number of common projects, insufficient understanding of the end result in joint activities, the fragmentation of professional interests and goals.

Worldwide, there is an acute shortage of qualified professionals with comprehensive training in the field of the Internet of Things and cyber-physical systems, as well as specialists, programmers and engineers capable of working at the interface of technologies in various fields of activity. In liberal education designers and curators of exhibitions must be also provided with structured knowledge about the possibilities of new technologies to expand the semantic field and enhance the emotional impact of the exhibits.

Obviously, it is needed to expand the possible applications of wireless technologies and sensor networks to create full-functional and actual objects of the cultural industry. There is a number of barriers to widespread the technology of the Internet of Things, including considerable financial and time expenditures in order to implement it in museums, there is also a lack of the amount of energy converted from the external environment which is essential for monitoring complex equipment and periodically send the information to the data center. Active development and practical application of the concept of the Internet of Things inside the cultural industry with the aim of commercialization of the objects become the leading driver of this process and will contribute to its rapid implementation.

6   Conclusion

Radical transformations of the cultural industry have influenced both the structure and form of the modern cultural product, in which today traditional and innovative approaches are organically combined. The spread of information and communication technologies has led to the need to change the traditional patterns of creating a cultural product, taking into account the peculiarities of the communication space in the cultural industries.

The instruments of enhancing participation and immersion become technologically more advanced and accessible, so it is natural that the space of socio-cultural practices will become more personalized and interactive.

The diversity of technological solutions increases the competitiveness of contemporary cultural objects in the market of cultural industries. Both entertainment and psychological aspects of the cultural object are becoming increasingly important regardless of its format.

Thus, we may conclude that the fundamental change in communication processes in cultural facilities has resulted in a substantial transformation that needs to be further integrated and interdisciplinary analysis.

In an attempt to frame the future of the research field, we identify a number of tasks that need to be addressed in the nearest future on the interdisciplinary basis.

1. Foresight - research of prospects of cultural industry development and transformation of its objects with the use of information and communication technologies in the digital economy;

2. Carrying out systematic research on the integration of new technologies in the cultural industry;
3. Comprehensive analysis of requests and visionary projects that humanitarian professionals can submit to scientists, engineers and programmers;
4. The creation of laboratories and research groups which will develop the collaboration of scientists, technicians, curators, artists, etc.;
5. Harmonization of terminology and lexical correspondences in order to improve professional and scientific communication;
6. Preparation and implementation of joint research in the field of digital Humanities (DIGITAL HUMANITY) and breakthrough interdisciplinary projects based on research data;
7. Development of classification of projects that can be implemented in the field of culture with the help of Internet of things technology, technological monitoring of this area;
8. Analysis of psychological and social impact of such projects, processing of feedback;
9. Quantitative evaluation of the IoT projects emerging in cultural industry;
10. Qualitative evaluation of the impact such projects have.

Together, these tasks become part of a unified communication process at the level of determining the directions of development of the cultural industry, education, science, art and design for the coming years. It seems to us that the use of the latest technologies in the sphere of both mass culture and contemporary art can go far beyond the simple "attraction" (which is the case today) and open up the possibility of creating new forms of communication and relations in the system of "artist-piece-viewer". We believe that the introduction of the Internet of things technology, used to expand art mediums, can be the key to the creation of impressive art projects, not only spectacular and entertaining, but also those that will strengthen the democratic component in contemporary art. The latest technologies may open the way not only to new messages, as we can assume recalling M. McLuhan, but also to new kind of experience. As we have argued, it can be the experience of understanding an audience as a "multitude", and each viewer – as an individual with its own creative initiative. In today's economy, where mechanical work is less needed and a creative approach to solving problems at all levels is more and more appreciated, the experience of transferring cultural industry projects to the tracks of co-creation and participation can become a way of radical transformation of society in the direction of more relevant life strategies. Solving this problem, of course, will require systematic work, including both relevant interdisciplinary re-search and the formation of creative laboratories where artists and technicians will be able to take the contemporary challenges together. We invite all interested parties to perspective interdisciplinary cooperation in the field of understanding the processes of practical application of information and communication technologies in the cultural industry.

Acknowledgments. The article was prepared within the framework of the Academic Fund Program at the National Research University Higher School of Economics (HSE) in 2017 — 2018 (grant № 17-05-0017) and by the Russian Academic Excellence Project «5-100».

References

1. Hesmondalsh, D.: Cultural industries/translation from English. I. Kushnareva under academic supervision of A. Mikhalev, National Research University Higher School of Economics. M.: Publishing House. Higher school of Economics (2014)
2. Gómez, J., Huete, J.F., Hoyos, O., Perez, L., Grigori, D.: Interaction system based on internet of things as support for education. Procedia Comput. Sci. **21**, 132–139 (2013)
3. Kawamoto, Y., Yamada, N., Nishiyama, H., Kato, N., Shimizu, Y., Zheng, Y.: A feedback control-based crowd dynamics management in IoT system. IEEE Int. Things J. **4**, 1466–1476 (2017). https://doi.org/10.1109/JIOT.2017.2724642
4. Ma, Y., Wang, Y., Yang, J., Miao, Y., Li, W.: Big health application system based on health internet of things and big data. IEEE Access **5**, 7885–7897 (2017). https://doi.org/10.1109/ACCESS.2016.2638449
5. Mohanty, S., Routray, S.K.: CE-driven trends in global communications: strategic sectors for economic growth and development. IEEE Consum. Electr. Mag. **6**, 61–65 (2017). https://doi.org/10.1109/MCE.2016.2614420
6. Li, J., Yu, F.R., Deng, G., Luo, C., Ming, Z., Yan, Q.: Industrial internet: a survey on the enabling technologies applications and challenges. IEEE Commun. Surv. Tutorials **19**, 1504–1526 (2017). https://doi.org/10.1109/COMST.2017.2691349
7. Chianese, A., Piccialli, F.: Designing a smart museum: when cultural heritage joins IoT. In: 2014 Eighth International Conference on Next Generation Mobile Apps, Services and Technologies, Oxford, pp. 300–306 (2014). https://doi.org/10.1109/ngmast.2014.21
8. Chianese, A., Piccialli, F., Jung, J.E.: The internet of cultural things: towards a smart cultural heritage. In: 2016 12th International Conference on Signal-Image Technology and Internet-Based Systems (SITIS), Naples, pp. 493–496 (2016). https://doi.org/10.1109/sitis.2016.83
9. Chianese, A., Benedusi, P., Marulli, F., Piccialli, F.: An associative engines based approach supporting collaborative analytics in the internet of cultural things. In: 2015 10th International Conference on P2P, Parallel, Grid, Cloud and Internet Computing (3PGCIC), Krakow, pp. 533–538 (2015). https://doi.org/10.1109/3pgcic.2015.56
10. Alletto, S., et al.: An indoor location-aware system for an IoT-based smart museum. IEEE Int. Things J. **3**(2), 244–253 (2016). https://doi.org/10.1109/JIOT.2015.2506258
11. Dao, V.-L., Hoang, V.-P.: A smart delivery system using Internet of Things. In: 2017 7th International Conference on Integrated Circuits Design and Verification (ICDV), pp. 58–63 (2017). https://doi.org/10.1109/icdv.2017.8188639
12. Mertz, J., Zapalowski, V., Lalanda, P., Nunes, I.: Autonomic management of context data based on application requirements. In: 43rd Annual Conference of the IEEE Industrial Electronics Society IECON 2017, pp. 8622–8627 (2017). https://doi.org/10.1109/iecon.2017.8217515
13. Al-Ruithe, M., Mthunzi, S., Benkhelifa, E.: Data governance for security in IoT and cloud converged environments. In: 2016 IEEE/ACS 13th International Conference of the IEEE Computer Systems and Applications (AICCSA), pp. 1–8 (2016). https://doi.org/10.1109/aiccsa.2016.7945737
14. Wu, P.-Y., Cheng, C.-W., Kaddi, C.D., Venugopalan, J., Hoffman, R., Wang, M.D.: Omic and electronic health record big data analytics for precision medicine. IEEE Trans. Biomed. Eng. **64**, 263–273 (2017). https://doi.org/10.1109/TBME.2016.2573285
15. McKee, D.W., Clement, S.J., Almutairi, J., Xu, J.: Massive-scale automation in cyber-physical systems: vision and challenges. In: 2017 IEEE 13th International Symposium on Autonomous Decentralized System (ISADS), pp. 5–11 (2017). https://doi.org/10.1109/isads.2017.56

16. Baskhar, M.: Curation: The Power of Selection in a World of Excess. Piatkus, London (2016). 368 p

17. Paul, C.: Digital Art. Thames & Hudson, New York (2003)

18. Bishop, C.: Radical Museology, or, What's "Contemporary" in Museums of Contemporary Art. Koenig books, London (2015). 80 p

19. Simon, N.: The Participatory Museum (2010). 388 p

20. Beacons: Exploring Location-Based Technology in Museums (2015). https://www.metmuseum.org/blogs/digital-underground/2015/beacons. Accessed 15 Jan 2018

21. Guggenheim App Adds Feature to Highlight Artworks Near Users (2015). https://www.guggenheim.org/news/guggenheim-app-adds-feature-to-highlight-artworks-near-users. Accessed 15 Jan 2018

22. iBeacon brings museum to life (2014). https://press.prophets.be/ibeacon-brings-museum-to-life. Accessed 15 Jan 2018

23. MAMM learned how to communicate with visitors via iPhone (2015). https://rg.ru/2015/06/02/mamm-site.html. Accessed 15 Jan 2018

24. IoE-Driven 'Gallery One' Boosts Attendance and Repeat Business for CMA – Cisco (2014). http://internetofeverything.cisco.com/sites/default/files/pdfs/Cleveland_Museum_Art_Jurisdiction_Profile_final.pdf. Accessed 15 Jan 2018

25. Hsy, S., Fait, H.: RFID enhances visitors' museum experience at the Exploratorium. Commun. ACM **48**(9), 60–65 (2015). https://doi.org/10.1145/1081992.1082021

26. For the first time the implementation of the RFID system took place in museum and display area (2015). https://www.popmech.ru/business-news/233965-vpervye-v-muzeyno-vystavochnom-prostranstve-realizovana-sistema-rfid-razmetki/. Accessed 15 Jan 2018

27. THE INTERNET OF THINGS (2012). https://australianmuseum.net.au/blogpost/museullaneous/the-internet-of-things. Accessed 15 Jan 2018

28. Rozenson, I.: Fundamentals of the Theory of Design. Peter, SPb. (2013)

29. Swedberg, C.: London history museum adopts technology of future. RFID J. (2011). http://www.rfidjournal.com/articles/view?8705. Accessed 15 Jan 2018

30. Alberro, A., Buchmann, S.: Art after conceptual art. In: Wien (eds.) Reihe Sammlung Generali Foundatio (2006)

31. Bishop, C.: Installation Art: A Critical History. Tate Publishing, London (2005). 144 p

32. Shuripa, S.: Things and objects. Polemics around minimalism. In: The oeuvres of ICA, Moscow, vol. 1, pp. 31–81 (2013)

33. Parfait, F.: Installations in Collections/Collection. New media installations. Editions du Centre Pompidou, Paris, pp. 33–63 (2006)

34. Virno, P.: A Grammar of the Multitude. For an Analysis of Contemporary Forms of Life. MiT Press (2004)

35. IEEE 802.15 WPAN Task Group 1 (TG1). http://www.ieee802.org/15/pub/TG1.html. Accessed 15 Jan 2018

36. IEEE 802.15 WPAN™ Task Group 4 (TG4). http://www.ieee802.org/15/pub/TG4.html. Accessed 15 Jan 2018

37. IEEE 802.11TM WIRELESS LOCAL AREA NETWORKS. http://www.ieee802.org/11/. Accessed 15 Jan 2018

38. IEEE 802.15 WPAN™ Task Group 6 (TG6) Body Area Networks. http://www.ieee802.org/15/pub/TG6.html. Accessed 15 Jan 2018

39. Dancing in the dark: a team from Dnepropetrovsk has created a computer show that amazed the world. https://ain.ua/2015/03/27/tancuyushhie-v-temnote-kak-komanda-iz-dnepropetrovska-sozdala-kompyuternoe-shou-porazivshee-mir. Accessed 15 Jan 2018

40. Light Balance. https://lightbalance.net/. Accessed 15 Jan 2018

41. Svetlisa: Theater of light performance. https://svet-litsa.com/performances. Accessed 15 Jan 2018
42. Shadow play. https://eurotour-group.ru/ekskursii/Czech-Republic/teatr-tenei.html. Accessed 15 Jan 2018
43. What is a Black Box Theater? http://www.wisegeek.com/what-is-a-black-box-theater.htm. Accessed 15 Jan 2018
44. Théâtre Le Comedia. https://www.le-comedia.fr/fr. Accessed 15 Jan 2018
45. Milgram, P., Kishino, A.F.: Taxonomy of mixed reality visual displays. IEICE Transactions on Information and Systems, **E77-D**(12), 1321–1329 (1994)
46. The TPCAST Wireless Adapter attaches to the top of the VR headset and replaces the need for cables. https://www.tpcastvr.com/product. Accessed 15 Jan 2018
47. Shuripa, S.: Action and meaning in the art of the second half of the twentieth century. In: Proceedings of IPSI, vol. 3 (2017)

The Integration of Online and Offline Education in the System of Students' Preparation for Global Academic Mobility

Nadezhda Almazova, Svetlana Andreeva, and Liudmila Khalyapina[✉]

Peter the Great Saint-Petersburg Polytechnic University, Saint-Petersburg, Russian Federation
almazovanadia1@yandex.ru, andreeva.teacher@gmail.com,
lhalapina@bk.ru

Abstract. In this study authors examined the growing role of the process of global academic mobility for the situation of professional staff preparation for the new conditions of growing global market. The requirements for the system of education lie in the field of searching and suggesting new approaches and technologies to solve this task, in particular, to find ways of teaching English academic discourse as it is the tool of communication in academic and professional environment. At the same time, nowadays many ESL teachers are studying the ways of implementing online courses into classroom education. This ongoing paper is devoted to the theoretical and experimental study of different models integrating online and offline education for this purpose and for students' preparation for effective usage of various genres of academic discourse in the English language as a language of global communication. The findings revealed the growth of various types of competences necessary for global academic mobility.

Keywords: Globalization · Internationalization · Global academic mobility
Digital education · Integration · Online education · Offline education

1 Introduction

With the development of the processes of globalization and the internationalization of the economy and business, a new goal has arisen before higher education - the training of professional staff able to work effectively in the changed conditions of the global market. For this purpose, the programs of global academic mobility are becoming more and more popular. Thus in 2017 the 39th Session of the General Conference of UNESCO adopted a document which gave the green light to continue the work on UNESCO's academic mobility convention - the Global Convention on the Recognition of Higher Education Qualifications, which underlined the importance of this idea for the whole world.

Despite the extensive growth of international students participating in the programs of academic mobility which is really becoming global, the scientific issue of this phenomenon has only started to be under detailed investigation. Different scientists try to analyze the importance of global academic mobility from various points of view. On

© Springer Nature Switzerland AG 2018
D. A. Alexandrov et al. (Eds.): DTGS 2018, CCIS 859, pp. 162–174, 2018.
https://doi.org/10.1007/978-3-030-02846-6_13

the one hand, they study the ways of expansion the curricula for effective training of students and professors in foreign partner universities improving the level of education in general. On the other hand, and this idea is very close to us, they try to investigate the process of global academic mobility as a factor eliminating the barriers to the international movement of educational services, intellectual products and human capital, strengthening international ties of universities, the formation of international consortia of universities, the growth of homogeneity and continuity of the world educational space [1]. The development of this interpretation of the phenomenon of global academic mobility is directly connected with the problem of searching for new approaches and methods of students' preparation for participation in these programs. Our suggestion is to come to the task of integration of online and offline education and to prepare student in various genres of academic discourse in the English language as a language of global communication, which in return will help them to become active participants of international programs and develop intellectual product of global society.

For this regard, a special research is needed. The first idea of the research is to evaluate the theoretical position of two interconnected scientific spheres: of global academic mobility and possible models of integration of online and offline education for students' preparation, and the second to observe the results of experimental teaching of students.

The paper presents preliminary results of the experimental research, aiming to develop the level of students' competences in different genres of academic discourse in the English language.

The paper's content is structured as follows. Firstly, we conduct a literature review, providing different viewpoints on global academic mobility and integration of online and offline education. Secondly, we describe the procedure and results of our experimental education and give their interpretation. In conclusion we summarize our findings.

2   The Global Academic Mobility and Integration of Online and Offline: How Are These Two Concepts Interconnected?

The concepts of academic mobility and global academic mobility are closely connected with such scientific idea as internationalization of education, which appeared in the 90s years of the XX century. Many scientists agreed that today the development of the internationalization of higher education is under the influence of objective trends of internationalization and globalization of economic life, implying the inclusion of an international intercultural and global dimension in the goals, methods and tools for the provision of higher education [2].

But at the same time it is quite clear that different approaches to the interpretation of the notion "internationalization of education" and even "international universities" could be found. Firstly, when this notion appeared (Scott, 1994, Sadlak, 1998) it was evidently used only for formal defining of the universities as more competitive than the others. The distinctive marks here were correlated with the number of foreign students or professors at this or that particular university. Further on the situation was radically changed. Nowadays the greater part of the researches in this sphere: Jane Knight [1],

Hans de Wit [3], Minna Söderqvist [4] and others are defining the internationalization as a change process leading to the inclusion of all international dimension in all aspects of its holistic management in order to enhance the quality of teaching and research and to achieve all the desired competences [3].

The role of internationalization of education is nowadays passing from its connection with contribution to country-wide growth and innovation to influence key areas of the world and global development [5].

These ideas of the final goal of internationalization to influence the global development through global and transnational projects where joined together intellectual, technological and economical recourses from different countries and universities are collaborating is a brand new trend in the development of this concept. There is an increase of transnational education institutions [6].

The forms of transnational education are becoming different and include various forms of academic mobility: from traditional mobility of students to a number of universities having very close net programs to 'any education delivered by an institution based in one country to students located in another' [7]. In this case more often we can speak about branch campuses, off-shore degrees, virtual universities, and distance education [1].

In other words, we speak about global academic mobility as an instrument or as a dimension of internationalization [8].

At the same time, we agree with those authors who argue about small enough amount of students participating in these programs in comparison with the whole number of students. "If internationalization is as high a priority as policy makers and higher education internationalization (HEI) leaders affirm, mobility must be greatly expanded" (Erasmus). In the report of this organization we found the following statistics: there is steady growth in mobility of degree seeking students: – In 15 years (1999–2014) the number of mobile students more than doubled from 2 to 4.3 million, but in the same 15 years total student numbers grew approximately equally: from 94 to 207 million, which gives the following conclusion: growth in absolute number of mobile students, no growth in proportion to all students – stable at 2%.

Due to these data and the results of theoretical investigations and empirical observation we came to the conclusion that one of the reasons of non-growing percentage of students participating in the programs of global academic mobility is not high enough level in their English language preparation in different genres of academic discourse which is used as a means of intercultural and professionally-oriented communication in foreign countries. The proof of this empirical research will be described in the next part of this article.

Our idea is supported by some European scientists who connect language teaching and learning with internationalization agendas because these are in turn driven by government agendas [9].

Now we would like to say some words about the so-called digital education which is becoming more and more popular in the digital society. Speaking about the wide development of distance education for the purposes of internationalization we should take into account the methodological advantages of integration of big opportunities of online education and our traditional ones which are still popular and demanded.

Recently a very famous phenomenon of mass open online courses (which in this work are designated by the term "online courses"), for which the set of platforms is created. There is an opinion that MOOC are the books of new generation containing videos, interactive tasks and a social component which at the same time "read" tens of thousands of people. It is obvious that online courses cannot replace completely classical higher education for a number of reasons. The task of the teacher consists in reasonable use of the Internet resources during the classroom work and to help students to organize their self-study minimizing or eliminating negative or useless role of incorrect integration of online courses into the process of traditional offline education.

Our research was concentrated on the development of possible models of integration of online and offline education for students' preparation in the sphere of English language academic discourse which will in its turn prepare students to participate in the programs of global academic mobility.

Many Russian and foreign scientists, such as N. Burmakina, L. Kulikova, A. Gillett, K. Hyland, and others, study the concept of academic discourse, its genres, constitutive and linguistic features of its oral and written forms. According to K. Tsymbal, academic discourse is a type of institutional status-role communication, participants of which are teachers and students, members of the scientific community, who must communicate in accordance with the norms of speech behavior adopted in it, in the educational and scientific situation of communication [10]. R. Patterson and A. Weideman state that academic discourse comprises all spoken and written activities connected with academia [11]. The functions of academic discourse are exposition, clarification and conclusion, therefore authors of academic texts are required to explain, define, compare and contrast, classify, conclude, etc. [12]. Spoken and written forms of academic discourse have their own features, which differ this type of discourse from others. Specific features of written academic discourse include:

1. Analytical, objective presentation of the material.
2. An impersonal, formal style of exposition.
3. Lack of colloquial and slang vocabulary (including, limited use of phrasal verbs).
4. Accurate use of vocabulary.
5. The use of specialized vocabulary, acronyms, abbreviations, passive verbal constructions, complex and impersonal sentences [13].

In terms of its distinctive characteristics, oral academic discourse has about the same set of linguistic and grammatical features as written form of academic discourse [14]. However, in oral speech there is a certain weakening of the formalities, which is caused by decrease in the number of complex sentences, more frequent use of first person pronouns, verbal constructions, lexical repetitions.

Among the genres of academic discourse is a lecture, seminar, essay, abstract, presentation and others. In this paper we dealt with those related to the curriculum of the discipline "English for Academic purposes" at the 1st and 2nd course of technical faculties at Peter the Great Polytechnic University, namely a comparative essay, lecture notes, a description of illustrative material (graph, chart or table) and an academic presentation. Each of the genres has its own structure and specific characteristics.

One of the most popular genres of written academic discourse is essay. In her paper A.G. Martynova defines the term essay as "an integral, coherent, logical, organized written work on a given topic in a limited scope, with the aim of demonstrating the competence of the author in a certain field by proposing a thesis and its proof/explanation". It is also added that ideas presented by the author in the essay are generated on the basis of the processed or reproduced information from the texts of other authors, and the style of presentation takes into account the lexical-grammatical and logical-compositional features of texts created earlier within a certain discipline [13]. One of the important characteristics of an academic essay is the type of rhetorical function that it performs. The rhetorical function specified in the writing instruction or in the title of an essay determines its purpose and type. Among the rhetorical functions (or cognitive genres) of an academic essay is description, narration, comparison and opposition, and others [15].

Academic essay is characterized not only by a clear structure, but also by a specific register, which is determined by the peculiarities of the context of this genre - the author demonstrating their competence, the discourse addressed to the teacher, and also the formality of writing [16].

The genre of a lecture in an academic environment is no less important than an essay. Characteristic features of this genre are the use of special lexical and grammatical elements that perform different functions, for example, encouraging students to talk, ask, confirm, disagree and others [17]. Each lecture has an organizing component, the share of which can vary. Among the organizing markers of the lecture N.G. Burmakina distinguishes the following: markers of expression of author's opinion, markers of citation and compositional signals, which help listeners orient themselves during the presentation [18]. Knowledge of the structure of the lecture and English-speaking identification phrases will help students navigate in the monologue of a teacher, highlighting the main points and predicting the content of a lecture.

One more genre of academic written discourse is the description of illustrative (graphical) material (describing visuals). The presence of tables and graphs in the text is one of the specific features of academic written discourse, and their description is characterized as a separate genre. The elements of the structure of this genre perform a number of functions, each of which is characterized by certain lexical and grammatical features: complex sentences, nominal constructions, use of times, vocabulary illustrating the growth and decline of tendencies, comparative and superlative degrees of adjectives. Knowing these features will allow students to choose the most relevant grammatical constructions and vocabulary to describe the schedule [19].

Academic presentation is used in almost all genres of oral academic discourse: a report at a conference, at a seminar, lecture, defense of a thesis, etc. In accordance with the classification of A.V. Oljanich, academic presentation combines the characteristics of scientific (information transfer), professional (the transfer of skills and abilities), pedagogical (the transfer of knowledge, historical, spiritual and cultural values) and presentation (self-presentation) discourses [20, 21].

3   Method and Procedure

3.1   Participants

The study was carried out at the premises of Peter the Great Polytechnic University in Saint-Petersburg. The participants of the experiment training were students of the 1st and 2nd courses of technical departments. The experiment had two groups at each course: (1) the control group without an intervention and (2) the experiment group, where participants were trained academic discourse with the integration of online and offline courses. On the 1st course the number of students participating in the experiment was 20 in the control group and 20 in the experiment group. The control group of the 2nd course comprises 22 students, while the number of students in the experiment group equals 24.

3.2   Procedure

We carefully studied the syllabus for these groups of students and identified the following genres of academic discourse covered during the Fall 2017 semester: comparative essay, covered at the 1st year at university, academic presentation and written description of visual information (for the students of the 2nd university year). Students were asked to complete the assignments in the above-mentioned genres before the experiment training, and therefore we were able to assess the initial level of their skills and abilities in academic written and spoken discourse. For the genres we developed assessment criteria, each of them rated 0–5: (1) Formality/Objectivity, (2) Cohesion and Coherence, (3) Argumentation, (4) Lexical Resource, (5) Grammatical Range and Accuracy.

Hereafter we describe the procedure of selection the materials of MOOCs based on the syllabus of the discipline "English for Academic Purposes". Interconnected use of online and offline courses in the framework of higher education means that the choice of the MOOCs materials is based on the university curriculum of the discipline "English for Academic purposes". We distinguished the two types of this curriculum-based model of integration of online and offline courses, which we called complementary model and supporting model. Complementary model is aimed at filling voids in the discipline's syllabus using MOOCs' materials to provide more information on the topic, which is not covered in the main textbook. Supporting model is used to provide more exercises and revision of the materials offered by the main textbook. These two models can be used in collaboration for more effective training. Table 1 gives the characteristic of the chosen MOOCs and the models of integration of online courses and the main textbook offline course.

Table 1. Models of integration of online and offline resources

Academic discourse genre	Integration model	Materials of the main textbook	MOOC	MOOC materials
Comparative essay	Complementary model	Example of a comparative essay; Exercises on structure of a comparative essay and linking words and phrases	Preparing for the AP English Language and Composition Exam (edX.org)	Video-lecture concerning proofreading, materials for outlining an essay
Academic presentation	Supporting and complementary model	Activities and advice on presentation preparation, structure, and delivery	Presentation Skills: Effective Presentation Delivery (coursera.org). Designed to help students develop their presentation skills	Video-lectures covering extralinguistic aspects of the genre: fear and anxiety, gestures, interaction with the audience and improvisation
Written description of visual information	Supporting and complementary model	Vocabulary describing trends, an example of written description of a bar chart, exercise on description structure	IELTS Academic Test Preparation (edX.org). Aims to provide students with the knowledge and practical skills required to successfully prepare to take the IELTS Academic Test	Video-lectures and practical exercises on types of visual information, description structure, vocabulary and grammar specific for the genre

The materials placed on the above-mentioned MOOCs were carefully selected in accordance with the syllabus so that they complement the main course book used at the lessons. Following methodological principles and taking into account the initial level of students' skills of academic discourse we identified the aspects of each of the discussed genres of academic discourse which were not sufficiently covered in the main textbook. The lacking materials were implemented in the educational process from MOOCs. In the classroom we introduced parts of MOOCs which were selected in accordance with goals and objectives of the lesson. Participants were also given tasks which they were to complete individually or in groups with the further presentation of the results of their work at the lesson. In the next sections we describe the process of implementation MOOC materials into the traditional classroom training in experiment groups of the 1st and 2nd courses.

Comparative Essay. The course "English for Academic Purposes" begins on the 1st year of training of students in technical faculties. The first genre with which freshmen work is a comparative essay. The main textbook presents exercises for developing the skills and abilities of this genre of academic discourse, which include an example of a comparative essay on the topic "Are women better drivers than men?", familiarizing with the structure of a comparative essay, exercises on linking words and phrases, and discussion of the topic of a comparative essay that students are to write as a homework. The analysis of initial level of skills and abilities of students in this genre of academic discourse, showed that the main problem faced by students in this type of work is inability to logically build their thoughts and ideas. Elements of the MOOC "Preparing for the AP English Language and Composition Essay" helped students cope with the aforementioned difficulties, and teachers were able to fulfill the deficit of class hours devoted to the discipline "English for Academic Purposes". The materials of the MOOC's section "Outline before you write" provide students with the materials to help them learn to structure their ideas in an essay logically and consistently.

Academic Presentation. Second-year students of technical faculties work with such genre of oral academic discourse as presentation. The sections of the Language Leader Intermediate textbook include topics that highlight the problems of preparing for an academic oral presentation (Preparing for a Talk), and delivering it (Delivering a Talk). The textbook offers exercises aimed at introducing students to the process of preparing their speech: students discuss their experience of speaking before an audience, listen to advice on arranging the preparation of an oral presentation, prepare mini-speeches on the given topics in pairs. They also work with the procedure of a presentation, as well as with typical mistakes of speakers. We deemed it necessary to supplement the material offered in the main textbook by the MOOC "Presentation Skills: Effective Presentation Delivery", presented on the coursera.org platform. The course covers such extralinguistic aspects of the genre as fear and anxiety, gestures, interaction with the audience and improvisation. The course consists of 4 sections which were studied by groups of students. Participants processed the information presented in the chosen part of the MOOC and prepared a presentation for their fellow students. Thus every participant got familiar with the essential extralinguistic aspects of academic presentation and also practiced delivering a talk.

Written Description of Visual Information. In this section we describe the process of implementation the materials from the MOOC "IELTS Academic Test Preparation" into the traditional classroom training. According to the syllabus of the discipline "English for Academic Purposes" on the 2nd course, students study the academic discourse genre "description of a chart" by the main textbook "Language Leader Intermediate". However, the results of the initial level of students' skills and abilities in this genre of written academic discourse show that training needs to be complemented by additional exercises and materials. To complete this type of assignment successfully students needed to get insight into the conventions of the genre, namely peculiarities of vocabulary and grammar use within the description of visual information. The exercises provided by the main textbook are aimed at students who have worked on this genre before, and therefore they do not cover all the aspects, such as specific vocabulary grammar constructions used to describe trends and changes in the graph or chart over time. Moreover, the vocabulary exercises are not developed in accordance with the main stages of working with vocabulary, which include familiarization and semantization, primary revision and development of skills and abilities of using vocabulary in various forms of oral and written communication [22]. For this reason, we chose the above-mentioned MOOC, which contains a wider range of vocabulary and more examples of its use in the description of a visual. The materials of the MOOC are presented in the form of video lectures, which were shown and discussed in the classroom, and interactive exercises completed by students individually. Interconnected application of online and offline courses in the course of academic discourse training facilitated students' interest and motivation to study English language.

3.3   Data Analysis

This section discusses the level of students' skills and abilities before and after experiment training. In Sect. 3.2 of this paper we described the assessment criteria for each of the chosen genres of academic written and spoken discourse. To get the most objective results we decided to assess the initial level of students' skills and abilities in each of the discussed genres of academic discourse. After the experiment training we again assessed students' skills and abilities in experiment and control groups to provide insight into the progress of participants of our experiment. At each stage we asked students to perform a task which we then assessed according to the criteria described in Sect. 3.2, the students' results were then summed up and arithmetical average in each group was calculated. These data were then compared to each other.

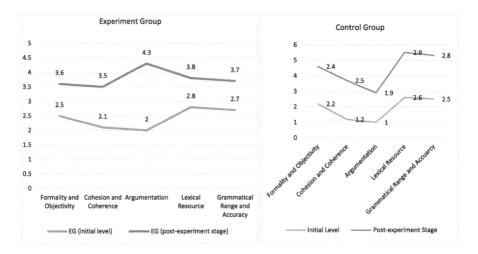

Fig. 1. The comparison of the initial level of students' skills and abilities with their skills and abilities at the post-experiment stage in the genre of a comparative essay.

This section discusses the level of students' skills and abilities before and after experiment training. In Sect. 3.2 of this paper we described the assessment criteria for each of the chosen genres of academic written and spoken discourse.

Comparative Essay. Figure 1 shows the comparison of the initial level of participants' skills and abilities in the genre of comparative essay in experiment and control groups and their level of skills at the final stage of experiment. The line graph indicates that the average initial level in the control group is slightly lower than that in the experiment group. While the progress in the development of students' skills and abilities in the experiment group is obvious, the students of the control group also made progress, but it remains insignificant. The average score for the first comparative essay in the experiment group was 10.1 points (out of 25 possible), in the control group - 9.8 points. After the experiment training, the students of the experiment group scored an average of 18.9 points, while in the control group the average score for the essay rose by 2.8 points and amounted to 12.6 points.

Academic Presentation. In this section we compare the initial level of students' skills and abilities in academic presentation with the level of their skills and abilities in this genre at the final stage of our experiment. Figure 2 demonstrates that students had the highest score on the criteria of "content" and "cohesion and coherence" (4.2 points). These aspects were partially covered in the materials of the offered MOOC. Moreover, the course video-lectures themselves are the genre of academic discourse, since the lecturer in the video, when explaining the rules and conventions of speaking to the audience, is himself a clear example of their compliance.

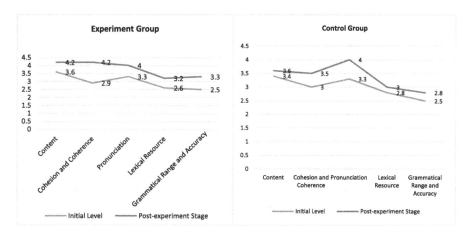

Fig. 2. The comparison of the initial level of students' skills and abilities with their skills and abilities at the post-experiment stage in the genre of an academic presentation.

Overall, the progress is observed in both groups, although the students of experiment group demonstrated the higher level of skills and abilities in the genre of academic presentation after the experiment training. We also carried out a correlation analysis of the data obtained during the determination of the initial level of skills and abilities, in order to identify the relationship between the individual components of academic presentation and to select the most suitable MOOC materials. High correlation was found between the scores obtained for the "Content" and "Cohesion and Coherence" criteria (the correlation coefficient is 0.7). We can conclude that students, increasing their level of skills and abilities by one of the above criteria, can also raise the level of the second. The same conclusion can be drawn regarding the criteria "Lexical Resource" and "Pronunciation", between these criteria high level correlation was also detected (correlation coefficient - 0.7).

Written Description of Visual Information. Figure 3 shows the comparison of the results of the assignment by the experiment and control groups at the beginning and in the end of the experiment training. A significant improvement in all criteria in the experiment group should be noted. On average, students increased the score for this type of work by 6.6 points. This shows that the materials we selected to work on the genre, effectively complemented the tasks presented in the main textbook. In the control group the progress is insignificant, students' works showed a certain stiffness in the choice of lexical means, because, in spite of the fact that the textbook offers expressions for describing trends on the charts, they were not enough to describe other kinds of illustrative material, such as diagrams and tables.

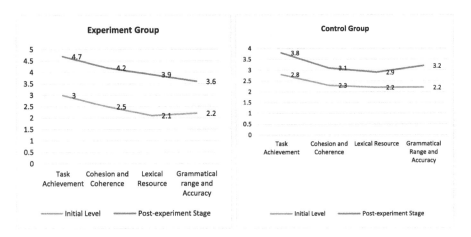

Fig. 3. The comparison of the initial level of students' skills and abilities with their skills and abilities at the post-experiment stage in the genre of written description of visual information.

4   Results

The purpose of this study was to develop possible models of integration of online and offline education which would help the educators in the process of students' preparation to the global academic mobility and in general to put our stone into the foundation of internationalization of the system of higher education.

For getting this goal suggested to implement online courses into traditional training to develop the level of students' competences in different genres of academic discourse in the English language. We analyzed scientific works dedicated to the study of the issues of academic mobility, academic discourse, and integration of online and offline education. We also described the procedure of the implementation of MOOCs into traditional English academic discourse training in the framework of higher education, and analyzed the data obtained during the experiment training.

We found, that despite the wide variety of MOOCs presented on various online platforms, none of them fully meets the goals and objectives of EAP training in a particular institution, if the course is not developed on the basis of this institution. Therefore, it is expedient to use separate elements of different courses in the educational process, which meet the goals and objectives of teaching a particular aspect of the English language. Methodically competently selected online and offline EAP training courses complement each other, solving a number of problems, such as a shortage of class hours devoted to the discipline "English for Academic Purposes", a lack of motivation for students and others. The selected MOOCs (IELTS Academic Test Preparation, Preparing for the AP English Language and Composition Exam - on the edX.org platform and Presentation Skills: Effective Presentation Delivery - on the Coursera.org platform) in combination with the offline course offered by the syllabus of the discipline "English for Academic Purposes" showed their advantages in such academic discourse genres as essay writing, description of illustrative material and academic presentation.

The growing level of skills in all genres of academic discourse which students demonstrated at the end of our experiment allows us to make a conclusion that our model will enlarge students' possibilities to become active participants of the programs of global academic mobility.

References

1. Knight, J.: Transnational education remodeled: toward a common TNE framework and definitions. J. Stud. Int. Educ. **20**(1), 34–47 (2016). https://doi.org/10.1177/1028315315602927
2. Mitchell, D., Nielsen, S.: Internationalization and globalization in higher education. business, management and economics "globalization - education and management agendas". In: Cuadra-Montiel, H. (ed.) Croatia, pp. 3–23 (2012). https://doi.org/10.5772/48702
3. De Vit, H.: What is an International University? - University World News, p. 359 (2015)
4. Söderqvist, M.: Internationalization and its management at higher-education institutions. Helsinki School of Economics (2007)
5. Henard, F., Diamond, L., Roseveare, D.: Approaches to Internationalisation and Their Implications for Strategic Management and Institutional Practice. OECD, pp. 1–46 (2012)
6. Bilecen, B., Van Mol, C.: Introduction: international academic mobility and inequalities. J. Ethnic Migr. Stud. **43**(8), 1241–1255 (2017). https://doi.org/10.1080/1369183X.2017.1300225
7. McBurnie, G., Ziguras, C.: Transnational Education: Issues and Trends in Offshore Higher Education. Routledge, New York (2007)
8. Egron-Polak, E.: Academic mobility in Higher Education worldwide - Where are we? Where might we go in the future? Brussels, pp. 1–26 (2017)

9. Byrom, M.: A note on internationalisation, internationalism and language teaching and learning. Lang. Learn. J. **40**(3), 1–7 (2012). https://doi.org/10.1080/09571736.2011.642528

10. Tsymbal, A.: Intonation-stylistic and communicative-pragmatic characteristics of oral presentations. – Ivanovo, pp. 1–220 (2015). (in Russian)

11. Patterson, R., Weideman, A.: The typicality of academic discourse and its relevance for constructs of academic literacy. J. Lang. Teach. **47**(1), 107–123 (2013)

12. Weideman, A.: Why is academic literacy important? Introduction to Weideman & Van Dyk (Editors). Academic literacy: test your competence. Geronimo, Bloemfontein, pp. 2–9 (2014)

13. Martynova, A.: Teaching academic discourse in the genre of the exposition essay: on the material of senior courses of a language university. Omsk, pp. 1–207 (2006). (in Russian)

14. Using English for Academic Purposes for Students in Higher Education. http://www.uefap.com. Accessed 6 Nov 2017

15. Tsirkunova, S.: On the peculiarities of the development of academic writing skills in the course "English for Academic Purposes" and "IELTS Academic Test Preparation", World of Linguistics and Communication. Tver State Agricultural Academy, Tver, vol. 38(1) (2015). (in Russian)

16. Mineeva, I.: Academic essay: theory and practice of the genre. Philol. Class **2**, 7–14 (2015). (in Russian)

17. Malavska, V.: Genre of an academic lecture. De Gruyter Open **3**(2), 56–84 (2010). https://doi.org/10.1515/llce-2016-0010

18. Burmakina, N.:The genre of the lecture and its culturally conditioned marking. Vestnik of the Leningrad State University named after I. Pushkin, vol. 1(1), pp. 185–191 (2013). (in Russian)

19. Khabarova, V.: Graph description as an issue in L2 academic english writing. J. Lang. Educ. **2**(4), 46–54 (2016). (in Russian)

20. Olyanich, A.: Presentation theory of discourse. Paradigma, Volgograd, pp. 1–507 (2004). (in Russian)

21. Ivanova, S.: Specificity of public speech. Delo, Moscow, pp. 1–124 (2001). (in Russian)

22. Galskova, N., Gez, N.: Theory of teaching foreign languages: linguodidactics and methodology, pp. 161–189 (2006). (in Russian)

LIS Students' Perceptions of the Use of LMS: An Evaluation Based on TAM (Kuwait Case Study)

Huda R. Farhan[(✉)] [iD]

Kuwait Authority for Applied Education and Training, Adailiyah, Kuwait
hudafarhan@gmail.com

Abstract. Including information technology within the learning and teaching environment, especially learning management systems (LMSs), has been under scrutiny for a long time, and studies in this area have focused on both teachers' and students' attitudes toward LMSs. In this paper, an evaluation of the development and application of students' acceptance of LMSs is presented. Public Authority of Applied Education and Training in Kuwait (PAAET) students' perceptions of the use of an LMS in a blended learning environment was investigated. The investigation was based on two methods: pre- and post-usage surveys and analyses of students' actual use of the system through system log mining. The findings show a significant improvement in students' computer skills after using an LMS. Students were willing to use the compulsory and optional components of an LMS if the motivation to do so was present. The nature of the subject being taught affected the students' intention and how they used the LMS. There was no significant difference in the students' perceptions of the use of LMSs before or after actual usage.

Keywords: Learning management systems · Technology acceptance model
Evaluation methodologies

1 Introduction

In the last two decades, more than ten major new technologies in the domain of Internet-based educational aids and tools for task collaboration have emerged (Singh and Reed 2001; [1]. The availability of such innovative approaches has resulted in twenty-first-century students having greater expectations regarding the form and quality of the instruction they receive. Given this demand for new forms and better quality of teaching, it is not surprising that educators and educational institutions are coming under increasing pressure to continually update their teaching methodologies and tools. Moreover, traditional face-to-face teaching and learning interactions within the confines of brick-and-mortar classrooms are incrementally being replaced by modalities of instructional delivery that rely on information and communication technology (ICT), such as distance learning, interactive e-learning, and blended learning. Singh and Reed (2001) noticed that although distance learning and e-learning are primarily ICT-based, blended learning represents a hybrid approach that seeks to

© Springer Nature Switzerland AG 2018
D. A. Alexandrov et al. (Eds.): DTGS 2018, CCIS 859, pp. 175–187, 2018.

combine the best features of ICT and classroom-based procedures for instructional delivery. To be more precise, Singh and Reed (2001, p. 2) define blended learning as an approach that "focuses on optimizing achievement of learning objectives by applying the right learning technologies to match the right personal learning style to transfer the right skills to the right person at the right time." However, whether the method relies exclusively on online learning or blended learning, there is a common component in both instructional methods: a learning management system (LMS).

Of course, the success of any instructional technique relies substantially on the attitudes of the learners [11], Al Fadhli (2009). Therefore, it is important for educators to adopt measures that motivate students' receptivity for non-traditional instructional methods. A number of researchers have tried to assess student acceptance of different LMSs based on various technology assessment models [3, 5, 11]. The findings of these studies emphasize the importance of future research in this field.

The main focus of this study is students' perceptions of the inclusion of LMSs within blended learning environments. Library and information studies students enrolled in two academic courses, database systems management and indexing and abstracting, were surveyed to assess their acceptance of LMSs. The technology acceptance model (TAM) was the essential element of assessment.

2   Related Work

The students' perceptions were assessed using TAM, which was developed by [2], with some modifications. TAM is widely used in the assessment of technology acceptance, whether in online learning or blended learning environments. However, few studies have analyzed students' perceptions toward LMSs in blended learning environments. One of these studies [11] used TAM to measure Greek university students' attitudes toward blended learning when the learning management system Moodle was used. Studying LMS acceptance from the learners' perspectives, the researchers found that perceived usefulness and ease of use had a positive effect on students' attitudes toward system usage. In an online learning environment, [1] used TAM to assess IDEWL, an online learning management system; the focus of the study was on the effect of interactivity on student usage, and the findings emphasized the influence of system interactivity on system effectiveness. [14] applied TAM to investigate higher education students' acceptance of LMSs compared to cloud technology; the findings indicated students have a higher comparative acceptance of cloud technology compared with LMSs. In this study, the three main TAM items—perceived ease of use, perceived usefulness, and attitude toward system usage—registered higher with cloud technology.

Some [10] used TAM to investigate pre-service teachers' attitudes toward the use of computers. Applying TAM, [6] investigated university students' acceptance of technology in an e-learning environment and found that TAM was effective in assessing the students' acceptance of e-learning. TAM's validity and reliability have also been tested in other studies [8, 12, 13]. In line with previous research, TAM was chosen for this study to scrutinize students' LMS acceptance.

It must be emphasized that there is ample choice of both commercial and free LMSs designed for online and blended learning environments. Some of the more well-known LMS packages include LMS ANGEL, BlackBoard, Desire2Learn, Pearson eCollege, Canvas, and Moodle. The effectiveness and suitability of these software application suites have been examined in prior investigations. For this research effort and because of financial and managerial constraints, the free version of BlackBoard was used. This application is available online and can be readily accessed from any computer that has an Internet connection. BlackBoard provides most of its available tools through the fee-based version of the system.

An LMS can improve a learning experience for both instructors and learners [14], and students may use LMSs for a variety of applications, such as acquiring and providing information (including downloading instructional material and assignments, posting and reading messages on online discussion boards, managing wikis, uploading assignments, exchanging files, and receiving announcements concerning upcoming events and grade notifications); accessing course materials (including YouTube videos, PowerPoint slides, and documents); participating in person-to-person communication, collaboration, and social learning via wikis and discussion boards (including communicating with instructors and classmates to receive relevant feedback concerning academic performance and sending e-mails and other messages); and taking online, open-book tests and measuring performance.

3   Methodology

The fundamental hypotheses for assessing students' acceptance of the chosen LMS are as follows:

H1: Utilization of an LMS has an impact on students' acceptance of LMSs because the LMS meets the students' learning preferences.

H2: Social factors affect students' acceptance of LMSs when they are applied in a blended learning environment.

H3: The nature of the courses taught using an LMS can strongly determine students' LMS acceptance.

This study was designed to answer several research questions: How did students perceive the difficulty or ease of use of BlackBoard? Does the actual usage of BlackBoard improve students' system acceptance? If so, to what degree? What is the effect of the nature of the course being taught (subjective vs. objective) on students' actual use of BlackBoard? How does students' acceptance of LMSs in a blended learning environment differ from expectations based on previous studies?

3.1   Instrument Development

Given that this study's principal focus is the evaluation of students' perceptions and attitudes toward the introduction of an online LMS, a questionnaire was developed and distributed before the start of the evaluation period to determine the students' pre-existing opinions about the effectiveness of online LMSs such as BlackBoard; this pre-evaluation also helped ascertain student readiness to accept and actively adopt blended learning approaches. A similar questionnaire was distributed at the end of the study, before the students were notified of their final grades; this was intended to measure any differences in attitudes resulting from interaction with the LMS.

This study's questionnaire uses instruments that focus on the measurement of the main TAM variables, namely perceived usefulness (PU) and perceived ease of use (PEOU) [2]. Attention was also devoted to the measurement of other variables operationalized in other TAM models that were modified after [2]' investigation. These variables are attitude toward use (ATT) [7, 14] and behavioral intention (BI) [12]. The questionnaire items used in this study were derived from previous studies and were only modified to accommodate any socio-cultural factors particular to Kuwait. A 7-point Likert scales ranging from strongly agree to *strongly disagree* was used to quantify user responses to the various questionnaire items.

To investigate students' stated perceptions of ease or frequency of use versus actual usage mentioned in Sect. 3, the following procedure was followed. Given the possibility that responses to survey items may not always be genuine, if the participants felt pressured to provide what they believed to be the "right" answer, the actual duration and frequency of the usage of the various BlackBoard services were logged to determine if the participants were utilizing the system optimally. Data mining methods were applied, and BlackBoard reporting tools provided data on the students' activities on the LMS. knowing that the usefulness of data mining in extracting "pedagogically meaningful information" was proven [4]. In addition, students were asked to perform information retrieval tasks related to the subjects taught. Participants conducted the tasks in one of the information studies department labs, and the class instructor and TA were available to resolve any technical problems students may have encountered while performing the tasks and to ensure that each participant performed the tasks without any external help. The computers used to perform the tasks were screen self-recorded, and participants were informed of the study and agreed to participate.

The design of the study adhered to factors affecting academic performance (Vector Mlambo 2011), and the design was considered when analyzing the data. Several factors can affect students' academic performance, including learning preferences; whether course materials have been designed to facilitate different learning styles (per Neil Flemming's (2001–2011) description); class attendance; interaction with course materials, instructors, and classmates through BlackBoard; entry qualifications and prerequisites (acquired via demographic survey); age; and gender.

3.2 Participants (Sampling)

Subjective convenience sampling was used to collect data because the implementation of LMSs is very limited, if not absent, at the college at which the current study took place. This explains the considerably small study population. Ninety-five undergraduate students (third- and fourth-year students) who were officially registered for two different library and information science classes were surveyed. A total of 27 students enrolled in a database management course and 68 students enrolled in an indexing and abstracting course were recruited. The two selected courses differ in nature. Indexing is based on subjective assessment (qualitative), and do not require or aims for the improvement of computer skills. Database management systems course based on objective assessment (quantitative) and require basic computer skills, introduction to computers is a course prerequisite for enrollment. The latter course also has the goal of students mastering database management related to libraries. Hence, the course supposedly improves computer skills. Both courses are optional and are taught as part of the undergraduate-level library and information science program at the Public Authority of Applied Education and Training (PAAET). The decision to recruit the study subjects from these two courses was made because these were the only courses at the department in which an LMS was utilized. The two courses were taught by the same instructor.

4   Results

Cronbach's alpha was used to measure the questionnaire's reliability coefficient. The overall reliability of the questionnaire is 0.96, indicating that the questionnaire is reliable. Ninety-five questionnaires were distributed at the beginning of the semester and again at the end of the semester, and the response rate of the pre-usage questionnaire was 81%. The total number of 77 participants was distributed over the two classes as follows: students enrolled in database management totaled 20 participants, or 74% of the total enrolled students. Students enrolled in the indexing and abstracting were 57 participants 83.8% of the total enrolled students. As for the participants' age ranges, in the pre-usage questionnaire, 63 participants were young students (18–25 years old), and 14 were mature students (26 years of age or older). Although in the post-usage questionnaire, 75 participants were young students (18–25 years old), and 11 were mature students (26 years of age or older).

The post-usage questionnaire response rate was higher than the pre-usage questionnaire. The total number of participants from both groups was 86 (90.5%), distributed as follows: For students enrolled in the database management class (objective), the participation rate was 88.8% (24 participants). For students enrolled in the indexing and abstracting class (subjective), the participation rate was 91.2% (62 participants). There was no significant difference between the young students' and the mature students' acceptance of the LMS. This is in line with [9], who found no differences in attitudes toward computer usage between different age groups. The participation rate was considerably high, particularly for the post-questionnaire in both classes. The response rate was also higher for the indexing class in both questionnaires.

Computer Skills (Pre-Questionnaire). More than 80% of the participants had e-mail addresses and used e-mail to communicate with instructors and classmates, and 95.3% stated that they had e-mail accounts that they used to send and receive e-mails and e-mail attachments. Most participants (87.5%) stated that they used their e-mail accounts to communicate with their teachers and colleagues. Almost all participants (90.6%) used social media to communicate with family and friends and send pictures and videos. It was also found that 32.8% of the participants were accustomed to downloading study materials available on teacher or college websites. Additionally, 42.2% of the students said that they did their homework through teacher or college websites. A small percentage of students (20.3%) indicated that they took exams online, and

26.6% used the messenger applications on teacher or college sites to communicate with classmates (23.4% did so to communicate with the teacher). Because the college provides students' final grades on the PAAET e-gate, most participants (73.4%) indicated that they got their grades online.

Less than 40% of the participants stated that they had used an online medium for learning purposes in prior classes, such as for exchanging documents or for exam administration. Notably, the student answers indicated that no one had previously used an LMS such as BlackBoard. In fact, only a small number of students used Internet facilities in their learning environments, though most of the students had used Internet facilities for communication purposes.

Computer Skills (Post-Questionnaire). A T test showed that, after using Black-Board, the computer skills of the students improved (Tables 1 and 2). This finding relates to [12] findings that "computer self-efficacy acts as a determinant of perceived ease of use both before and after hands-on use." Furthermore, there was no significant difference in the students' perceptions of BlackBoard before and after use. Therefore, only the post-questionnaire will be analyzed in detail.

Table 1. T test results according to course variable (pre-test sample).

Course		N	Mean	Std. deviation	t	df	Sig. (two-tailed)
Computer skills	Database management systems	16	1.52	0.194	.659	62	0.51
	Indexing and extraction	48	1.48	0.184			
Usefulness of the system	Database management systems	16	3.83	0.607	.026	62	0.98
	Indexing and extraction	48	3.82	0.844			
Ease of use	Database management systems	16	3.33	0.669	−1.368	62	0.18
	Indexing and extraction	48	3.62	0.748			
Student perception	Database management systems	16	3.52	0.727	−1.801	62	0.08
	Indexing and extraction	48	3.96	0.888			

Stated Perception of Effectiveness of Use

1. On the questionnaire, 58.2% of the participating students stated that they agreed/totally agreed that the use of BlackBoard enabled them to accomplish tasks more quickly.
2. On the questionnaire, 62.8% of the participating students stated that they agreed/totally agreed that using BlackBoard enhanced their performance during the practical sessions.

Table 2. T test results according to course variable (post-test sample).

Course		N	Mean	Std. deviation	t	df	Sig. (two-tailed)
Computer skills	Database management systems	24	1.66	0.210	−.433	84	0.67
	Indexing and extraction	62	1.68	0.216			
Usefulness of the system	Database management systems	24	4.08	0.883	2.124	84	0.04
	Indexing and extraction	62	3.58	1.032			
Ease of use	Database management systems	24	3.83	0.976	1.490	84	0.14
	Indexing and extraction	62	3.49	0.934			
Student perception	Database management systems	24	3.90	1.115	.475	84	0.64
	Indexing and extraction	62	3.78	0.945			
Expected behavior toward the program	Database management systems	24	3.69	1.007	.967	84	0.34
	Indexing and extraction	62	3.45	1.090			

3. On the questionnaire, 62.8% of the participating students stated that they agreed/totally agreed that using BlackBoard made it easier for them to follow and study the course material.
4. On the questionnaire, 66.3% of the participating students stated that they agreed/totally agreed that by using BlackBoard, they were sure that they would be informed of announcements and tasks on time.
5. On the questionnaire, 65.1% of the participating students stated that they agreed/totally agreed that they found BlackBoard useful in learning experiences.
6. On the questionnaire, 63.9% of the participating students stated that they agreed/totally agreed that using BlackBoard enhanced their participation in the practical sessions (self-developed).
7. On the questionnaire, 60.4% of the participating students stated that they agreed/totally agreed that using BlackBoard in their studies increased their productivity/increased.

Stated Perception of Ease/Frequency of Use

1. .On the questionnaire, 50% of the participating students stated that learning to operate BlackBoard was easy for them.
2. On the questionnaire, 56.9% of the participating students stated that navigating BlackBoard was easy for them (T).
3. On the questionnaire, 56.9% of the participating students stated that their interactions with BlackBoard were clear and understandable (D).
4. On the questionnaire, 41.9% of the participating students stated that they thought that BlackBoard was flexible to interact with (T).
5. On the questionnaire, 59.3% of the participating students stated that it was easy for them to become skillful at using BlackBoard.
6. On the questionnaire, 59.3% of the participating students stated that they found BlackBoard easy to use (D).

Attitude Toward Use

1. On the questionnaire, 62.8% of the participating students stated that learning on the web-based system was fun.
2. On the questionnaire, 66.3% of the participating students stated that using the web-based system was a good idea.
3. On the questionnaire, 64% of the participating students stated that the web-based system was an attractive way to learn.
4. On the questionnaire, 62.8% of the participating students stated that overall, they liked using the web-based system.

Behavioral Intention

1. On the questionnaire, 50% of the participating students stated that, assuming they had access to BlackBoard, they would use it, and 15.1% did not agree.
2. On the questionnaire, 48.8% stated that they intended to use BlackBoard frequently in other courses that implement blended learning, and 12.8% did not agree.
3. On the questionnaire, 52.3% of the participating students stated that they intended to choose more courses using BlackBoard in the following semesters, and 11.6% did not agree.
4. On the questionnaire, around 50% of the participants stated that they intended to use BlackBoard, use it frequently, and choose more courses using BlackBoard (50%, 48%, and 52.3%, respectively). This finding reflects the actual usage of BlackBoard when compared with the number of students who participated in the voluntary tasks conducted after the students finished their courses.

An ethnographic content analysis was performed to analyze the content optionally generated by the students via Blackboard. The students' optional activities on Black-Board were analyzed exclusively because assessing a tool that students were obliged to use would not give an indication of the students' willingness to use said tool. Students were asked to submit a number of assignments and create wikis, as well as develop individual projects, to pass the courses. The students' participation in these activities was 100% and hence is not included in this analysis. The optional tools assessed in this part of the study are the discussion board, journal, and subject-related questionnaire.

Students were anonymized, so it was not possible to look at the correlation between involvement in BlackBoard and academic performance. This possible correlation is a potential avenue for future research. There was not a significant difference in the perception of LMSs across age groups in dealing with technology.

Discussion Boards. The number of participating students from the database management course totaled 27, and they collectively generated 55 posts. In the indexing class, the number of participants totaled 68, and they collectively generated 51 posts.

Uses of the Discussion Boards. Students taking the objective course were more active on the discussion board than students taking the subjective course. The 27 students in the database management course (objective) created 55 posts. The 68 indexing students produced 51 posts. The average entries per student for the database class was 2.03, and the average for the indexing class was 0.75.

Personal Journals. There were 49 participating students from the indexing class, and they generated 179 entries. The 23 participating students from the database management course generated 107 entries.

Uses of the Journals. The database class students were slightly more active with their journals. The average number of entries was 4.1 and 3.6 for database and indexing students, respectively. Similar to the discussion boards, both classes posted learning and teaching materials in their journals rather than using them for other activities (communication, socialization, or scheduling and administration). This is in contract to the findings from the questionnaire, which indicates that students use the Internet mainly for communication. It seems that the LMS shifted the students' focus toward learning activities.

5   Discussion

From the results of the two questionnaires, students showed high expectations regarding BlackBoard usage in the learning environment, whether before or after using the system, which contradicts [11] findings emphasizing the effect of actual usage on students' attitudes toward system use. This might be because of the nature of the description and orientation of the system given to students before conducting the study. This could also be due to the differences in the features and interface design between the systems used in the two studies (Moodle vs. BlackBoard). In addition, this finding might be related to social factors not within the scope of this study. More than 60% of the students believed that they would find BlackBoard effective for accomplishing tasks and enhancing performance. The same percentage agreed that this remained true after actually using BlackBoard. Additionally, more than 62% of the participants believed that web-based learning systems were fun to use, an attractive way to learn, and a good idea. In general, more than 60% of students had a good impression of web-based learning systems and would be willing to use this type of system if it were made available by their learning institution. Further, there was no significant difference in the students' attitudes toward frequency of use, effectiveness, or ease of use before or after using the system.

Stated perceptions of the effectiveness of using the LMS versus actual performance were also investigated. In the indexing class, 49 students actually used the system for journals, and 68 students used the discussion boards. In the database class, 27 students used the discussion board, and 23 students used the journals. Both services were optional. Moreover, these activities were ungraded, with the students only gaining virtual badges after inputting a certain number of journal entries. These badges might have encouraged the students to use BlackBoard's optional services.

Based on the pre-assigned themes (Tables 3 and 4), the posts on the discussion boards that were generated by both groups were mainly for learning and teaching activities. In fact, 61.8% of the database course students' posts and 58.8% of the indexing course students' posts were related to learning and teaching activities. The percentage of participants who created posts related to scheduling and administration was 32.7% for the database students and 25.5% for the indexing students. Other themes (including communication and socializing) were represented in 1.65% of posts for the database students and 25.5% of posts for the indexing students.

Table 3. Example of vocabularies based on the themes of the discussion boards.

Theme	Example concepts
Scheduling and administration	tomorrow, delay, Monday, inquiry, Wednesday, date, test, Sunday Tuesday, hours, hand over
Learning and teaching activities	project, assignment, database test, term, indexing, periodicity, study, midterm, assignment, brochure, internet indexing, course description, access, import, subscript, form, password, contact, slides, document
Other	benefit, discussion, hello, thank you, diary, yes, finish

Table 4. Examples of vocabularies based on the themes of the personal journals.

Theme	Example concepts
Scheduling and administration	Monday, class, today, late, Tuesday, Sunday, week, attend, tomorrow
Learning and teaching activities	midterm, exercise, lecture, project, practical, essay, indexing, subject summary, system, book, class, database, computer
Other	sick, clean, sleep, like, good morning, remember, salaam, situation, miss you, hi, evening, mother, friend, God, mother day, happy people, about me, thanks, information, bad day, good evening, admits library, bad, sorry, teacher hour, good year, useful, finish, finally, important, try, good

Like the discussion board posts, the journal posts were related to learning and teaching activities for both groups, at 67.3% for the database students and 72.6% for the indexing students, followed by other themes at 29.9% for the database students and 26.3% for the indexing students. Finally, scheduling posts accounted for 10.2% of posts for the database students and 6.7% for the indexing students. Both classes used the journals for communication and socialization more than the discussion boards.

In addition, the actual performance of the students was assessed by observing student interactions with BlackBoard as they performed pre-assigned tasks. Applying HCI measurements, students' screen motions were self-captured as they performed subject-related IR tasks. Students were asked to find a specific page on a website to review and summarize and then input information into a suitable place in BlackBoard. The tasks were performed in a computer lab with the class instructor and TA, who both resolved any technical problems and ensured the tasks were performed individually by each student without any external help. The assigned tasks were designed to measure the following dimensions: ease of finding information, quantified by the number of attempts before finding the needed information; ease of uploading homework assignments and practical tasks, as well as preferred tools for uploading (written or computer submissions); the time students took to understand the task; the time students needed to complete the task; and the number of completed tasks and uncompleted tasks, as well as whether students asked for help regarding system issues while performing the task.

Students were asked to click the "help" button at the top of the page and pause the recording if they wanted to ask for help from either the instructor or the TA. The time a student spent on the screen before starting a task was considered the time a student needed to understand the task. The number of clicks on different icons not related to a task was considered attempts to do the task. Clicking on the back button or erasing and rewriting answers were considered attempts to do a task. Students taking the objective class were more willing to participate in the tasks. This was noticed also with the number of posts generated for the journals and discussion boards. This might be due to the fact that the database management course had slightly more access to the computer lab.

Fourteen students from the database management course participated in this task, and only one student did not complete the first task. Three students sought help while doing the first and second tasks. Most students correctly finished the tasks in a reasonable amount of time without any difficulties. Most students used one attempt to finish the tasks. Eighteen students from the indexing course participated in the tasks, and all the students completed both tasks. Four students sought help for the first task, and three students sought help for the second. Most students needed one attempt to finish the tasks. The considerably low participation rate in performing the tasks for both groups contradicts the participants' perception of their willingness to use the LMS, especially knowing that the tasks were optional. This emphasizes the need for further investigations of the effects of social factors on LMS usage.

6 Conclusion

Low computer skills do not need to hinder the implementation of LMSs in the learning process; the findings of this study show a significant improvement in students' computer skills after using an LMS. Students were willing to use the LMS for both social and learning purposes. However, learning purposes pre-dominated use. Students were willing to use the compulsory and optional components of an LMS if the motivation was present. Students considered the LMS easy to use and beneficial, and they were willing to use it in future classes if available. The nature of the subject being taught did not have an effect on the students' intensity and form of LMS usage. Hence, the possible effects of the design of the classes and delivery methods need to be investigated in future studies. Pre-usage orientation might also affect students' acceptance of an LMS. Therefore, a well-designed implementation program should focus not only on students' computer skills but also on the class design and instructors' abilities.

There was no significant difference in the students' perceptions of the use of an LMS before or after actual usage; therefore, future studies should focus on hands-on usage of an LMS. However, a limitation of this study is the relationship between students' academic performance and their use of an LMS, which requires further investigation as well. Future study should analyze the effects of system usage on students' academic performance in a more controlled environment to eliminate any other factors that might influence students' performance. Finally, the ability to compare students' general performance with their final scores in the courses incorporating an LMS would be a beneficial measure for future research.

Acknowledgment. The researcher would like to acknowledge the managerial and financial support provided by Public Authority of Applied Education and Training PAAET.

References

1. Baharin, A.T., Latehb, H., Nathan, S., Nawawi, H.M.: Evaluating effectiveness of IDEWL using technology acceptance model. Soc. Behav. Sci. **171**, 897–904 (2015). https://doi.org/10.2305/iucn.ch.2000.pag.6.en
2. Davis, F.D.: Perceived usefulness, perceived ease of use, and user acceptance of information technology. Manag. Inf. Syst. Q. MIS **13**(3), 319–340 (1989). https://doi.org/10.2307/249008
3. Horvat, A., Dobrota, M., Krsmanovik, M., Cudanov, M.: Student perception of Moodle learning management system: a satisfaction and significance analysis. Interact. Learn. Environ. **23**(4), 515–527 (2015). https://doi.org/10.1080/10494820.2013.788033
4. Leah, P., Macfadyen, S.D.: Mining LMS data to develop an "early warning system" for educators: a proof of concept. Comput. Educ. **54**(2), 588–599 (2010). https://doi.org/10.1016/j.compedu.2009.09.008
5. Orfanou, K., Tselios, N., Katsanos, C.: Perceived usability evaluation of learning management systems: empirical evaluation of the system usability scale. Int. Rev. Res. Open Distrib. Learn. **16**(2), 227–246 (2015). https://doi.org/10.19173/irrodl.v16i2.1955
6. Park, S.Y.: An analysis of the technology acceptance model in understanding university students' behavioral intention to use e-learning. Educ. Technol. Soc. **12**(3), 150–162 (2009)
7. Sánchez, R.A., Hueros, A.D.: Motivational factors that influence the acceptance of Moodle using TAM. Comput. Hum. Behav. **26**(6), 1632–1640 (2010). https://doi.org/10.1016/j.chb.2010.06.011
8. Park, S.Y., Min-Woo, N., Seung-Bong, C.: University students' behavioral intention to use mobile learning: evaluating the technology acceptance model. Br. J. Educ. Technol. **43**(4), 592–605 (2011). https://doi.org/10.1111/j.1467-8535.2011.01229.x
9. Teo, T.: Pre-service teachers' attitudes towards computer use: a Singapore survey. Aust. J. Educ. Technol. **24**(4), 413–424 (2008). https://doi.org/10.14742/ajet.1201

10. Teo, T., Lee, C.B., Chai, C.S.: Understanding pre-service teachers' computer attitudes: applying and extending the technology acceptance model. Comput. Educ. **52**(2), 302–312 (2009)
11. Tselios, N., Daskalakis, S., Papadopoulou, M.: Assessing the acceptance of blended learning university course. Educ. Technol. Soc. **14**(2), 224–235 (2011)
12. Venkatesh, V., Davis, F.D.: A model of the antecedents of perceived ease of use: development and test. Decis. Sci. **27**(3), 451–481 (1996). https://doi.org/10.1111/j.1540-5915.1996.tb00860.x
13. Viswanath Venkatsh, F.D.: A theoretical extension of the technology acceptance model: four longitudinal field studies. Manage. Sci. **46**(2), 186–204 (2000). https://doi.org/10.1287/mnsc.46.2.186.11926
14. Stantchev, V., Colomo-Palacios, R., Soto-Acosta, P., Misra, S.: Learning management systems and cloud file hosting services: a study on students' acceptance. Comput. Hum. Behav. **31**, 612–619 (2014). https://doi.org/10.1016/j.chb.2013.07.002

International Workshop on Internet Psychology

Big Data Analysis of Young Citizens' Social and Political Behaviour and Resocialization Technics

Galina Nikiporets-Takigawa[1,2(✉)] ⓘ and Olga Lobazova[2]

[1] University of Cambridge, Sidgwick Avenue, Cambridge, CB3 9DA, UK
gn254@cam.ac.uk, nikiporetsgiu@rgsu.net
[2] Russian State Social University, Vilgelma Pika St. 4, 129226 Moscow, Russian Federation
lobazovaof@rgsu.net

Abstract. The paper is based on the experience of an ongoing project 'Monitoring and prevention of antisocial behaviour of the young people based on Big Data and communication in social networks' which is created and implemented by the authors with the aim to prevent politically and socially destructive behaviour of the Russian adolescences based on Big Data and the mediation in social networks. We describe the three stages of the project: educational (preparatory-organizational), analytical (information-prognostic) and mediating (social-pedagogical); and introduce the system 'Social-political insider' for the collecting, processing and the sentiment analysis of the data about the sphere of the interests, the interest groups, subcultures that are in high demand among the young people. Then we discuss the preliminary results of the first, educational, stage of the project and the special knowledge and skills which are essential for the project team. The importance of such projects which can create the well-trained teams of the professionals and to organize monitoring and modification of the social and political behaviour of young people on a systematic basis is emphasized at the state level as one of the tasks of the state youth policy.

Keywords: Big data · Youth social and political behaviour
Social-Political insider system · Youth policy · Russia

1 Introduction

We live in the political and socio-cultural space, which is characterized by the impact of the cyber information and cyber communication, new technologies and networks to the all aspects of the human behaviour [6, 7, 17, 19, 21]. Information appears as an important resource for the government, business, civil society. The formation of a global information world is one of the positive consequences of the active use of the information technologies and the networks. But there are also negative sides: a person with unformed ideological views, civic identities and limited knowledge and understanding of the current political issues at times cannot make the right choice struggling to distinguish between the fake and reality in the information and communication flow in the internet. In addition, in the internet various political forces, including radical, extremist, compete

© Springer Nature Switzerland AG 2018
D. A. Alexandrov et al. (Eds.): DTGS 2018, CCIS 859, pp. 191–201, 2018.
https://doi.org/10.1007/978-3-030-02846-6_15

for such individuals trying to recruit them as the potential followers of their parties[1], movements, associations. The virtual 'market of ideas' offers the whole range of possible options, including the antisocial ones. The market is the most active in the social network sites [2]. Social networks are a platform where, apart from everything that is undoubtedly positive because it is associated with the rapid provision of new knowledge and information; the personal safety is treated and the ethical norms and the personal rights of the individual are distributed and violated.

The internet is also a 'global matrix of war' [10], a digital war zone [11, 12], information war [22, 29] because "the digital technologies are used by both state and non-state actors, often blurring the line between military and entertainment and between control and resistance. Attacks on the ground and in the air now go hand in hand with information warfare, propaganda and racist attacks in blogs and YouTube vlogs; conflicts between states or stateless groups reverberate in cyberattacks by hackers from each side' (Ibrahim 2009; Kaplan 2009; Reading 2009)" [15: 2]. Victory or defeat in the war affect the whole society and senior civil servants and other policy decision-makers globally think how to tackle online harms and wider international internet governance, to determine how to take forward the response to the current public consultation on internet safety, try to gain a wider perspective on the range of policy options for managing online harms; understand better the impact of policy in this space; determine measures for success, given increasing volumes of both illegal and 'legal but harmful' material that is online, to find the right balance between freedom of expression and protecting vulnerable individuals online; to ensure privacy online whilst maintaining security; to find who is responsible for online content, whether it is the internet companies, particularly social media companies and whether they should be required through legislation to ensure their platforms are free from (a) illegal and (b) legal but harmful material, etc.[2] The Russian Doctrine of Information Security of the Russian Federation sets the task of "developing effective means of countering the technologies of manipulation of consciousness and the development of methods of public administration aimed at protecting the information space of the country and the security of its citizens." [28].

The answers to such questions are often seen in the need for the content filtering and unwanted content clipping. These goals are pursued by some state programs around the world. In Russia, for example, the Federal Agency for Youth Affairs in 2015 announced a competition for the development of technology for filtering internet content under a state contract [14].

However, since filtering and especially blocking of the internet and its various technologies and products is recognized worldwide as ineffective practice[3], the monitoring,

---

[1] See the success of the Labour party at General election in the UK in 2017 [3, 16], which was caused among other factors by the effective tactics of the recruiting the young voters via the internet.

[2] Questions, addressed to one of the authors of this paper by the Director for Security and Online Harms in the Department for Culture, Media and Sport of the British government, Sarah Connolly, during the meet-up and the expert interview in the University of Cambridge in March 2018.

[3] On the experience of Tunisia, Egypt, Iran, China see [1, 8, 23, 26].

the prevention, the prophylactics and other preliminary measures can be more productive.

Obviously, the first step is to monitor the dissemination of information and communication channels in the social networks. We based this monitoring and analysis of information and communication flows in various internet segments on the technologies that are, in fact, an "organic" part of cyberspace – Big Data. The many various studies have been devoted to the promises of the Big Data to study consumer behaviour and appropriate targeting of advertising in trade, economy, banking, public administration, transport and medical services, actively promoting the idea of the positive impact of monitoring demand and supply in the relevant sectors of the economy [9, 13, 18, 24, 25, 27]. Using Big Data which localized in open media sources and social networks for political purposes [4, 5], however, is a much more recent trend. Nowadays obviously, the Big Data is gradually entering the Russian political space [20], but the number of applications for political analysis and forecast is disproportionate to the demand that is formed in the field of the politics and the political science by the politicians, political scientists and political technologists. The more noticeable is the publication on the use of Big Data in monitoring and managing political issues.

The growing activity of all groups and strata of the population in social networks affects not only the political space, but all those territories, where the personal interests are tightly involved and a person is most responsive to the signals of the ongoing justice/injustice, truth/lie, morality/immorality. Here, cyberspace enhances the ability to manipulate the consciousness and behaviour of a user of the social networks, and this ability is increasing even more if the person is young. The question which can be asked is whether the social networks themselves can be used as a counterweight to manipulation. The project 'Monitoring and the prevention of the antisocial behaviour of the youth based on Big Data and communication in social networks', which is introduced and discussed in the paper answers this question positively.

1.1 The Work-in-Progress 'Monitoring and the Prevention of the Antisocial Behaviour of the Youth Based on Big Data and Communication in Social Networks' Project Objectives

This project aims to monitor and prevent politically and socially destructive behaviour among the young people in Russia based on the Big Data, the 'Socio-Political Insider' system, and the mediation in the social networks; and for this requires:

- To create a team of professionals who can monitor an antisocial behaviour and mediate it
- To teach them how to use Big Data, the Social-Political Insider System, and the mediation practices
- With help of this team and Social-Political Insider System conduct monitoring and mediation

The project is created by the authors of this paper in response to the request of the practitioners, who express concern about the fact that the main place where adolescences and young people communicate is cyberspace, and the social networks predominately, as well as about the fact that the young people are exposed to the illusions of a virtual 'market of the ideas' and emotions, risking their health and lives because many adults manipulate them. February 2017 marked a tragic record in the number of the suicides of the adolescents who were victims of a suicidal movement "Blue Whale". There is controversy about the existence and the extent of this movement, but there was a "moral panic" associated with a sharp increase in the number of the teens' suicides, primarily in the Russian province. The cases of the withdrawal from the life under the influence of virtual communities "Blue Whale" revealed to the public and the state the degree of danger of uncontrolled network communication, an indisputable fact of the teenagers' vulnerability and the need to prevent them from the antisocial, extremist, deviant behaviour. Then March and June of 2017 brought the evidences of the political oppositional activism among the youth who was mobilized via YouTube. Again, and at the new level, this raised the task of the monitoring and prevention of the political and social behaviour of the adolescences and the young people. Many practitioners, with the head of several schools, the officials in the Ministry of Education and Science of the Ulyanovsk Oblast and the Department of Education of Moscow among them, rushed to ask the researches to develop the security measures against the so-called "death groups", or virtual communities "Blue Whale", as well as against the other cases of the involvement of the young people in the destructive activity, and to arm the teachers and those who are involved in the problems solving with the young individuals at the various levels, with the means to counter such phenomena.

2   Method

Since the process of the research of the information influence on the political and social behaviour comes across the intersection of psychology, sociology, information technology, political science and other disciplines, and at the same time new technics and methods are constantly emerging, the methodology of the project is syncretic.

The methodology of the main part of the project is based on the understanding of Big Data as a source of information about the sphere of interests, groups of interests, communities, subcultures that are in high demand among the young people. In addition to the material for monitoring, social networks are also a platform and mechanism for the preventive measures that limit or minimize negative impact of the internet. Big Data of the social media is our main empirical source, and we collect them, select, quantify and qualitatively analyze using a system called the Socio-Political Insider. Choosing Big Data of the social networks as a material, we built a methodology based on our understanding of the essence of the information influence, which is implemented in the space of social media and within the theoretical approach of the theory of mass communication. The concept of the information war as a "communicative technology for impact on the mass consciousness" gives the researcher operational possibilities for investigating not only the results of information influence, but also the ways to counter this influence.

Using the methods of social media mining, online community detection, social network analysis, opinion mining, leaders' detection, quantitative and qualitative methods of the content-analysis, the central part of the 'Monitoring and prevention of the antisocial behaviour of the young people based on the Big Data and communication in social networks' Project allows to gather a variety of the information about the main interests of the most popular online communities of the adolescents and the young persons and the key patterns that ideologically form these communities. On the base of the results of the monitoring, the degree of radicalization of the most popular communities is calculated, and those that requires the preventive work, which limits or minimizes their negative informational and ideological impact on youth and adolescents, are determined. The mediating, as the third stage of the project, is based on the data which are obtained and analyzed during the second stage about the characteristics of self-organization of the youth and adolescents in the social networks.

The methodology of the final part of the project is based on socio-psychological approaches to the implementation of the resocialization of young people who have fallen under destructive influence. At the same time, we refer to the destruction as follows: political extremism (quasi parties, NGOs, associations and associations); religious extremism (sects and cults); mental extremism (death group); social extremism (sports fans and mass idols). The ways and methods of communication in social networks, proposed by the authors for the resocialization of adolescents and young people, are based on the results of monitoring and are the mediation of a potential or actual conflict (internal or external). We also used the author's experience and data obtained by the authors as a result of the study of the archetypes of the youth of modern Russia (based on domestic products of mass culture), commissioned by the Federal Agency for Youth Affairs, as well as recommendations on the system of information filters developed within the framework of the same FADM project, minimizing the negative impact of information flows on young people, preventing radicalization of youth and involving youth in antisocial activities.

2.1 'Social-Political Insider System'

The system is created by a research team including Galina Nikiporets-Takigawa (PI), Andrei Koniaev, Artemi Krasheninnikov and Anna Larionova, and tested for various government and commercial customers. It allows monitoring of specified sources in all social networks and messengers by key words, sentiment analysis and an opportunity for the mapping via the social networks of the 'social well-being' and political attitudes of the various strata of society, for further analysis and forecasting of the political and social trends in real time. The advantage of the interface of our system which significantly exceeds those available on the market, is that it is customized for the needs of any task and any customer and is adjustable together with the customer in the process of the monitoring. Our web interface consists of an authorization module, a personal cabinet that stores themes, keywords and reports, as well as functions for downloading reports in csv, xlsx format and contains messages with selected parameters for the specified period. Reports in the format doc, docx, pdf are formed from pre-created templates and, if necessary, adjusted manually, based on the uploaded data in csv, xlsx format.

3   Results

The project 'Monitoring and prevention of the antisocial behaviour of the youth based on Big Data and communication in social networks' includes three successively implemented stages:

The first stage – *educational (preparatory-organizational)* aims to build a team of professionals who can monitor an antisocial behaviour and mediate it and to teach them how to use Big Data, the Social-Political Insider System, and the mediation practices. We find three categories of professionals involved in the monitoring and prevention of antisocial behaviour:

- *Customers* (who set up the task to monitor, analyze, forecast and prevent the antisocial behaviour of the young people) – from the state higher management who are involved in the youth policy, as well as the regional leaders and the practitioners in the educational and other institutions that are directly involved in the work with young people at risk.
- *Analysts*, among whom there are representatives of educational institutions (pedagogical workers who are engaged in educational work and can implement the functions of an analyst and current monitoring supervisor with appropriate training), but mainly researchers interested in skills of working with large data and implementation various projects and orders with their help.
- *Mediators* - the young journalists and PR specialists, teachers, local community leaders, psychologists and political scientists including the students who can work as a mediator under the supervision of the head of the project and other more experienced members of the project team.

For the three groups the authors of this project and this paper created and designed the three formats of the training as the programmes of the CDE (continuing professional development).

The second stage – *analytical (information-prognostic)*, implying the collection in real time of data from social networks to identify and analyze areas of antisocial behaviour. It involves collecting data based on keywords in social networks in the aggregate of selected sources, further shipment of data for a certain period, clustering and systematization, depending on the task, and the creation of analytical and forecast reports. Analysts, whom we train with monitoring techniques, learn how to use the system "Socio-political insider".

And third stage – *mediating* means the communication at the online areas at risk, that are identified by the analysts; the implication of the technics of the re-socialization of the young people, involving them to the prosocial activity and prosocial modification of their behaviour as the final result.

4   Discussion

The *customers* come to this training to know how to apply Big Data as an instrument of socio-political analysis. For this, the customer needs to form knowledge on:

1. The opportunities, limitations, and struggles in working with Big Data as a material and method;
2. The applications of the Big Data and the variety of the tasks which can be possibly proceeded with the help of Big Data (based on the various case studies);
3. The various projects involving Big Data analysis and the results, the drawbacks and achievements;
4. The principles of a team building for the Big Data monitoring and mediation based on the social media and the members' essential skills and specific knowledge;
5. And the financial issues with the central idea that the analytics based on the Big Data is highly time, labour and money-consuming as it requires an extensive effort of a team of the customers-analysts-mediators together.
6. Further and based on 1–5 knowledge, it is necessary to teach to the *customers* how to set the realistic goals for the monitoring and analysis and how to formulate the requests to avoid an ambiguous interpretation among the analysts.

Even though working with large data sets makes it possible to talk about the direction of trends with greater certainty, it should be understood, which tasks Big Data solve better and what is the limitations of such data. It is also useful to know the data market, because the customer should secure a relevant payment for the project team including not only analysts and mediators, but also IT specialists and programmers.

For the *analysts,* we deliver the knowledge of:

1. The social media mining, monitoring and social network analysis;
2. The methods of online community detection; the analysis of the online identity
3. The opinion mining and sentiment analysis
4. The identification of trends and forecasting the further development of the trends;
5. How to trace a degree of marginality and possible radicalization of the interest groups, movements and subcultures and how to identify cases of manipulation of the behaviour of the young people
6. And how to prepare an analytical report for the 'customer' with customer-friendly description and visualization, and the clear recommendations for the mediators.

The fact is that IT people are very good with dealing with Big Data but they do not know and do not care what to collect, and normally, it is senseless to seek a creative attitude to the data from them. And the analyst should set the task to the programmer. Here the analysts' knowledge of the field, and intuition, and the theoretical and methodological basis, that is, all knowledge and skills which an analyst should have, will come in handy. The accuracy in the setting the task, which an analyst poses to the programmer and to the system, results in the validity of the data. The analysts should know not only the techniques of working with Big Data, but their applications, how to identify the manipulation of the young people in somebody's interests, tracking such intentions and posing the task for the mediators to start a conversation with young people. The important skills for this group is also to constantly refine their knowledge about the characteristics of the network communication of the young people, the preferences of the young people regarding various internet platforms, the archetypes of the modern youth, the sphere of their interests, the platforms where the young people self-organize, the movements and subcultures which attract the young people the most, about the leaders of the youth movements and subcultures, as well as about the system of the information filters and other measures taken by the state to manage online harms and undesirable impact on the youth, and ways to counteract these protections which are practiced by the leaders of various movements in the internet. They should be trained to carry the monitoring on different internet platforms and to change the set of sources for monitoring regularly considering their popularity among the young people.

The third group – the group of the *mediators* – are trained to work to overcome the negative processes in the process of the communication directly with the individual users. A very advisable is to have the young persons as the mediators. Such specialists with the proper qualifications and high competence obtained due the training within our project form a highly effective 'rapid reaction force', capable to be engaged promptly in network communication with young individuals and groups at risk. We train them to know:

1. The subcultures, interest groups, political and religious communities and movements;
2. Formats which are comfortable and preferred by the young people for the communication;
3. Technics to be implemented (to subscribe and start to communicate there) in groups with the aims of explanatory work and prevention of manipulation of the consciousness and behaviour of young people, as well as of the dialogue with young people on popular platforms for young people, aimed at developing pro-social behaviour of young people;
4. Physiological issues and archetypal patterns of the youth in Russia based on the recent updates;
5. The variety of the groups at risk and the technics of the communication which are adapted for the specific needs and interests of all of them.
6. Specific characteristics of the online communication and leadership in the networked social movements

The main competencies necessary for the mediators are knowledge of the norms of communication with deviant personalities and the ability to apply methods of psychological self-defense. After obtaining this education programme, they can be introduced into groups at risk, interact with the young people through networks and organize a dialogue with groups at risk online deliberately engage them to active communication, becoming the opinion-makers and leaders in the online discussions, or supporting the existing discussions and conducting the explanatory tasks aimed at exposing the drawbacks of the ideology and activity which unite young people in the online group. They offer a prosocial alternative to a person and invite him or her to face-to-face communication to discuss further their political and social demands and their struggles in political, social identity construction, and about the subcultures and movements they are involved by means of the internet.

5   Conclusions and Recommendations

As the result of the first *educational* stage we have a team of the researches, students and practitioners from various fields which are related to the new technologies, youth politics, the dealing with the various issues of young and, especially, vulnerable individuals; who are familiar with Big Data and sufficiently trained and ready to work together on the task of monitoring and resocialization of the young people at risk. They can maintain a constant monitoring of the networked activity of the young people and identify areas of their interests, preferences, inclinations, including suicidal, extremist and other, about the degree of involvement of the young people in various subcultures, movements, associations, including destructive ones, and, in general, to have an updated portrait of youth at hands.

To use Big Data for all these purposes, a set of skills, competencies and knowledge is needed and a team of a competent customer, purposely trained analysist, professional programmers and experienced mediators is crucially demanded. The effectiveness of the well trained team and the adequate training for this team is proven during the approbation of the educational stage of our project with the Ministry of the education and science of the Ulyanovsk Oblast and two partner schools.

The training of the customers and other project members (the analysts, mediators) can be implemented in the administrative and educational institutions routine and the teams of the 'rapid reaction' can be formed everywhere on a regular basis. In the case of mastering the above competencies among all members of the team and in the case the many schools and other administrative and education bodies which are struggling to understand how to help young people and to prevent them from dangerous involvement in various activities – have the team of the trained professionals, it is very likely:

(a) for a young audience exposed to the risk of destruction:
- to escape the influence of antisocial virtual and real actors.
- to be involved in the prosocial virtual and real communication and activities online and offline.

(b) for participants of the project:
- to acquire the experience of positive influence on the audience of social networks.
- to develop further the skills in the application of education and political technologies in the online space.
- to gain the experience of monitoring of the archetypes of Russian youth, their interests, preferences, inclinations, including suicidal, extremist and other, to obtain reliable information about the degree of involvement of young Russians into the various subcultures, movements, associations, including destructive, and, in general, to create a portrait of contemporary youth in Russia.

(c) for society:
- to strengthen the social unity – as the value system of the youth which is aligned with the other strata in the society can lead to a better cross generational communication. The stabilized and fully developed value system of the young people is the main counteraction to the involvement of the young people in the radical subcultures, to the social activity that is harmful and undesirable for the society, and serves as the basis for educating the prosocial behaviour of the young people.
- to ensure the national security – as the focus on increasing the level of consolidation among the young people of the country is the task for ensuring national security. To accomplish this task, it is necessary to investigate the value

consciousness of the young people, the value orientations, dynamics and trends of their changes.

Social networks are, apart from the reach and useful space of free communication and information, a place where the destructive effect is not only on the youth, but on all users, while being conducted individually, anonymously, beyond the ethical and aesthetic framework and norms, and therefore the most productive. The community, interested in the development of the information technologies and their better influence on the political and socio-cultural processes, needs to focus on the interdisciplinary analysis of the information flows reflecting the social reality through the internet that have a huge potential for ideological influence. Our project allows to identify and resist the themes and areas of destructive influence.

References

1. Ackerman, S.: Egypt's internet shutdown can't stop mass protests. Wired (2011). http://www.wired.com/dangerroom/2011/01/egypts-internet-shutdown-cant-stop-mass-protests/
2. Boyd, D., Ellison, N.: Social network sites: definition, history, and scholarship. J. Comput.-Mediated Commun. **13**(1) (2017). Article 11. http://onlinelibrary.willey.com/doi/10/1111/j.1083-6101.2007.00393.x/full
3. General election 2017: what caused Labour's youth vote surge? BBC News Online, 16 June 2017 (2017). http://www.bbc.com/news/uk-politics-40244905
4. Gorbachev, A.M.: Big Data kak instrument protivodejstvija ugrozam jekstremizma. Mezhdunarodno-pravovye sredstva protivodejstvija terrorizmu v uslovijah globalizacii. Problemy terroristicheskogo naemnichestva sredi molodezhi i puti ih preodolenija. Sbornik materialov vserossijskoj konferencii. Stavropol'skij gosudarstvennyj pedagogicheskij institute, pp. 15–17 (2016)
5. Gricenko, R.A., Prokopchuk, D.D., Tancura, M.S.: Ispol'zovanie «Big Data» v prikladnom politicheskom analize. Voprosy nacional'nyh i federativnyh otnoshenij **2**(37), 85–92 (2017)
6. Habermas, J.: Political communication in media society: does democracy still enjoy an epistemic dimension? The impact of normative theory on empirical research. Commun. Theory **16**(4), 411–426 (2006). https://doi.org/10.1111/j.1468-2885.2006.00280.x
7. Heverin, T., Zach, L.: Microblogging for crisis communication: examination of Twitter use in response to a 2009 violent crisis in the Seattle-Tacoma, Washington area. In: Proceedings of the 7th International ISCRAM Conference, Seattle, USA (2010). https://doi.org/10.1002/1944-2866.POI335
8. Huang, E.: What you need to know about China's VPN crackdown Quartz, 12 July 2017 (2017). https://qz.com/1026064/what-you-need-to-know-about-chinas-vpn-crackdown/
9. Il'jasova, NJu, Kuprijanov, A.V., Popov, S.B., Paringer, R.A.: Osobennosti ispol'zovanija tehnologij Big Data v zadachah medicinskoj diagnostiki. Sistemy vysokoj dostupnosti **12**(1), 45–52 (2017)
10. Kaplan, C.: Twitter terrorists, cell phone jihadists and citizen bloggers: the "global matrix of war" and the biopolitics of technoculture in Mumbai. Theory Cult. Soc. **26**(7–8), 1–14 (2009). http://journals.sagepub.com/doi/abs/10.1177/0263276409349281
11. Karatgozianni, A., Kuntsman, A. (eds.): Digital Cultures and the Politics of Emotion: Feelings, Affect and Technological Change. Palgrave Macmillan, Basingstoke and New York (2012)
12. Karatzogianni, A.: The Politics of Cyberconflict. Routledge, London and New York (2006)

13. Kazakov, R.I.: Tehnologii Big Data v upravlenii krupnymi bankami. Biznes-obrazovanie v jekonomike znanij **2**(2), 19–22 (2015)
14. Kolpakova, M., Nikiporets-Takigawa, G.: Archetypal patterns of youth in Russia in the continuum of socio-political formations. Contemporary Probl. Soc. Work **1**(4), 76–80 (2015). https://doi.org/10.17922/2412-5466-2015-1-4-9-12
15. Kuntsman, A.: Online memories, digital conflicts and the cybertouch of war. Digital Icons: Studies in Russian, Eurasian and Central European New Media 4, 1–12 (2010). http://www.digitalicons.org/wp-content/uploads/issue04/files/2010/11/Kuntsman-4.12.pdf
16. Labour is winning the election on social media. Campaign, 07 July 2017. https://www.campaignlive.co.uk/article/labour-winning-election-social-media/1435748. Accessed 07 June 2017
17. Livingstone, S.: On the mediation of everything. J. Commun. **59**(1), 1–18 (2009). https://doi.org/10.1111/j.1460-2466.2008.01401.x
18. Mal'ceva, A.V., Mahnytkina, O.V., Shilkina, N.E.: Izuchenie povedencheskih patternov pol'zovatelej social'nyh setej: vozmozhnosti Big Data. Zhizn' issledovanija posle issledovanija: kak sdelat' rezul'taty ponjatnymi i poleznymi. VI Sociologicheskaja Grushinskaja konferencija. Moskva. Izd-vo RANHiGS, pp. 988–991 (2016)
19. Mercea, D.: Digital prefigurative participation: the entwinement of online communication and offline participation in protest events. New Media Soc. **14**, 153–169 (2010). http://journals.sagepub.com/doi/10.1177/1461444811429103
20. Meshherjakov, I.S.: Tehnologii Big Data v dejatel'nosti organov gosudarstvennoj vlasti. Obrazovanie i nauka kak strategicheskie resursy razvitija sovremennogo gosudarstva: sbornik nauchnyh trudov. Saratov: Povolzhskij in-t upr. im. P. A. Stolypina - fil. RANHiGS, pp. 123–125 (2017)
21. Nikiporets-Takigawa, G., Pain, E.: Internet and Ideological Movements in Russia. NLO, Moscow (2016)
22. Pochepcov, G.G.: Informacionnye vojny. Refl-buk, Moskva (2000)
23. Shveits, M.: Internet: konets svobody i anonimnosti (2012). http://www.polit.ru/article/2012/12/04/www_blackout/
24. Sizov, I.A.: Big Data - bol'shie dannye v biznese. Ekonomika. Biznes. Informatika **3**, 8–23 (2016). https://doi.org/10.19075/2414-0031-2016-5-36-44.4
25. Smirnov, V.A.: Kontury novoj modeli sociologicheskogo analiza jeffektivnosti molodezhnoj politiki s ispol'zovaniem «Big Data». Aktual'nye problemy sociologii kul'tury, obrazovanija, molodezhi i upravlenija. Materialy Vserossijskoj nauchno-prakticheskoj konferencii s mezhdunarodnym uchastiem (Ekaterinburg, 24–25 fevralja 2016 g.) S. 918–923 (2016)
26. Telegram tells Rudaw it has not closed any accounts, including KDPI on Iran's demand. Rudaw.net, 02 January 2016 (2016)
27. Terent'eva, E.I., Morbah, E.S., Vozgrina, A.V.: Sposoby primenenija Big Data dlja PR-zadach. Nauka-Rastudent.ru **3**(17) (2016)
28. Ukaz Prezidenta RF «Ob utverzhdenii Doktriny informacionnoj bezopasnosti Rossijskoj Federacii» № 646 ot 05.12.2016. Sobr. zakonodatel'stva Ros. Federacii. № 50 (2016)
29. Vasil'eva, E.N., Cynarjova, N.A.: Informacionnaja vojna v kontekste teorii massovoj kommunikacii. Vestnik Tverskogo gosudarstvennogo universiteta. Serija: Filologija, N. 3, S. 117–122 (2017)

"I Am a Warrior": Self-Identification and Involvement in Massively Multiplayer Online Role-Playing Games

Yuliya Proekt$^{(\boxtimes)}$ [ID], Valeriya Khoroshikh [ID],
Alexandra Kosheleva [ID], Violetta Lugovaya [ID], and Elena Rokhina [ID]

Herzen State Pedagogical University of Russia,
Saint-Petersburg, Russian Federation
proekt.jl@gmail.com, vkhoroshikh@gmail.com,
alkosh@inbox.ru, violetta_lugovay@mail.ru,
elenarokhina@yandex.ru

Abstract. The aim of this paper is to explore the connections among the indicators of involvement in massively multiplayer online role-playing games (MMORPG) of Russian-speaking adult gamers. The measurement model of involvement in MMORPG playing that includes motivation to play, engagement on a gamer's level, identification on a game construction level, and presence on a life environment level, were considered. The findings revealed the statistically significant correlation among all the indicators of involvement that fit the proposed model. Results indicate that gamers have seen more possibilities for social integration in the game rather than in day-to-day life. Two key tendencies were revealed: intentions to prosocial and competitive game behavior.

Keywords: MMORPG · Involvement · Engagement · Motivation to play
Self-Identification · Presence · Gamer

1 Introduction

The psychological meaning of online games has been one of the most discussed topics in cyber psychology for the last thirty years [2, 35]. Many studies explored and evaluated different aspects of human behavior in online games such as video game addiction [14, 17, 20]; gamer's experience and flow [9, 31, 36]; involvement, engagement, and immersion [4, 6, 18, 24, 25]; outcomes of participation in games [12, 19]; social structure of game communities [11, 38] and others. Despite the growing interest of researchers, not enough attention has been paid to the issue of the connection between a perception of a fictional game world and the game behavior of adults.

The essential feature of MMORPG is its role-playing constituent which means using an avatar by gamers. The interaction of a gamer with the gaming world and other gamers is realizing only through avatars which have been chosen and equipped by the people who created them [2]. Most of MMORPGs have been based on a fantasy plot that makes a player possible to take part in collective narrative construction. A.E. Voiskounsky et al. suggest that game users evolve at the same time as their characters and gain the experience that cannot be useful in any other part of their life [37]. Thus, the game reality has become an independent part of the gamer's living environment with its rules, values and social interactions. We believe that the game narrative components' contribution to a player's experience is an important research issue. Therefore the main questions of this study are how the gamers' identification with their avatar contributes to involvement in MMORPG and the consequences of the game – such as satisfaction and excessive usage. The paper also explores the concept of involvement and its empirical filling. Thus the aim of the study is to explore the connection among the indicators of involvement in MMORPG of adult gamers including specific associations between self-identification and involvement in the game.

2   Involvement in MMORPG

Involvement is one of the key characteristics to have been studied in media research. Media psychologists have considered three components of involvement (emotion, cognition and behavior) through which various forms of interaction with the virtual environment could be represented [39]. Whereas this concept is traditionally explained as motivational continuum towards particular situations or objects, involvement rather denominates the game's meaning for its users [32].

Involvement in online games is often associated with immersion. Although immersion hasn't got enough empirical interpretations [16] this concept is very popular in game studies. Brown and Cairns found that immersion can be defined as the degree of involvement with a game [6]. They distinguished three levels of immersion depending on the types of barriers that could interrupt involvement (gamer preference, game construction, environmental factors). This concept is also close to flow experience [9, 31, 36]. They are both connected with a distorting sense of time, a loss of self-awareness, challenge, and controllable tasks [9, 31]. Yet the specifics of immersion are revealed through the concentration of mental resources of a player on game tasks (cognition, emotional dispositions, motivation, and etc.) without demands of an experienced gamer or quick feedback [16].

Another concept closely connecting with involvement is presence. Presence is defined through depiction of a gamer's experience of fictional world as real [22]. Researchers have developed different typologies of presence [see e.g., 23]. There might be three spaces where gamers could experience this state of mind. The first space is depicted the physical environment of game world. A gamer hasn't felt that this space is artificial entity but perceived it as real. The second space is defined as social and it means that gamer believes that other game actors are real people. And the third dimension reflects the self-presence attitudes of gamers according to their virtual self. This concept is close to one of the components of transportation into narrative world in terms of Green and Brock [13]. Authors argue that mental imagery is needed for readers' ability to imagine events, places and themselves in the narrative. Researchers found that similar processes may be observed in online games [5, 9, 31].

Engagement is the next concept that is often overlapped with involvement in many ways. So Brockmyer et al. have defined engagement as "a generic indicator of game involvement" [4, p. 624]. Often engagement is defined as the state of "being there" during the game experience [18] that makes this concept very closely linked to presence. In some studies engagement was considered through the players' activity or their behavior in the game including pleasure, attention, interactivity, perceived user control, challenge, and other states [24, 25]. We suppose that the useful definition is given by Brown and Cairns who marked it as the first stage of immersion on the level of gamer's preference [6]. This provides an opportunity to analyze engagement as a characteristic of player's inner states.

2.1 Motivation to Play MMORPG

An increasing number of game researches are devoted to the study of motivation to play video games. Most of the studies have shown that researchers tried primarily to portray types of motives that explain why users play MMORPG and what they look for in this form of pastime. A widely accepted Yee's model of game play motivation includes three main motives. The first is achievement as a possibility to reach high level of competence in game. The second is social interaction as a possibility to establish pleasant and lasting relationships with other people. And the third is immersion as a possibility to resettle into the virtual world of game [40]. Przybylski et al. proposed the motivational approach to video games engagement. They have demonstrated how video games may provide opportunities for satisfaction of the universal human needs (competence, autonomy and relatedness) [27]. We could say that both taxonomies are closely related to each other as long as achievement is relevant competence; and social interaction is closely to relatedness. Immersion is relevant to autonomy because the concepts are both about the possibilities to find the well-controlled and comfortable reality.

2.2 Identification

Researchers have been studying identification with game characters (avatars) in order to find out how narrative elements of the plot structure of game and game tasks encourage players to give voices to their multiple selves in virtual game worlds. If identity is defined as a gamers' mental model of themselves, identification is interpreted in process terms. The processes may include experimentation with multiple selves and the construction of possible new identities [29]. MMORPG's construction is built in such a way that it makes strong links between the player and the avatar through creating a joint performance. The player's interactions with the game environment have a strong semiotic nature. Human bodies, actions, interplay and communication are given as interactive texts and icons. The integration of social and mechanic components of the game is named by Gee as Social Semiotic Space (SSS) [10]. Its content is defined as "a characteristic set of multimodal signs to which people can give specific sorts of meanings and with which they can interact in various ways" [10, p. 218]. Such an approach lets us to consider identification in the context of media research. There are two different points of view on identification with fictional characters. The main focus of the first point of view is a similarity between the person and the character [1, 21]. Another approach emphasizes the positions of the recipients in a character perception. So Oatley [26] supposed that identification occurs when a person takes on character's goals. Such the position makes possible for the person to embody an illusion of experiencing a character's life.

Based on the definitions of identification in media studies researchers differentiate various identities of gamers. Gee [11] analyzed three existing in game identities (virtual (character), real (gamer) and projective (interacting real and virtual identities). Tronstad [33] proposed to differ "empathic" and "sameness" identities. A gamer is able to empathize with through the understanding of the avatar's experience. Sameness identity comes from the gamers' experience of game world as if they were an avatar. Other taxonomies are based on an analysis of gamer selves' interplay [7, 34]. Avatar identities associated with game environment whereas player and user identities designated personal and social components of persistent self-model of gamers.

In summary, most reviewing studies indicated that involvement is a generic concept of understanding personal processes during playing in MMORPG and specifics of interaction between gamers, game environment, and other social agents (see the theoretical proposed model of involvement on Fig. 1).

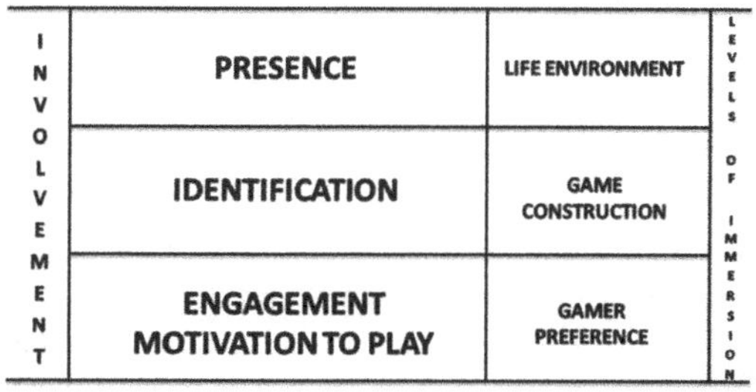

Fig. 1. The measurement model of involvement in MMORPG playing

3 Method

3.1 Participants and Procedures

Russian-speaking MMORPG players were invited to take part in this study through word-of-mouth on social media and also mailing lists of our university students. 130 participants were recruited as volunteers without any financial compensation. Data from 11 participants were removed as a result of problems with completing their forms. 77 participants were male (64,2%), and 42 were female. Their ages ranged from 16 to 50 years ($\bar{x}$ = 24,2; Me = 23; SD = 5,2). 64,2% of participants have played MMORPG for more than 5 years, while 22,5% have played for 3 to 5 years, and 13,3% have played no more than 3 years. Participants were asked to fill on online form which contained a set of questionnaires.

3.2   Measures

Motivation to Play MMORPG

"Goals and Principles of Avatar" Scale (GPAS)
The Scale involves 18 items that were based on descriptions of the video games narrative plots and typical actions of characters. Participants were asked "How often does your avatar act in accordance with following principles and goals?" A five-pointed Likert scale (1-almost never; 2-rarely; 3-sometimes; 4-often; 5-constantly) was used (see all items in Table 1). The Cronbach's alpha for this and other forms are given below.

Table 1. Factor loading and factor structure of an avatar's motivation (Note. SH = social harmony, MP = magic power, C = conquest, A = achievement)

Items	SH	MP	C	A
The victory of the good over the evil	**0,85**	–0,18	0,05	0,12
Triumph of justice	**0,83**	–0,07	–0,02	0,21
Protect the weak and vulnerable	**0,79**	–0,06	–0,13	0,01
Protection of the world	**0,76**	0,10	–0,01	–0,29
Love and acceptance from others	**0,71**	0,14	–0,31	0,09
The maintaining balance of nature	**0,70**	0,03	0,00	–0,34
Being the good knight without fear and without reproach	**0,68**	–0,24	0,17	0,01
Comprehension of the dark magic depths	–0,12	**0,83**	0,13	0,10
Passion for dark mysticism	–0,18	**0,82**	0,23	0,14
Magic strength management	0,24	**0,81**	0,04	0,05
Conquest of new territories, suppression of enemies	0,02	0,12	**0,79**	0,08
Contempt for other races and clans	–0,02	0,09	**0,74**	0,09
Freedom from morality and conscience	–0,08	**0,43**	**0,57**	0,10
Independence from rules and taboos	–0,17	**0,47**	**0,54**	0,12
Power over the world	–0,06	**0,42**	**0,49**	0,20
Aspiration to constantly improve own playing skills	0,12	0,30	0,03	**0,79**
Improving the game ranking by any means	–0,13	–0,04	0,19	**0,63**
Aspiration to improve own skills in the Art of Battle	0,22	0,13	0,37	**0,59**
Explored variance	4,88	3,08	2,68	1,95
Unique Variance Accounted for by Factors	23,24	14,67	12,78	9.29

Gamer's Motivation Evaluation Questionnaire (GMEQ)
The form involves the next statement "People usually play MMORPG because they want…" and 7 suggestions about personal reasons including social motives (social interaction/relatedness/) like "to meet new people"; immersion/autonomy motives like "to feel uniqueness"; and achievement/competence motives like "compete with other people". Participants were asked to evaluate to what extend items resemble their

motivation to play using a ten-pointed scale (where 1-Absolutely distinctive from me; 10 – extremely similar to me). The overall scores were calculated as a common variable.

3.2.1 Game Experience Questionnaire Form (GEQF)

The multi-item self-report questionnaire GEQF was developed by the authors. The questionnaire involves questions about life-time gaming habits (period of playing; frequency and duration of playing; and others), satisfaction of game playing and experience of playing MMORPG. 3 scales for dimensions excessive usage, engagement and presence were included as parts of the questionnaire. The excessive usage scale was based on the Chen Internet Addiction Scale (CIAS) [8] which was adopted for the criteria of MMORPG playing. The scale included 6 items that corresponded with Internet addiction, such as salience, tolerance, mood modification, and others. A five-pointed Likert scale was used to evaluate the frequency of the participants' symptoms (1-almost never; 2-rarely; 3-sometimes; 4-often; 5-constantly). The Engagement scale included 10 items. These items were devoted to reveal in which extend players feel an inclusion in the game world (e.g. "I really live in this game world", or reverse item "I found my mind wandering while playing the game"). The Presence scale contained 5 items that presented the experience of being in the game world (e.g. "How often did you feel as if you walked through the game area while you were playing the game?). A seven-pointed Likert scale was used (1- absolutely disagree; 7 - completely agree; 1-almost never; 7-constantly).

3.2.2 Identification

3.2.3.1 Twenty Statements Test (TST)

The Twenty Statements Test is well known as an instrument of identities study. This test was used for the dimension of attitudes towards one's own self [19]. We used a modification of TST to adopt the objectives of this study. Participants were asked to provide ten statements responding to the question "Who am I when I'm playing MMORPG?" and other ten different statements in response to the question "Who am I when I'm out of the game?" For coding the free responses of participants we used Hartley's recommendations [15]. The three indicators were personal, interpersonal and collective self-descriptions.

3.2.3.2 Identification Questionnaire Form

The Form includes 5 questions such as "How often did you perceive your game avatar as yourself?", "To what extent can you use the word "I" for your game avatar descriptions?" Participants respond on a seven-pointed Likert scale. The overall score was calculated as common variable.

3.3 Data Analysis

We conducted a descriptive and comparative analysis to evaluate gender and age differences through analysis of variance. The total sample was divided into young (range = 16–22; n = 49) and adult (range = 23–50; n = 70) groups. The next step was performed by analyzing a correlational structure of all the study variables. Exploratory factor analysis was conducted for developing GPAS. Finally we conducted the median split to divide participants into high and low involvement in MMORPG playing groups using the overall scores of 5 scales (Motivation, Engagement, Identification, Presence, and Excessive usage) (mdn = 114, range = 55–193) to compare the self-descriptions of gamers. All analyses were calculated in Statistica v. 6.1 (StatSoft Inc.).

4   Results

The results of revealing essential characteristics of involvement in MMORPG according to the gender and age of gamers are shown in Table 2.

Table 2. Means and SD of involvement characteristics

Involvement indicators	GENDER				AGE			
	Male		Female		Young		Adult	
	Mean	SD	Mean	SD	Mean	SD	Mean	SD
Motivation	30.58	14.33	32.44	13.08	37.24	14.56	27.90	12.05
Engagement	35.00	8.85	39.91	8.87	40.67	8.88	34.34	8.62
Identification	21,04	8,71	23,39	8,36	25.20	8.46	19.76	8.06
Presence	16.05	8.11	19.40	7.63	19.54	7.81	15.90	8.03
Excessive usage	9.00	3.47	9.27	2.90	9.84	3.42	8.66	3.12

It was assumed that age, gender and the parameters of game behavior might moderate involvement in MMORPG playing. The results demonstrated that female gamers were significantly more engaged (F = 8.46, p < 0,01) and felt presence (F = 4.89, p < 0,05) than male gamers. Significant age differences were revealed for almost all indicators of involvement: motivation to play MMORPG (F = 14.12, p < 0,01), engagement (F = 14.65, p < 0,01), identification (F = 13.55, p < 0,01), and presence (F = 14.12, p < 0,01). Significant differences in excessive usage did not appear for both dimensions.

Table 3. Correlation coefficients between indicators of involvement in MMORPF, satisfaction of game playing, means, SD and Cronbach α (Note: *p < 0.05; **p < 0.01)

Variables	1	2	3	4	5	Mean	SD	α
1. Motivation	–					31.25	13.87	0.82
2. Engagement	0.45**					36.76	9.13	0.76
3. Identification	0.39**	0.47**				21.88	8.63	0.82
4. Presence	0.41**	0.62**	0.69**			17.25	8.07	0.84
5. Excessive usage	0.22*	0.55**	0.13	0.24*		9.10	3.27	0.66
6. Satisfaction	0.31**	0.16	0.12	0.27*	-0.12	7.47	1.13	–

The findings of this study revealed a statistically significant correlation among all the indicators of involvement (Table 3). This fits our model of involvement in MMORPG playing. Furthermore the results showed a connection between satisfaction of game playing, presence and motivation. Engagement, identification and excessive usage didn't contribute to satisfaction of game playing.

We further conducted exploratory factor analysis for developing GPAS. Principal components analysis with varimax raw rotation was computed for 18 items (see Table 1). We supposed that the representations of goals and principles of an avatar were the connecting links between player motivation and identification with a game character. We analyzed the Scree plot and Eigenvalues greater than one that let us extract four factors which accounted for 60.0% of the variance in item scores. There were only three cases with cross loading of items across factors with items loaded above 0.40 on two factors. See Table 4 for reliability, means, and standard deviations of the scale items.

Table 4. Descriptive statistics and intercorrelations among GPAS subscales (Note: **p < 0.01)

Subscale	Mean	SD	Cronbach α	Factor intercorrelations		
				1	2	3
1. Social Harmony (9 items)	23.82	9.47	0.88	–		
2. Magic Power (6 items)	13.24	6.22	0.83	−0.05	–	
3. Conquest (6 items)	13.60	5.69	0.80	−0.13	0.79**	–
4. Achievement (3 items)	9.95	2.99	0.61	0.09	0.39**	0.48**

As shown in Table 4 factor structure included one independent and three consistent factors. The independent factor captured different aspects of intentions to prosocial behavior. Other factors were closer to competitive behavior and its variations in fictional worlds of MMORPG.

Table 5. Correlation coefficients between indicators of involvement in MMORPF, satisfaction of game playing and GPAS subscales (Note: *p < 0.05; **p < 0.01)

Indicators of involvement	GPAS subscales			
	SH	MP	C	A
1. Motivation	0.20	0.10	0.22*	0.34**
2. Engagement	0.23*	0.19	0.31**	0.30**
3. Identification	0.26*	0.29**	0.17	0.14
4. Presence	0.32**	0.28**	0.18	0.13
5. Excessive usage	−0.04	0.13	0.16	0.27**
6. Satisfaction	0.25*	0.04	0.05	0.18

Table 5 shows the relationships between GPAS subscales and the key indicators of involvement in MMORPG playing. Social harmony is positively associated with identification, engagement, presence and satisfaction of game playing. We could say that intentions to prosocial behavior contribute to positive experiences and involvement in the game. Magic power is positively linked with identification and presence. This connection would be expected. The participants who accept the fictional game world's rules lead towards feeling of presence in this world. Conquest is positively correlated with motivation and engagement. Achievement is positively associated with motivation, engagement and excessive usage of game. This also could be expected because those players who have a need for game skills improvement and competition elevate their online activity up to excessive usage of the game.

Table 6. Descriptive statistics of the types of self-descriptions

Types of self-descriptions	Means (standard deviations)					
	Total sample		High involvement		Low involvement	
	Virtual	Real	Virtual	Real	Virtual	Real
Personal	3.44 (2.17)	5.40 (2.54)	3.66 (2.16)	5.56 (2.61)	3.27 (2.20)	5.27 (2.51)
Interpersonal	3.13 (2.08)	1.84 (1.61)	3.54 (2.24)	1.74 (1.23)	2.79 (1.89)	1.91 (1.88)
Collective	2.61 (2.13)	2.16 (1.87)	1.80 (1.66)	1.98 (1.61)	3.29 (2.25)	2.31 (2.08)

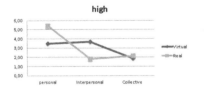 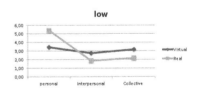

Fig. 2. Plots of self-descriptions profiles in high and low involved gamers' groups

The final step of study was performed by analyzing self-descriptions of participants that are indicators of identification in and out of the game. See Table 6 for means, and standard deviations of the types of participants' self-descriptions; see Fig. 2 for plots of self-descriptions profiles.

The comparative analysis indicated significant differences in the self-description profiles that revealed more personal self-descriptions in real identification (total: $t = 7.72$, $p < 0.001$; high: $t = 4.78$, $p < 0.001$; low: $t = 6.08$, $p < 0.001$), and more interpersonal self-descriptions in virtual identification of gamers (total: $t = 5.69$, $p < 0.001$; high: $t = 4.81$, $p < 0.001$; low: $t = 3.29$, $p < 0.01$). Significant differences in collective self-descriptions appeared only for low involved gamers ($t = 2.64$, $p < 0.05$). They tend to approve their virtual self through social categorization.

5   Discussion

The paper's findings support the theoretical proposed model of involvement in MMORPG. In this context involvement might be considered as a multifaceted and multidimensional construct that comprises distinct indicators and reflect rich and meaningful experiences of playing MMORPG. It is important to note that the interaction between the game mechanics and gamer personality provides possible interpretations of the game fictional world and player's role models. The gamers could choose who they want to be and how they behave in the game. These possibilities are more attractive for young gamers who have greater involvement in MMORPG playing. Altogether, gamers have seen more possibilities for social integration in the game rather than in day-to-day life which confirms the greater interpersonal self-descriptions in virtual self.

We couldn't agree with the point of view that refers to MMORPG as the space of ideal selves' realization by gamers [3, 28] because these selves are not under total control by users but they reveal themselves in social interaction and depend on many situational and interpersonal factors. Many gamers have been staying in the same game and the same game community for years [38]. They tend to change avatars and their roles from normative to margin and vice versa. Gamers are more likely to look for an ideal reality where their selves could enact in more suitable and comfortable conditions where they could do things that couldn't be conducted in the real world. Our findings confirm this position. Two key tendencies in identification with an avatar were revealed as intentions to prosocial and competitive behavior. In general these tendencies might correspond to two types of culture (collective and individualistic) that are connected to the oppositions of self-transcendence to self-enhancement values [30]. Typical MMORPG mechanics suppose a lot of possibilities for gamers to get drawn into communities and collaborations that might explain the link between Social Harmony preferences and satisfaction of game playing. At the same time competitive behavior connected with excessive usage that might show just a suitability of the game world for self-affirmation and the need to prove the importance of own self.

6   Conclusion

People always need both stories and social games. Before the digital era there were tournaments, masquerades, carnivals, and other forms of massive playing' encouragement. Today we have Internet as a space of collective creativity. Modern game dynamics and mechanics like persistence, avatar-mediated play, perpetuity and social interaction [2] let invert MMORPG playing in a never ending story. In this study we examined connections among self-identification and key indicators of involvement in MMORPG playing. The results highlight the value of social integration in a fantasy game world where every person could be a part of heroes' teams which pursue noble goals and form own sodalities.

The findings of this study may serve as a framework for the development of the gamification of learning and psychotherapeutic processes. We also suppose that this study may be a starting point in the exploration the role of cultural and social factors in involvement of Russian gamers in MMORPG playing.

References

1. Andsager, J.L., Bemker, V., Choi, H.-L., Torwel, V.: Perceived similarity of exemplar traits and behavior: effects on message evaluation. Commun. Res. **33**, 3–18 (2006). https://doi.org/10.1177/0093650205283099
2. Barnett, J., Coulson, M.: Virtually real: a psychological perspective on massively multiplayer online games. Rev. Gen. Psychol. **14**(2), 167–179 (2010). https://doi.org/10.1037/a0019442
3. Bessiere, K., Seay, A.F., Kiesler, S.: The ideal elf: identity exploration in World of Warcraft. Cyber Psychol. Behav. **10**, 530–535 (2007). https://doi.org/10.1089/cpb.2007.9994
4. Brockmyer, J.H., Fox, C.M., Curtiss, K.A., McBroom, E., Burkhart, K.M., Pidruzny, J.N.: The development of the Game Engagement Questionnaire: a measure of engagement in video game-playing. J. Exp. Soc. Psychol. **45**, 624–634 (2009). https://doi.org/10.1016/j.jesp.2009.02.016
5. Brookes, S.: Playing the story: transportation as a moderator of involvement in narratively based video games. MA thesis. The Ohio State University, Columbus, Ohio (2010)
6. Brown, E., Cairns, P.: A grounded investigation of game immersion. In: CHI, pp. 1279–1300. ACM Press (2004). https://doi.org/10.1145/985921.986048
7. Carter, M., Gibbs, M., Arnold, M.: Avatars, characters, players and users: multiple identities at/in play. In: Proceedings of the 24th Australian Computer-Human Interaction Conference, pp. 68–71. ACM, New York (2012). https://doi.org/10.1145/2414536.2414547
8. Chen, S., Weng, L., Su, Y., Wu, H., Yang, P.: Development of a chinese internet addiction scale and its psychometric study. Chin. J. Psychol. **45**(3), 279–294 (2003). https://doi.org/10.1037/t44491-000
9. Cowley, B., Charles, D., Black, M., Hickey, R.: Toward an understanding of flow in video games. Comput. Entertain. **6**(2), 1–27 (2008). https://doi.org/10.1145/1371216.1371223
10. Gee, J.P.: Semiotic social spaces and affinity spaces: from the age of mythology to today's schools. In: Barton, D., Tusting, K. (eds.) Beyond communities of practice: Language power and social context. Cambridge University Press. Cambridge (2005). https://doi.org/10.1017/cbo9780511610554.012
11. Gee, J.P.: What Video Games Have to Teach us About Learning and Literacy. Palgrave Macmillan, New York (2003)
12. Granic, I., Lobel, A., Engels, R.: The benefits of playing video games. Am. Psychol. **69**, 66–78 (2013). https://doi.org/10.1037/a0034857
13. Green, M.C., Brock, T.C.: The role of transportation in the persuasiveness of public narratives. J. Pers. Soc. Psychol. **79**(5), 701–721 (2000). https://doi.org/10.1037/0022-3514.79.5.701
14. Griffiths, M.D., Kuss, D.J., King, D.L.: Video game addiction: past present and future. Curr. Psychiatry Rev. **8**, 308–318 (2012). https://doi.org/10.2174/157340012803520414
15. Hartley, W.S.: Manual for the twenty statements problem. Department of Research, Greater Kansas City Mental Health Foundation. Kansas City (1970)
16. Jennett, Ch., et al.: Measuring and defining the experience of immersion in games. Int. J. Hum Comput Stud. **66**, 9 (2008). https://doi.org/10.1016/j.ijhcs.2008.04.004

17. Kardefelt-Winther, D.: A conceptual and methodological critique of internet addiction research: towards a model of compensatory internet use. Comput. Hum. Behav. **31**, 351–354 (2014). https://doi.org/10.1016/j.chb.2013.10.059

18. Klimmt, C., Vorderer, P.: Media psychology "is not yet there": Introducing theories on media entertainment to the presence debate. Presence Teleoperators Virtual Environ. **12**, 346–359 (2003). https://doi.org/10.1162/105474603322391596

19. Kuhn, M.H., McPartland, T.S.: An empirical investigation of self-attitudes. Am. Soc. Rev. **19**, 68–76 (1954). https://doi.org/10.2307/2088175

20. Kuss, D.J., Griffiths, M.D.: Internet gaming addiction: a systematic review of empirical research. Int. J. Mental Health Addict. **10**(2), 278–296 (2012). https://doi.org/10.1007/s11469-011-9318-5

21. Larsen, S.F., Seilman, U.: Personal reminding while reading literature. Text **8**, 411–429 (1988). https://doi.org/10.1515/text.1.1988.8.4.411

22. Lee, K.M.: Presence, explicated. Commun. Theory **14**, 27–50 (2004). https://doi.org/10.1111/j.1468-2885.2004.tb00302.x

23. Lombard, M., Ditton, T.: At the heart of it all: the concept of presence. J. Comput.-Mediated Commun. **3**(2), 20 (1997). https://doi.org/10.1111/j.1083-6101.1997.tb00072.x

24. Mikeal Martey, R., et al.: Measuring game engagement: multiple methods and construct complexity. Simul. Gaming **45**(4–5), 528–547 (2014). https://doi.org/10.1177/1046878114553575

25. O'Brien, H.L., Toms, E.G.: What is user engagement? A conceptual framework for defining user engagement with technology. J. Am. Soc. Inform. Sci. Technol. **59**, 938–955 (2008). https://doi.org/10.1002/asi.20801

26. Oatley, K.: Meetings of minds: dialogue, sympathy, and identification, in reading fiction. Poetics **26**(5–6), 439–454 (1999). https://doi.org/10.1016/S0304-422X(99)00011-X

27. Przybylski, A.K., Rigby, C.S., Ryan, R.M.: A motivational model of video game engagement. Rev. Gen. Psychol. **14**(2), 154–166 (2010). https://doi.org/10.1037/a0019440

28. Przybylski, A.K., Weinstein, N., Murayama, K., Lynch, M.F., Ryan, R.M.: The ideal self at play: the appeal of video games that let you be all you can be. Psychol. Sci. **23**(1), 69–76 (2012). https://doi.org/10.1177/0956797611418676

29. Reid, E.: The self and the Internet: variations on the illusion of one self. In: Psychology and the Internet. Academic Press, San Diego (1998)

30. Schwartz, S.H.: An overview of the Schwartz theory of basic values. Online Readings Psychol. Cult. **2**(1), 1–20 (2012). https://doi.org/10.9707/2307-0919.1116

31. Sweetser, P., Wyeth, P.: GameFlow: a model for evaluating player enjoyment in games. Comput. Entertain. **3**(3), 1–24 (2005). https://doi.org/10.1145/1077246.1077253

32. Takatalo, J., Häkkinen, J., Kaistinen, J., Nyman, G.: Presence, involvement, and flow in digital games. In: Bernhaupt, R. (eds.) Evaluating User Experience in Games. Human-Computer Interaction Series, pp. 23–46. Springer, London (2010). https://doi.org/10.1007/978-1-84882-963-3_3

33. Tronstad, R.: Character identification in world of warcraft: the relationship between capacity and appearance. In: Corneliussen, H.G., Rettberg, J.W. (eds.) Digital Culture, Play, and Identity: A World of Warcraft Reader, pp. 249–263. MITPress. Cambridge (2008)

34. Van Looy, J., Courtois, C., De Vocht, M., De Marez, L.: Player identification in online games: validation of a scale for measuring identification in MMOGs. Media Psychol. **15**, 197–221 (2012). https://doi.org/10.1080/15213269.2012.674917

35. Voiskounsky, A.E.: Group Playing Activity in Internet. Psikhologicheskii Zhurnal **20**(1), 126–132 (1999). (in Russian)

36. Voiskounsky, A.E.: On the Psychology of Computer Gaming Psihologiya. ZHurnal Vysshej shkoly ehkonomiki **12**(1), 5–12 (2015). (in Russian)

37. Voiskounsky, A.E., Mitina, O.V., Avetisova, A.A.: Obshcheniye i "opyt potoka" v gruppovykh rolevykh Internet-igrakh. Psihologicheskii zhurnal **25**(5), 47–63 (2005). (in Russian)
38. Williams, D., Ducheneaut, N., Xiong, L., Zhang, Y., Yee, N., Nickell, E.: From tree house to barracks: the social life of guilds in World of Warcraft. Games Cult. **1**(4), 338–361 (2006). https://doi.org/10.1177/1555412006292616
39. Wirth, W., Hofer, M., Schramm, H.: The role of emotional involvement and trait absorption in the formation of spatial presence. Media Psychol. **15**(1), 19–43 (2012). https://doi.org/10.1080/15213269.2011.648536
40. Yee, N.: Motivations for play in online games. CyberPsychol. Behav. **9**(6), 772–775 (2007). https://doi.org/10.1089/cpb.2006.9.772

Development of the Internet Psychology
in Russia: An Overview

Alexander Voiskounsky[✉]

Lomonosov Moscow State University, 11/9 Mokhovaya street,
Moscow 125009, Russia
vaemsu@gmail.com

Abstract. A brief description of prehistory, history and current state of the art in the development of the Internet psychology, or cyberpsychology in Russia is presented. Prehistory refers to a "non-meeting" stage: unsuccessful enthusiasts of computer networking in Russia (then the USSR) have not had recourse to social science researchers, including psychologists, who might have provided computer scientists with valuable "human factor" reasons to make the innovative appeal fully argumentative. The second period, named "culture psychology", refers to the beginning of the appropriate studies which happened to start earlier than the public access to the Internet became available. The main theoretical platform of the culture psychology studies was the Vygotskian paradigm in psychology. The dominant characteristic of the third period was multitheoretical approach, and this phase got an ad hoc name "positive psychology" – simply due to the fact that a series of cyberpsychogical studies was performed within a positive psychology paradigm. The last enlisted period refers to the "current studies": it includes both multi-theoretical works and diverse projects targeted on numerous Internet mediated activities such as interaction, cognition (e.g. learning), video game playing and various online entertainments.

Keywords: Culture psychology · Cyberpsychology · Positive psychology
Mediation · Flow · Digital risks · Local area network

1 Introduction

The origin of the Internet (initially ARPANET) dates back to 1969. This particular date in the world-wide Internet history refers primarily to the USA. Other countries may have their own important dates related to the beginning of network transactions and the start of computer-mediated connections. Similarly, the starting points of studies in the field of the Internet psychology vary in different countries, sometimes independently of a wide or narrow access to computer networking.

The latter case characterizes Russia: the earliest publications in the (future) field of cyberpsychology happened to precede both mass and even selective access to global computer networking. Phases, or periods of the growth of the Internet psychology in Russia seem to deserve special discussion. The Internet psychology is rather a new research field everywhere, including Russia. This paper covers all the periods starting

© Springer Nature Switzerland AG 2018
D. A. Alexandrov et al. (Eds.): DTGS 2018, CCIS 859, pp. 215–226, 2018.

from prehistory (it may be called a "Jurassic" period) and up to the current works in the field; the earliest periods, being the least known, are discussed in full, while much less space is given to the modern state of the art. When possible, the development of the Internet psychology in Russia is compared to parallel processes in other countries; nevertheless, any approach towards a cross-national perspective, such as [3], needs to be addressed to the future. Though the materials on which the paper is based include the whole scope of studies done during all the periods, the references to the paper are limited to the works which are *available in English*; hopefully, such an abridgment may be helpful for all those who do not follow the sources published in Russian.

2 Non-meeting

The Internet psychology has grown from several disciplines, including first of all psychology with its diverse fields such as social, organizational, clinical, cognitive, educational psychology, and computer science – the latter provided networks and connection protocols, mobile communications, browser programs, software for gaming, for real-time and delayed interactions, including audio and visual presentations, etc. All the elements worked out within the computer science field may be jointly named as a channel, or a medium, while particular services based upon the channel provide multiple activities such as online and mobile interactions, learning, gaming and entertainment, shopping, sexting and consuming porn, etc. Social scientists and psychologists feel professionally interested in the numerous ways the Internet services are included into everyday life. This is rather a new research field accelerating since 1990s, although the earliest works have been published before 1980 [16].

It is doubtful that "the medium is the message," the famous Marshall McLuhan's formula, is still correct. Instead, the abovementioned medium serves as a frame for enriching the world-wide web: the Web 2.0 idea is that the users themselves create content within the given broad frames. While the computer science practitioners constantly update the online behavior medium, psychologists do their best to unblock the message of the universally accepted medium – namely, the content which people insert into learning, entertainments, interactions, etc. while they stay online. Younger generations of psychologists are quite fond of doing studies related to the Internet psychology, though the latter is rarely taught. To the best of the author's knowledge, there is only one textbook in the field [7].

Nowadays, psychology and the Internet are closely connected; decades ago few, if any people would foresee the current liaison. For example, the initiators of an elegant project which was worked out within the computer science, namely the idea of building telecommunication networks to connect mainframe computers, have lamentably disregarded a chance (probably, questionable) to get support and possibly benefit from collecting psychological data. The event of non-meeting psychology and computer science dates back as long as 1959; since that time it lasted until mid-1980s.

The story of the computer networking project initiated by Anatoly Kitov, a Colonel in the Russian (then Soviet) Army and a visionary, is thoroughly presented in several books written mainly by the US scholars. First to be named is a book by Gerlovich [12]; more details are presented in a recently published book by Peters [23].

Earlier than the famous Internet pioneer, the MIT professor psychologist Joseph Licklider, Kitov explicitly formulated the actual ideas of computer networking. In 1959 he applied to Nikita Khruschev proposing to start computer networking within the whole Soviet Union. (Licklider's project "Intergalactic Computer Network" appeared in 1962). Though the Kitov's project has been estimated positively, there was no order to perform it immediately in practice; the second appeal to the state leader (this time it was Leonid Brezhnev) ruined Kitov's military career. He managed nevertheless to build a new career as a consultant and professor of cybernetics, and even more, he found a strong sympathizer and upholder, Viktor Glushkov – the founder and director of the Institute for Cybernetics (Kiev, the Ukraine, then USSR) and an influential member of the Academy of Sciences.

Glushkov's project called OGAS (acronym, stands in Russian for All-State Automated System), close to Kitov's project, has been presented as a pioneering innovation, highly helpful in collection of unbiased managerial data related to the economic growth, or maybe unwanted recession. Since Glushkov was decades-long insistent in making the project known to the highest-status decision makers, the OGAS project has several times been close to being approved, but it has never been funded and Glushkov has never enjoyed an expected success. When computer networks became available in the late 1980s, the project was based on the Western connection protocols and Western hardware and had little if any common with the Glushkov's or Kitov's layout.

Both Gerlovich and Peters provide many views, details and rumours explaining the "uneasy" [23] story of non-networking the vast territory of the USSR with many thousands computerized enterprises – industrial, agricultural, etc. Leaving aside possible or hardly possible explanations, we may dare to throw a reproach to the late networking activists, namely Anatoly Kitov and Viktor Glushkov. Neither of them made an attempt to promote at least a moderate experience of computer networking in order to lean afterwards upon the collected evidences of the expected mass enthusiasm towards being networked; the documented evidences might be attached to the claims shipped to the state leaders. Surely such a non-existent experience can hardly be called otherwise than totally virtual: at that time nobody have had any experience in global telecommunications.

But not so with local area networks (LANs). Here starts the "non-meeting" moment when the proponents of computer networking disregarded any psychological data, and the "human factors" data collected within the organizational psychology in particular. Being the director of the Kiev Institute of Cybernetics with up to 2,000 employees, Glushkov was more than able to give his subordinates an access to computer networks, such as LANs. But this time he underestimated what might be called a sort of a human factor in computer telecommunications.

The LAN technologies have been rapidly developing during 1970s. All those who have had a chance to use the most advanced LANs, got a feeling close to being able to meet a sensational miracle. Upon the author's experience, the LAN users combined professional and friendly talks, discussed organizational problems, shared news and humor: some of them from time to time filled in various psychological and/or sociological questionnaires. The author's experience hints that enough LAN related material was collected by Moscow University psychologists by mid-1980s to write the first research papers on psychology of computer-mediated communication; publications in English followed a decade later [1, 33, 34].

Glushkov was far from considering evidences related to the human factor. He was well aware that the best way to achieve the desired goal was to get an approval from the highest-rank decision makers. Probably this is the best way indeed, but not the only one. His supposed would-be OGAS users were intended to be assigned bureaucrats – very likely, differing from networking enthusiasts. Glushkov was not sensitive to any possible support which might come from ordinary well-educated people who might acquire a genuine interest in the use of local or global computer networks, e.g., for interaction and information retrieval.

It is more than likely that Glushkov had little or no information concerning the ways the ordinary specialists use LANs. He could nevertheless collect data referring to the mass access and exchange of information via Videotex systems – a combination of simple software, rapidly updating databases, phone channels, TV screens and special keyboards. Such systems – named Prestel in Great Britain and several other North-European countries, or Minitel in France – were in heavily exploit during 1970s, when Glushkov was alive and active. Commercial success of the Videotex networks, as well as personal attachment of millions phone subscribers to these systems might give Glushkov and his collaborators an alternative view on the way to "network a nation": the "uneasy" work might go both bottom up and top down. Ordinary people (the bottom) stayed neglected in the descriptions of the networking projects presented by Kitov and Glushkov. The early Soviet visionaries of computer networking were in their own way the products of the command system of decision making. The early proponents of computer networking made their best trying to impact the top officials (the tops); at the same time they never applied to the bottom, i.e. all those who may feel genuine interest in the use of the would-be medium.

This is the story of a prolonged period during which the specialists in the two disciplines, namely computer science and psychology, did not take a chance to collaborate in an innovative initiative of building global computer networks in Russia (then USSR) for exchange of data and – we suggest – interaction. The result of the "non-meeting" story was the loss of priority in computer telecommunications.

3 Culture Psychology

The next episode is the story of meeting. As it was mentioned in several works in Russian and in English [36, 39, 41], psychologists turned to be among the first group of Russian scholars who initiated investigations of human behavior mediated by computer networks. The studies started in 1980s, about a decade before the Internet became widely available in the USSR. Two research projects need to be mentioned.

One project was partly based first, on interviews with Soviet scientists who happened to participate in a Delfi procedure of expert evaluations made online on assignment of international professional committees; second, on analysis of Usenet posts; lastly, some findings were obtained by a participant observation of the LAN users' interactions. The results were published in 1980s in Russian and over a decade later – in English [33, 34]. The second study – the Soviet/Russian-American project

(1985–1994) "Cognition and Communication" – was done in cooperation by research groups from the University of California at San Diego and the Moscow Institute of Psychology, Academy of Science; the former provided the hardware and telecommunication facilities, the latter introduced computer networking to the Soviet/Russian academic institutions, libraries, primary and high schools. The main results [15] were obtained by observing and testing schoolchildren in educational computer-rich environments, such as the one called the 5th Dimension [6]. Besides, researchers observed and described some particular patterns referring to computer-mediated interactions of adult novice users, mainly researchers in humanities. Both projects were carried out in the Vygotskian paradigm [1, 6, 14].

This is an important point. At that time and afterwards, neither communication science nor human-computer interaction were developed enough to be in the list of academic disciplines in Russia. While it is a common opinion today that the current civilization is a network society, the genesis of this particular idea in 1980s did not attract due attention of scholars in social sciences and humanities, at least in the Soviet Union, later Russia. At that time philosophers, economists, linguists, sociologists, or historians did not express strong professional interest in the studies of computer mediation of numerous human behaviors. Not so with psychologists, as it was said above. The reason is that the Vygotskian perspective turned to be among the most promising platforms in the studies of computer (later – Internet) mediation.

Indeed, the introduction of networking ideas into psychology, unlike many other academic disciplines, met no conceptual problems due to the theory which is popular in Russia as well as abroad: namely, the socio-historic theory of psychic development introduced by Lev Vygotsky [43]. From this theoretical background it is emphasized that higher mental processes are of social origin, their development is based on joint child-adult actions (particularly within the zone of proximal development) and interpersonal communication, and presumably on mediated forms of behavior. Within this theory, called also as "culture psychology," mediation is a fundamental principle in mental development; it includes the acquisition and usage of practical and psychological instruments: material tools, signs, and semiotic systems. According to Vygotsky, genuine human forms of behavior are mediated by culture-related sign systems. Thus, the acquisition of personal skills and social norms through signs is the mainstream of psychic development. The developing sign systems often remediate behaviors which have already been mediated, and probably not once. Remediation refers to significant changes in the selection of newly-mediating semiotic systems; an evident historic example is a transfer from syllabic to alphabetic writing systems [6].

Investigations of mediated and remediated forms of behavior are traditional for the Vygotskian approach. In the 1980s and later psychologists in Russia first discovered that by using networks, computer facilities and software, individuals could mediate and remediate processes of cognition and interaction. They realized as well that a software program that runs on a computer is, in its essence, a semiotic system; information technologies represent quite complicated sign systems. Being universal mediators in almost any type of activity, the developing information technologies impact (e.g., remediate) human behaviors; this process is of primary interest to psychologists who practice the Vygotskian approach.

4 Positive Psychology

After the prolonged period called above "non-meeting" of psychologists and computer scientists, the next period may be characterized by a priority interest of psychologists in the problems of computer mediation and the peculiarities of the new-born network communities. Soon after the regular access to the Internet became widely available in Russia, scholars in diverse disciplines joined psychologists in carrying out, from their professional perspectives, multiple studies of the Internet use. Nowadays, on a par with all the other researchers, psychologists go on investigating human behavior mediated by the Internet, computers, smartphones, gadgets, ipad tablets, etc. from diverse theoretical platforms, including those differing from the culture psychology. By 2000 a certain theoretical diversity became apparent in the development of the Internet psychology in Russia: current newborn psychological methods and concepts have been widely applied. As an example, it is reasonable to mention positive psychology which has been developed during the last decades, mainly in the works of M. Seligman and M. Csikszentmihalyi [26], and their collaborators and followers.

Among the most promising platforms within the positive psychology are the Self-Determination theory developed by Deci and Ryan [9, 25] and the optimal experience (flow) theory developed by Csikszentmihalyi [8]; within the both approaches well-established research methods have been developed. Positive psychology is being extensively applied in numerous studies of human behavior mediated by the Internet, computers, smartphones and gadgets, such as problem solving, group and dyadic interaction, learning and exploration, gameplaying and entertainment, shopping, etc. Numerous publications and practical work reports in the field are known, as it is evident from analytical reviews and popular science books which have been issued during the last decade [25, 37, 40]. While both of the approaches are being intensely used in the world-wide studies of computers/Internet mediated behavior, the optimal experience (also known as flow experience) methodology has been quite intensely applied in Russia in the context of studies of the Internet related environments.

The notion of a flow has grown from the materials Csikszentmihalyi got hold of: his respondents systematically descried their sensational, though possibly repeated experience for which one could have hardly chosen a name other than flow – a sort of an optimal level of their experience when people control themselves and the environment, concentrate over the processes they are in and little care of the would-be results, have a distorted sense of time or lose the very idea of time management, find that their skills match the task challenges, often feel themselves as creative creatures. Flow, both habitual and rarely occurring, was reported to happen irrespectively of the type of the work being performed: be it spiritual or mundane, creative or routine, unique or known to almost everyone, individual or team-work, rarely or regularly performed. Flow experience, as well as self-determination tendencies, is a well-grounded prospective model for designers and managers to plan and construct digital projects which are going to be a long time in demand.

The scope of works dealing with the positive psychology approach to the Internet mediated behavior, unlike the works done during other periods, is fully enough available in English and reviewed in a number of publications, such as [39, 40].

Optimal experience very often takes place when human beings use digital technologies; indeed, flow has been described and measured in numerous Internet related behaviors, e.g. online types of instruction, entertainment, interaction, explorative behavior, usability testing, web marketing and shopping, psychological rehabilitation, etc. (for an overview see [37, 40]). Optimal experience was first investigated as an important component of the motivation of Russian computer hackers [42]; data close to that were reported as well outside Russia [18]. The main result is that the computer hackers who are highly or poorly qualified, report of flow experience; the moderately qualified hackers report it less often, and when they do, they admit facing problems in keeping this experience when they update their "professional" knowledge and renew qualification. At the time when the study was done, the meme "Russian hackers" was not yet well-known to the ordinary people as it is nowadays, though as long back as two decades ago and earlier the computer hackers from Russia enjoyed a high enough reputation among specialists – both criminals and security experts, including proponents of non-regulated use of computer networks.

The most extensive pool of a flow related research has been devoted to an empirical study of motivation of video game players (gamers). Online studies were carried out using the same methodology within the Russian, French, US, and Chinese samples of gamers: the initial questionnaire was worked out and after it proved to be reliable taken Russian-speaking gamers, it was adapted (back-translated and corrected) and performed within the non-Russian samples. The results of the factor study show that almost all the gamers report they experience flow; this factor was found to be no less (in fact, usually more) important than other factors, such as achievements, enjoyment of having interactions or wish to cognate navigating the net. Thus flow experience is among the major factors attracting gamers and pushing them to play their favorite games and/or mastering new ones. While the results of the studies related to each ethnic group have been published both in Russian and in English, comparative multi-ethnic results of the confirmatory factor analysis use are fully presented in [38]. Flow is the first factor in an almost every model; the single exception is the factor structure characterizing the French gamers. The latter model can be called minimalistic since it contains only three factors, while the factor structure characterizing Russian gamers is rather complex, it contains as many as six factors; the factor structures characterizing American and Chinese gamers lay in between. To make it short, it has been shown that the flow experience is a common motivating element for online gamers in four different ethnic groups such as Chinese, Americans, French and Russians. This proves the fact that optimal experience is a basic element explaining the world-wide attractiveness of playing computer video games [38, 40, 41].

At the positive psychology period and up to now free choice to navigate through diverse Web sources began to stumble upon restrictions entrusted by Russian security forces, ministries and Parliament: an online "struggle between dictators and revolutionaries" [28] is being more and more fierce. Various evidences say though that this sort of "struggle" is activated elsewhere, on the agenda of anti-terrorist tactics [22].

5 Current Studies

All the studies performed within the abovementioned stages (such as non-meeting, culture psychology and positive psychology) have at least one important consideration in common: they refer to small or very small samples, since the number of the Internet users in Russia was not too great prior to the advance, in 2007 and subsequent years, of social networking media, such as VKontakte, Odnoklassniki, Twitter or Facebook. After that the coverage of the Internet users increased and approached to be representative. That means, mass studies became possible. At the same time, local social media (such as VKontakte: vk.com) effectively compete with the most acknowledged international media and messengers; this competition is a probable reason that in cross-cultural studies of the intrusion and use of the major social networking sites, such as Facebook, the Russian population has not been investigated [3].

A mass study is being conducted for over a decade by a team headed by Soldatova [29–31] who started her career in cyberpsychology at the stage mentioned above as 'culture psychology.' Many thousands children, parents and teachers habituating in diverse regions of Russia fill in each year the questionnaires developed by the team. The main themes include the types of computers'/Internet phobias, anxiety or stress, digital competence of children and adults, varieties of a possible gap between the generations of the Internet users, motivations towards improvement of competence, attitudes and the likely preoccupation of the grown-ups towards the kids' overuse of new technologies, likelihood of becoming a victim of cyberharassment, cyberbulling or cybermobbing, expectations of schoolchildren referring to the future challenges connected with the development of digital technologies, etc. Besides, the team is responsible for the HelpLine phone channel and gives recommendations to kids in emergency cases. The continuous research project is partly affiliated with the work project EUKidsOnline and commissioned by the European Union department which is responsible for the kids' safety in the digital world – the current and especially the future world. The methodology of the studies held in Russia systematically corresponds, for about ten years, to the methodology of the EUKidsOnline II project (http://detionline.com/assets/files/helpline/Russian_KidsOnline_Final_Report_2013.pdf). The use of common instruments suggests perspectives of comparing the readiness of Russian and European children (as well as their parents and teachers) to meet and withstand numerous risks inseparable from the contemporary digital world.

For example, Soldatova and her team investigated the active and passive strategies of coping behavior related to the risks which include negative online content, such as bulling, sexual offers, fraud, theft of personal information, insistent appeals to meet online acquaintances, etc. [29–31]. The likelihood of becoming a victim significantly depends on the preferred types of activity: for example, those who are focused on learning while they are online, are less exposed to the risks listed above. Moreover, they can easier apply for assistance to their significant adults to get what is called parental mediation [30]. To assess the digital competence in adolescents and parents in Russia the researchers started with a description of digital competence which was considered by being dependent for making "confident, effective, critical and safe choices" [29, p. 66] when online. After that a specially developed Digital Competence Index was introduced, which consists of such components as knowledge, skills, motivation, responsibility and safety. The Index includes information and media competence, communicative competence, technical competence and consumption competence. To assess digital competence, a 52-item instrument was worked out; the studies confirmed reliability and usefulness of the Digital Competence Index [31].

Among the most widely accepted risks is Internet addiction, also known as overuse or pathological use of the Internet. The addictive phenomenology is being intensely discussed and investigated from medical and physiological [21, 24], psychological [19, 20], educational [27] and technological [35] perspectives. Diverse views on the theme were introduced and debated, mostly in publications in Russian. In 2009 and 2015 several translations into Russian have been published, including papers by M. Griffiths, K. Young, A. Weinstein, Z. Demetrovics and other respectable psychologists. The papers written in English on the theme are not numerous. Voiskounsky [35, 40] discussed the tendencies to interpret repetitive actions either as Internet addiction or as flow (optimal experience), and declared that optimal experience has rather little in common with addictions. Many authors avoid the term "Internet Addiction Disorder" and prefer to consider "pathological" or "excessive" use of the Internet, smartphones, and gadgets. For example, the reasons for the excessive use are believed [10, 11] to be twofold: first, the changes in human needs such as the need for convenience and functionality (for example, a "cool", stylish or expensive gadget), and second, a transformation of a human body in its psychological boundaries (subjective extension and subjective violation), since the human boundaries seem to extend nowadays as long in distance, as their gadgets may reach; thus, human beings may attain elements of a so-called universal or world-wide consciousness. We may conclude that in Russia, as well as outside Russia, many professionals in human health and well-being find rather ambiguous both a theory of Internet addiction and its treatment.

A great deal of studies was devoted to personality traits and cognitive abilities of video gamers [4, 17] but we will not follow this theme in any detail since this voluminous task deserves a special paper. More and more work is being done on Internet-mediated communications via social networking sites, though most of the papers stay unavailable for those who read only in English.

It was reasonably observed that the newborn virtual identities are being built via the use of blogs and social networking sites [2]; later it was stated that reputations are being leveled up, or probably, and unhappily, down [41]. A detailed chronological description enlists numerous virtual fictitious personae which were created and promoted on the Russian-language web-sites (up to early 2000s), all being creative projects in literature and journalism; it is specially noted that multiplayer role gamers are not taken into account since all the gamers accept certain virtual statuses [13]. A study of harmful online behavior [5] is quite fundamental and may be thought of as an approach to a Big Data technology: 6724 participants filled in the Dark Triad scale; the researchers analyzed the personal data of the participants plus as many as 15,281 posts placed upon the Facebook walls. Over 25% of the participants reported of being engaged earlier in some types of harmful online behaviors. Males are found more likely to send out insulting or threatening messages, to post aggressive comments. Meanwhile, dissemination of other people's private information does not correlate to gender. Not surprisingly, researchers have found (1) psychopathy and (2) male gender to be the unique predictors of engagement in harmful online behaviors. Besides, numerous parameters of the Dark Triad scale are shown to be significantly correlated to the linguistic patterns of the participants' posts [5].

6   Conclusion

The current paper presents a brief review of the development of the Internet psychology, or cyberpsychology in Russia; the work is restricted to publications available in English. A systematic review of the sources published in Russian would be a no less important task for scholars in the future. The Russian perspective is in a way peculiar. On one side, the Internet access in Russia is highly intrusive, and originally national sources effectively compete the most acknowledged sources such as Google or Facebook: the browser yandex.ru competes the former, the social networking site vk.com – the latter. On the other side, sufficiently wide use of various Internet sources is becoming restricted in Russia due to unfriendly decisions of legislative and administrative governance which now and then demands to block up the access to the sources initiated by liberal, or pretending liberal political activists within Russia and abroad.

The Internet psychology has a long enough history in Russia, in the sense that the relevant studies started as early as in 1980s. Russian specialists participate in international projects related to the Internet psychology, though probably less actively than their qualification deserves. Since Russia has stepped into a group of countries which regulate the public access to various Internet sources, we may expect that ongoing and future psychological studies will show numerous peculiarities. The Internet psychology will be prospering, since it attracts young psychologists and college students.

Acknowledgment.   The study was supported by the Russian Science Foundation, project # 18-18-00365.

References

1. Arestova, O., Babanin, L., Voiskounsky, A.: Psychological research of computer-mediated communication in Russia. Behav. Inf. Technol. **18**(2), 141–147 (1999)
2. Asmolov, A.G., Asmolov, G.A.: From We-media to I-media: identity transformations in the virtual world. Psychol. Russia State Art **2**, 101–123 (2009). http://psychologyinrussia.com/volumes/pdf/2009/05_2009_asmolovi.pdf
3. Błachnio, A., Przepiorka, A., Benvenuti, M., et al.: Cultural and personality predictors of Facebook intrusion: a cross-cultural study. Front. Psychol. **7**, 1895 (2016). https://doi.org/10.3389/fpsyg.2016.01895
4. Bogacheva, N.V.: Cognitive styles specifics of adult computer gamers. Ann. Rev. CyberTherapy Telemed. **14**, 84–88 (2016)
5. Bogolyubova, O., Panicheva, P., Tikhonov, R., Ivanov, V., Ledovaya, Y.: Dark personalities on Facebook: harmful online behaviors and language. Comput. Hum. Behav. **78**, 151–159 (2018). https://doi.org/10.1016/j.chb.2017.09.032
6. Cole, M.: Cultural Psychology: A Once and Future Discipline. The Belknap Press of Harvard University Press, Cambridge and London (1996)

7. Connolly, I., Palmer, V., Barton, H., Kirwan, G.: An Introduction to Cyberpsychology. Routledge, London (2016)
8. Csikszentmihalyi, M.: Beyond Boredom and Anxiety: Experiencing Flow in Work and Play. Jossey-Bass, San-Francisco (2000)
9. Deci, E., Ryan, R.: Intrinsic Motivation and Self-Determination in Human Behavior. Plenum Press, New York (1985)
10. Emelin, V.A., Rasskazova, E.I., Tkhostov, A.S.: Technology-related transformations of imaginary body boundaries: psychopathology of the everyday excessive Internet and mobile phone use. Psychol. Russia State Art **10**(3), 178–189 (2017)
11. Emelin, V., Tkhostov, A., Rasskazova, E.: Excessive use of Internet, mobile phones and computers: the role of technology-related changes in needs and psychological boundaries. Procedia Soc. Behav. Sci. **86**, 530–535 (2013)
12. Gerlovich, S.: From Newspeak to Cyberspeak: A History of Soviet Cybernetics. The MIT Press, Cambridge (2002)
13. Gorny, E.: The virtual persona as a creative genre on the Russian Internet. In: Schmidt, H., Teubener, K., Konradova, N. (eds.) Control + Shift. Public and Private Usages of the Russian Internet, pp. 156–176. Books on Demand, Norderstedt (2007). http://www.katy-teubener.de/joomla/images/stories/texts/publikationen/control_shift_01.pdf
14. Griffin, P., Belyaeva, A.V., Soldatova, G.U.: Socio-historical concepts applied to observations of computer use. Eur. J. Psychol. Educ. **7**, 269–286 (1992)
15. Griffin, P., Belyaeva, A.V., Soldatova, G.U., & the Velikhov-Hamburg Collective: Creating and reconstituting contexts for educational interactions, including a computer program. In: Forman, E.A., Minnick, N., Stone C.A. (eds.) Contexts for Learning: Sociocultural Dynamics in Children's Development, pp. 120–152. Oxford University Press, New York (1993)
16. Hiltz, S.R., Turoff, M.: The Network Nation. Human Communication via Computer. Addison-Wesley, Reading (1978)
17. Kulikova, T.I., Maliy, D.V.: The correlation between a passion for computer games and the school performance of younger schoolchildren. Psychol. Russia State Art **8**(3), 124–136 (2015). https://doi.org/10.11621/pir.2015.0310
18. Lakhani, K.R., Wolf, R.G.: Why hackers do what they do. Understanding motivation and effort in free/open source software projects. In: Feller, J., Fitzgerald, B., Hissam, S.A., Lakhani, K.R. (eds.). Perspectives on Free and Open Source Software, pp. 3–22. The MIT Press, Cambridge (2005). http://mitpress.mit.edu/books/chapters/0262562278chap1.pdf
19. Malygin, V.L., Merkurieva, Y.A., Iskandirova, A.B., Pakhtusova, E.E., Prokofieva, A.V.: Specific features of value orientations in adolescents with Internet addictive behaviour. Med. Psihol. Ross. **4**(33), 9 (2015). http://mprj.ru/archiv_global/2015_4_33/nomer02.php
20. Malygin, V.L., Merkurieva, Y.A., Krasnov, I.O.: Neuropsychological peculiarities as risk-factors of Internet addictive behaviour in adolescents. Med. Psihol. Ross. **4**(33), 12 (2015). http://mprj.ru/archiv_global/2015_4_33/nomer04.php
21. Mendelevich, V.: Psychiatry during the era of addiction medicine: modern diagnostic and therapeutic realities. Psychopathology and Addiction Medicine 1.1 (2016). http://pam-eng.ruspsy.net/article.php?post=486
22. Morozov, E.: The Net Delusion: The Dark Side of Internet Freedom. Perseus Books, Cambridge (2011)
23. Peters, B.: How Not to Network a Nation: The Uneasy History of the Soviet Internet. The MIT Press, Cambridge, London (2016)
24. Rabadanova, A.I., Abacharova, Z.S.: Comparative study of bioelectric brain activity in drug and internet addicts. Hum. Physiol. **40**(3), 252–257 (2014). https://doi.org/10.1134/S0362119714020133

25. Rigby, C.S., Ryan, R.M.: Glued to Games: How Video Games Draw us in and Hold us Spellbound. Prager, New York (2011)
26. Seligman, M.E.P., Csikszentmihalyi, M.: Positive psychology: an introduction. Am. Psychol. **55**(1), 5–14 (2000)
27. Shubnikova, E.G., Khuziakhmetov, A.N., Khanolainen, D.P.: Internet-addiction of adolescents: diagnostic problems and pedagogical prevention in the educational environment. EURASIA J. Math. Sci. Technol. Educ. **13**(8), 5261–5271 (2017). http://www.ejmste.com/Internet-Addiction-of-Adolescents-Diagnostic-Problems-and-Pedagogical-Prevention,76427,0,2.html
28. Soldatov, A.A., Borogan, I.P.: The Red Web: The Struggle between Russia's Digital Dictators and the New Online Revolutionaries. Public Affairs, New York (2015)
29. Soldatova, G.V., Rasskazova, E.I.: Assessment of the digital competence in Russian adolescents and parents: digital competence index. Psychol. Russia State Art **7**(4), 65–73 (2014). http://psychologyinrussia.com/volumes/pdf/2014_4/2014_4_65-74.pdf
30. Soldatova, G.U., Rasskazova, E.I.: Adolescent Safety on the Internet: risks, coping with problems and parental mediation. Russian Educ. Soc. **58**(2), 133–162 (2016)
31. Soldatova, G., Zotova, E.: Coping with online risks: the experience of Russian schoolchildren. J. Child. Media **7**(1), 44–59 (2013)
32. Tikhonova, M.N., Bogoslovskii, M.M.: Internet addiction factors. Autom. Documentation Math. Linguist. **49**(3), 96–102 (2015)
33. Voiskounsky, A.E.: The development of external means of communicative orientation. J. Russian East Eur. Psychol. **33**(5), 74–81 (1995)
34. Voiskounsky, A.: Investigation of Relcom network users. In: Sudweeks, F., McLaughlin, M., Rafaeli, Sh. (eds.) Network and Netplay: Virtual Groups on the Internet, pp. 113–126. AAAI Press/The MIT Press, Menlo Park, Cambridge, London (1998)
35. Voiskounsky, A.E.: Two types of repetitive experiences on the Internet. INTERFACE J. Educ. Commun. Values **7**(6), 1/6–6/6 (2007). http://bcis.pacificu.edu/journal/2007/06/voiskounsky.php
36. Voiskounsky, A.E.: Cyberpsychology and computer-mediated communication in Russia: past, present and future. Russian J. Commun. **1**(1), 78–94 (2008)
37. Voiskounsky, A.E.: Flow experience in cyberspace: current studies and perspectives. In: Barak, A. (ed.) Psychological Aspects of Cyberspace: Theory, Research, Applications, pp. 70–101. Cambridge University Press (2008)
38. Voiskounsky, A.E.: Positive psychology centered online studies. In: CENTRIC 2011: The Fourth International Conference on Advances in Human-Oriented and Personalized Mechanisms, Technologies, and Services, Barcelona, Spain, 23 October 2011–28 October 2011, pp. 8–14. XPS Publ. (2011)
39. Voiskounsky, A.: The origin and current status of cyberpsychology in Russia. In: Yan, Z. (ed.) Encyclopedia of Cyber Behavior, pp. 1328–1338. IGI Global, Hershey (2012)
40. Voiskounsky, A.E.: Flow experience in Internet-mediated environments. In: Leontiev, D. (ed.) Motivation, Consciousness and Self-Regulation, pp. 243–269. Nova Science Publishers, N.Y. (2012)
41. Voiskounsky, A.: Online behavior: interdisciplinary perspectives for cyberpsychology. Ann. Rev. CyberTherapy Telemed. **14**, 16–22 (2016)
42. Voiskounsky, A.E., Smyslova, O.V.: Flow-based model of computer hackers' motivation. CyberPsychol. Behav. **6**(3), 171–180 (2003)
43. Vygotsky, L.S.: Mind in Society: The Development of Higher Psychological Processes. Harvard University Press, Cambridge (1978)

Problematic Internet Usage and the Meaning-Based Regulation of Activity Among Adolescents

O. V. Khodakovskaia[1]([✉]) [iD], I. M. Bogdanovskaya[2] [iD],
N. N. Koroleva[2] [iD], A. N. Alekhin[2] [iD], and V. F. Lugovaya[2] [iD]

[1] Saint - Petersburg State University of Culture,
Dvortsovaya emb. 4, 191186
St. Petersburg, Russia
olga-khodakovskaya@yandex.ru
[2] Herzen State Pedagogical University of Russia,
emb. of the Moika River, 48, 191186 St. Petersburg, Russian Federation

Abstract. This paper explores the relationship between problematic internet usage and the meaning-based regulation of activity among adolescents. Participants were 77 adolescents (36 males, 41 females; M = 15.16 years, SD = 1.1) in grades 9–10 of two secondary schools predominantly for middle and lower-middle socioeconomic-status families in St. Petersburg (Russian Federation). Personal meaning-based regulation of adolescents' activity can be defined as a structure connected with various aspects of the adolescents' inner world and behavior. The data obtained make it possible to identify the personality-meaning-based preconditions for PIU in adolescence: difficulties in modelling the conditions for activity and programming behavior to achieve goals; a pronounced tendency to independent activity; a high level of susceptibility to psychological problems. The findings revealed that PIU may combine with a tendency to frequent usage of various electronic devices and a desire to acquire expensive technical novelties. The results given may be use in the development of psychological prophylaxis and correction of PIU in adolescence.

Keywords: Problematic internet usage · Adolescents
Meaning-based regulation of the activity

1 Introduction

In this paper terms of basic approaches to the determination of problematic internet usage are considered. PIU main symptom as being the reduction or loss of control over behavior in the Internet, the reduction of awareness and self-regulation in the process of Internet communication are discussed [7, 15, 23, 37–39]. Next, outline the important positions of the relationship between PIU and the meaning-based regulation of activity in adolescence. Personal meaning-based regulation of adolescents' activity defined as a structure connected with various aspects of the adolescents' inner world and behavior. Style of individual self-regulation, its individual profile, including cognitive, volitional, personality characteristics, adolescents' problematic feelings in various spheres of life and specific characteristics in their use of digital electronic devices were considered part of this structure [2, 32]. Paper concludes with some possible ideas and directions for future research.

2 Main Approaches to an Understanding of Problematic Internet Usage

Basic Approaches to the Determination of Problematic Internet Usage. In the opinion of Robert S. Tokunaga [37, 38], currently three basic approaches to the determination of inappropriate Internet usage have formed: clinical, cognitive-behavioristic and socio-cognitive. Accordingly, in the first instance PIU is viewed as a type of addiction and its manifestations display a similarity to other non-chemical types of addiction [41–44]. Due to the appearance and spread of mobile devices and the expanding opportunities to access Internet resources, the intensity and length of their usage in the contemporary world is rather the norm than a symptom of dependent behavior [19]. Cognitive-behavioral conceptions explain PIU in terms of cognitive distortions, insufficient self-control over behavior online, a low level of development of social skills, frequent experience of stress in interactions [8, 11, 14, 16, 21]. The socio-cognitive interpretation of PIU also associates it with the loss of conscious self-regulation of time spent on line, but to a greater degree in connection with the existence of psycho-social problems (loneliness, shame, social anxiety) [10, 17, 18, 25, 26, 37].

Meaning-Based Regulation of Behavior and Activity. In theory of the meaning-based regulation of behavior and activity (L.S. Vygotskii, A.N. Leont'ev, A.G. Asmolov, D.A. Leont'ev, et al.) [4, 27, 40], the leading role in the regulation of activity is played by personality meanings that reflect the significance for the person of this or that set of objects or phenomena in the life-world, their connection with needs, motives and goals. Meaning-based structures determined the impulse to action, the content and direction of activity, its beginning, maintenance and cessation, possible changes of goals in the course of activity, the connection between actions and personality values and convictions. According that theory PIU is defined as an increase in the personality significance of Internet communication, the transformation of an online presence into the leading meaning-forming motive.

3 Problematic Internet Usage and the Shaping of Meaning-Based Regulation of Activity Among Adolescents

In Russia the number of teenage active Inter-net users is soaring. According to long-term studies carried out by the Internet Development Foundation, in the period between 2009 and 2016, the number of adolescents spending a significant amount of time online (more than five hours a day) increased sevenfold. In 2009, only 8% of teenagers used the Internet intensively, but by 2016 that figure was 52%. In addition PIU may be increased by the appearance of new varieties of addiction: "gadget addiction", "digital electronic devices addiction" and a passion for technological novelties [12, 28, 30].

Teenagers' experience connected with the intensive development of self-awareness along with the incomplete formation of personality mechanisms of self-regulation becomes a precondition for a "withdrawal into virtual reality", the formation of patterns of Internet addiction [2, 34]. A number of studies have shown that adolescents with PIU typically display abnormalities in the formation of the meaning-based sphere of the personality: reduced self-control and a high level of emotional tension coupled with a tendency to avoid problems and failure to accept responsibility [17, 36]. This testifies to a need to examine the role of meaning-based regulation of behavior and activity in the shaping of teenagers' problematic behavior in cyberspace, which determines the relevance of the present work. As personal meanings regulating adolescents' activity we examined the style of individual self-regulation and adolescents' problematic feelings in various spheres of life (relations with parents, peers, in school, with re-gard to health, the future, social life, leisure, oneself).

The specific research questions include: (1) if there is a significant relationship between PIU with characteristics of personal meaning-based regulation of activity among adolescence? (2) what are the purpose and destination of digital electronic devices in adolescents with PIU? Whether the PIU could combine with a tendency to usage of various electronic devices?

4   Methods

4.1   Participants

Participants were 83 pupils in grades 9–10 of two secondary schools predominantly for middle and lower-middle socioeconomic-status families in St. Petersburg (Russian Federation). All the pupils were invited to participate in the study. Of these participants, 6 were dropped from the analysis because they scored more than 65 on the Chen Internet Addiction Scale (CIAS). According to the report by Malygin et al., the cutoff point at a score of 65 to determine Internet addiction via CIAS gave a good performance in respect of reliability and validity [47]. In the current study participants with scores in the range 27–42 and 43–64 were studied further. Following Malygin et al., we adopted this cutoff value to classify normal internet use (27–42) and problematic internet use (43–64). The final sample consisted of 77 adolescents (36 males, 41 females; M = 15.16 years, SD = 1.1).

4.2   Measures

Problematic Internet Use. The adolescents completed a CIAS – the self-rating questionnaire comprising 26 items, with a four-point Likert's scale ranging from 1 (Does not match my experience at all) to 4 (Definitely matches my experience). The questionnaire was specially developed for assessing internet addiction [9]. The scale (IA) is made up of five subscales: (1) compulsive use (5 items); (2) withdrawal symptoms (5 items); (3) tolerance (4 items); (4) interpersonal and health-related problems (7 items); (5) time management problems (5 items) and two integral indicators: (6) key symptoms of IA (IA-Sym = (Com + Wit + Tol)); (7) negative effects of Internet use (IA-Rp = (In + Tm)). The CIAS was adapted for use in Russia by Malygin et al. [29]. According to their report, the cutoff point at a score of 65 was used to define IA; ranges of 27–42 and 43–64 respectively were classified as normal internet use and problematic internet use (PIU). Cronbach's alpha fell in the range of 0.757 the scale of compulsive use to 0.9 on the scale of time management problems. IA test/re-test correlation on all subscales showed a good performance on reliability (a Pearson's correlation coefficient not less than 0.7–0.75).

Style of Behavioral Self-regulation [31]. The statements in the questionnaire are based on typical everyday situations and are not connected with the specifics of any particular work or study activity. The methodology is aimed at a diagnosis of conscious individual self-regulation and its components: planning, modelling, programming, evaluation of results, and personality characteristics – flexibility and independence. The questionnaire consists of 46 statements and functions as a single scale "Overall level of self-regulation", assessing the general level of formation of the individual system of conscious self-regulation of a person's voluntary activity on a four-point Likert's scale ranging from 1 (Does not match my experience at all) to 4 (Definitely matches my experience).

The questionnaire has six sub-scales (each containing 9 statements): Planning – the conscious planning of activity; Modelling – the level of development of concepts of external and internal significant conditions; the degree of conscious awareness, detailing and appropriateness of them; Programming – the person's conscious programming of his or her action; Evaluation of results –the adequacy of the assessment of oneself and of the results of one's activity and behavior; Flexibility – the ability to adjust, to make corrections to one's system of self-regulation when external and internal conditions change; Independence – level of development of autonomous behavior. The test/re-test reliability for the various scales has values between 0.583 and 0.78. The internal reliability was determined by a method of split-halfs (Cronbach's alpha = 0.437 to 0.675)

Psychological Problems of Adolescents [33]. The questionnaire consists of 93 statements and comprises 8 subscales: Problems related to the future (13 items), Problems related to the parental home (13 items), Problems related to school (13 items), Problems associated with peers (10 items), Problems with oneself (13 items), Problems related to leisure activities (9 items), Problems related to health (10 items), Problems related to the development of society (12 items). The summarizing scale of the questionnaire "Overall preoccupation with problems", Cronbach's alpha: 0.746.

Demographic Items and Digital Electronic Devices Usage. A questionnaire was used that incorporated demographic questions, and others measuring digital electronic device (DED) usage.

Demographic questions: (1) Gender: Male/Female; (2) Age, (3) Parental education: Mother/Father Secondary Education/Vocational School or College/Higher Education; (4) Parental employment status: Mother/Father in work/not working.

DED usage items included questions about the number of smartphones, tablets, smart watches owned, total number of digital electronic devices and the cost of DEDs.

The survey also included questions on types of DED personal meanings

(5) Signs of changing psychological boundaries (10 items), as described by Emelin et al. [13] e.g. "If I forget my smartphone/tablet/smart watch at home I feel uncomfortable (agree/disagree)." Kuder-Richardson reliability = 0.588;

(6) DED and interpersonal communication (4 items), Kuder-Richardson reliability = 0.582; e.g. "I like it when real-life communication goes hand in hand with communication in social networks (agree/disagree)";

(7) DED usage and maintenance of health (4 items), Kuder-Richardson reliability = 0.516; e.g. "I like to look after my health using a smartphone app (agree/disagree)";

(8) DED usage and self-development (5 items), Kuder-Richardson reliability = 0.733; e.g. "My self-development depends directly on information accessible via the Internet (agree/disagree)";

(9) DED usage and education (6 items), Cronbach's alpha = 0.695; e.g. "Do you use educational app/services on your electronic devices? From 1 (Does not match my experience at all) to 4 (Definitely matches my experience)."

4.3 Data Analysis

In this study the chi-square test was used for determining whether level of PIU were independent of adolescent gender, parental education, mother's and father's employment status.

We conduct the ANOVA to examine the CIAS scores in different groups of independent variables (normal/problematic internet use), the Scheffe test to correct alpha for to account for multiple comparisons and Pearson's correlation to examine the associations between CIAS and other variables. Statistical significance was set at a level of $p < 0.05$. The Statistica 10.0 software package was used for analyses in this study.

5 Results

In the first stage of the study, we tested the dependence of PIU level on the influence of different demographic variables. There was no correlation with the sex of the adolescents ($\chi^2(1) = 2.62$, $p = 0.11$), the mother's level of education ($\chi^2(2) = 0.67$, $p = 0.72$), the father's level of education ($\chi^2(3) = 2.23$, $p = 0.52$), the mother's employment status ($\chi^2(1) = 0.06$, $p = 0.93$), the father's employment status ($\chi^2(1) = 0.83$, $p = 0,36$), the level of the family's income ($\chi^2(2) = 2.27$, $p = 0.32$). In the present study girls detected signs of PIU more frequently than boys (78.05% vs 61.11%), while normal Internet usage was to a somewhat greater degree characteristic for the boys (38.89% vs 21.95%). Boys predominate (5.1% vs 2.6%) among the adolescents who displayed signs of Internet addiction (PIU $\geq$ 65).

Table 1 shows the statistically significant differences in the parameters studied between the groups of adolescents with normal and problematic Internet usage. The PIU group have more pronounced compulsive symptoms, withdrawal symptoms, a higher tolerance to being online, difficulties with management of time spent on line, key symptoms and issues associated with Internet addiction.

Table 1. Means and SD of subscales CIAS for normal and problematic Internet Use

Variables	Normal internet use		Problematic internet use		Adjusted p-value
	Mean	SD	Mean	SD	
Compulsive use	6.74	2.00	10.28	2.71	<0.001
Withdrawal symptoms	7.30	1.77	11.30	2.57	<0.001
Tolerance	6.04	1.52	8.91	2.15	<0.001
Interpersonal and health-related problems	9.74	1.51	12.20	3.04	<0.001
Time management problems	7.13	1.94	9.98	2.89	<0.001
Key symptoms of IA	20.09	3.15	30.50	5.27	<0.001
Negative effects of Internet use	16.78	2.58	22.19	4.03	<0.001
Overall level of IA	36.74	3.86	52.69	7.38	<0.001

Table 2 Shows the Statistically Significant Differences in the Parameters of Behavioral Self-regulation.

Table 2. Means and SD of parameters behavioral self-regulation

Variables	Normal internet use		Problematic internet use		Adjusted p-value
	Mean	SD	Mean	SD	
Modelling	6.35	1.61	5.24	1.73	0.01
Independence	3.91	2.09	5.02	2.18	0.04

Significant differences were revealed between the groups of study participants in two parameters of behavioral self-regulation: "Modelling" and "Independence" (Morosanova's "Style of Behavioral Self-Regulation" questionnaire). Among the teenagers with PIU we observe a lower value on the "Modelling" scale that reflects the consciousness and adequacy of conceptions about the external and internal conditions for the achievement of the goals of activity. However, they have higher values on the "Independence" scale. This is indicative of greater autonomy by the adolescents with a high level of PIU in the organization of their own activity, a capacity to independently set goals, find means of achieving them, monitor the course of the realization of an activity and asses its result. The results of the correlation analysis showed the presence of a statistically significant link between the characteristics of behavioral self-regulation and symptoms of PIU. The parameter "Modelling" (the capacity to model the conditions of activity) has a negative connection with such symptoms of PIU as tolerance ($r = -0.35$ $p = 0.002$); time-management ($r = -0.26$ $p = 0.025$); key symptoms of Internet addiction ($r = -0.23$ $p = 0.047$). "Flexibility" (of the regulatory processes) has an inverse connection with the scale of intrapersonal problems and health issues ($r = -0.32$ $p = 0.005$). Meanwhile, the self-regulation parameter "Independence" is positively connected to the majority of PIU characteristics: compulsive symptoms ($r = 0.31$, $p = 0.006$), withdrawal symptoms ($r = 0.23$ $p = 0.047$), tolerance ($r = 0.27$ $p = 0.018$), key symptoms of Internet addiction ($r = 0.34$ $p = 0.003$); problems connected with Internet addiction ($r = 0.30$ $p = 0.008$), the overall score ($r = 0.37$ $p = 0.001$). Table 3 shows the statistically significant differences of subscales PPA (Psychological Problems of Adolescents).

Table 3. Means and SD of subscales PPA (Psychological Problems of Adolescents)

Variables	Normal internet use		Problematic internet use		Adjusted p-value
	Mean	SD	Mean	SD	
Problems related to the future	2.43	0.63	2.91	0.66	0.0041
Problems with oneself	2.49	0.72	3.01	0.73	0.005
The overall level of problems	19.65	3.38	22.36	3.68	0.003

Adolescents with pronounced PIU also differ from their peers with normal Internet usage in the characteristics of problematic feelings. There is a statistically significantly higher level in this group for the parameters reflecting concern over problems connected with the future and with oneself, and a general intensity of psychological problems. The integral indicator of problematic feelings in the adolescents is positively connected with the general level of PIU ($r = 0.40$ $p = 0.0001$), compulsive symptoms ($r = 0.35$ $p = 0.002$), withdrawal symptoms ($r = 0.24$ $p = 0.036$), intrapersonal problems and health issues ($r = 0.45$ $p = 0.0001$), key symptoms ($r = 0.35$ $p = 0.002$) and problems connected with Internet addiction ($r = 0.38$ $p = 0.001$).

Adolescents with PIU are marked by specific characteristics in their use of various digital electronic devices (smartphones, tablets). In comparison to adolescents without pronounced symptoms of PIU, they display clear signs of a change of psychological boundaries in interaction with the devices. They make more frequent use of various technological aids in the course of their studies. The parameters for adolescents' use of various digital electronic devices present a large number of statistically significant correlations with characteristics of PIU. PIU is determined to the highest degree by the change in psychological boundaries in the process of using digital electronic devices.

There are links between this parameter and such characteristics of PIU as compulsive symptoms ($r = 0.52$ $p = 0.0001$), withdrawal symptoms ($r = 0.51$ $p = 0.0001$), tolerance ($r = 0.23$ $p = 0.041$), time-management ($r = 0.28$ $p = 0.015$), key symptoms of Internet addiction ($r = 0.55$ $p = 0.0001$) and the overall score ($r = 0.45$ $p = 0.0001$). There is also a positive connection between key symptoms of Internet addiction and the average cost of the tablet that the adolescents use ($r = 0.23$ $p = 0.047$), the intensity of use of electronic devices for self-development ($r = 0.24$ $p = 0.038$), for maintaining health ($r = 0.28$ $p = 0.015$) and for studies ($r = 0.35$ $p = 0.002$).

Table 4. Means and SD of DED (digital electronic devise usage)

Variables	Normal internet use		Problematic internet use		Adjusted p-value
	Mean	SD	Mean	SD	
Total number of digital electronic devices	1.57	0.90	1.91	1.19	0.22
The total cost of digital electronic devices	17.74	12.19	21.24	12.95	0.27
DED usage and signs of changing psychological boundaries	2.30	1.36	4.15	1.94	0.0001
DED usage and interpersonal communication	1.48	0.85	1.67	0.85	0.37
DED usage and maintenance of health	0.26	0.45	0.28	0.45	0.88
DED usage and self-development	0.65	0.88	0.96	1.06	0.22
DED usage and education	6.71	1.68	9.41	3.13	0.0002

The use of electronic devices with the aim of self-development or in educational activities also both display a direct correlation with the overall level of PIU ($r = 0.25$ $p = 0.027$; $r = 0.28$ $p = 0.013$) (Table 4).

6 Discussion

As the data obtained indicates, socio-demographic and socio-economic factors, such as an adolescent's sex, the education and employment status of the parents or the material prosperity of the family, do not exert a significant influence on the level of PIU. The results obtained on the relationship between an adolescent's gender and PIU do not entirely accord with the study made by Jeong et al. [22], which showed that PIU is more typical for males. Possibly the results that we obtained can be explained by the greater frankness of Russian girls compared to boys participating in the study, their preparedness to admit to showing symptoms of PIU. With regard to the remaining socio-demographic and socio-economic determinants, it is possible to point to the review by Lai and Kwan [24], which notes the indeterminacy of their influence on PIU. There are some studies that indicate a role for the parents' level of education, the mother's employment status, the family's prosperity and social status as factors that raise the risk of Internet addiction [1]. At the same time, a number of works display the opposite tendency, suggesting that the socio-economic factors listed act to reduce the risk of Internet addiction [18].

In adolescents with PIU there is a problem in the sphere of the meaning-based behavioral self-regulation, which manifests itself in the pursuit of independence, simultaneously with a reduced ability in modelling - understanding or awareness of their own needs and resources that might accompany the achievement of desired goals. The data that we obtained accord overall with the results of non-Russian studies, indicating that PIU presupposes poorly developed self-control mechanisms, connected with impulsive behavior [3, 5, 6].

Adolescents with a pronounced concern about psychological issues are more inclined to PIU. They are most worried about the future and their own self. There is a fairly large number of works that indicate the psychological problems of individuals with PIU that include a tendency to depression, loneliness, a low level of social skills development, dissatisfaction with relations with other people, but the link between PIU and problematic feelings in adolescents is the least researched.

It is a noticeable fact that PIU combines with adolescents' intensive and frequent use of digital electronic devices, which leads to an alteration of psychological boundaries in the process of interaction with them, to a failure to distinguish the real from the virtual. The link between PIU and other forms of technology addiction is confirmed by other studies [35]. However, it should be noted that the specifics of interaction with digital electronic devices and PIU among Russian adolescents remains a little-studied field. The use of devices to maintain health to all appearances aggravates PIU as it forms a compulsive urge to constantly consult gadgets to keep track of the parameters for physical activity, nutrition. Such adolescents try to acquire expensive tablet computers, which are the most popular among Russian schoolchildren. At the same time teenagers with PIU often use electronic devices as a study aid. Perhaps this comprises an inner resource of their self-realization in the Internet.

7 Conclusion

The present work shows that the phenomenon of PIU among adolescents can be examined from the position of meaning-based regulation of the activity of the individual. According to this position, PIU is to a greater degree characteristic of teenagers with a low level of meaning-based regulation and pronounced problematic feelings.

Independence in the performance of activities among adolescents is coupled with a reduced ability to model the conditions for activity, to choose a programme of action to achieve particular goals, to correct and regulate that programme in the course of the activity. Adolescents with PIU use the Internet and various digital electronic devices to tackle tasks relating to their studies and healthy living. The findings revealed that it possible to identify the personality-meaning-based preconditions for PIU in adolescence: difficulties in modelling the conditions for activity and programming behavior to achieve goals; a pronounced tendency to independent activity; a high level of susceptibility to psychological problems. PIU may combine with a tendency to frequent usage of various electronic devices and a desire to acquire expensive technical novelties. The results given may be use in the development of psychological prophylaxis and correction of PIU in adolescence.

Prospects for further research might be connected with the identification of the content of meaning-based attitudes, motives and goals as regulators of activity in the virtual environment; with the study of personal meanings to the use of digital electronic devices among teenagers with PIU; with cross-cultural studies would make it possible to establish culturally determined manifestations of PIU and its personal meaning-based factors.

References

1. Ahmadi, K.: Internet addiction among Iranian adolescents: a nationwide study. Acta Med. Iran. **52**(6), 467–472 (2014)
2. Alekhin, A.N., Koroleva, N.N., Ostasheva, E.I.: Semantic structures of world image as internal factors in the self-destructive behavior of today's teenagers. Psychol. Russ. State Art. **8**(1), 125–138 (2015). https://doi.org/10.11621/pir.2015.0111
3. Anderson, E.L., Steen, E., Stavropoulos, V.: Internet use and problematic internet use: a systematic review of longitudinal research trends in adolescence and emergent adulthood. Int. J. Adolesc. Youth **22**(4), 430–454 (2017). https://doi.org/10.1080/02673843.2016.1227716
4. Asmolov, A.G.: O predmete psihologii lichnosti. Voprosy psihologii **3**, 118–125 (1983). (in Russian)
5. Billieux, J., Van der Linden, M.: Problematic use of the internet and self-regulation: a review of the initial studies. Open Addict. J. **5**, 24–29 (2012). https://doi.org/10.2174/1874941001205010024
6. Błachnio, A., Przepiorka, A.: Dysfunction of self-regulation and self-control in facebook addiction. Psychiatr. Q. **87**, 493–500 (2016). https://doi.org/10.1007/s11126-015-9403-1
7. Cao, F., Su, L., Liu, T., Gao, X.: The relationship between impulsivity and Internet addiction in a sample of Chinese adolescents. Eur Psychiatry. **22**, 466–471 (2007). https://doi.org/10.1016/j.eurpsy.2007.05.004
8. Caplan, S.E.: Theory and measurement of generalized problematic Internet use: a two-step approach. Comput. Hum. Behav. **26**, 1089–1097 (2010). https://doi.org/10.1016/j.chb.2010.03.012
9. Chen, S.H., Wen, L.C., Su, Y.J., Wu, H.M., Yang, P.F.: Development of Chinese internet addiction scale and its psychometric study. Chin. J. Psychol. **45**(3), 279–294 (2003). https://doi.org/10.3390/brainsci2030347
10. Craparo, G.: The relationships between self-efficacy, internet addiction and shame. Indian J. Psychol. Med. **36**(3): 304–307 (2014). https://doi.org/10.4103/0253-7176.135386
11. Davis, R.A.: A cognitive-behavioral model of pathological internet use. Comput. Hum. Behav. **17**, 187–195 (2001). https://doi.org/10.1016/S0747-5632(00)00041-8
12. Egorov, A.Y.: Nekhimicheskie zavisimosti. Rech', Sankt-Peterburg (2007). (in Russian)
13. Emelin, V.A., Rasskazova, E.I., Tkhostov, A.S.: Development and validation of the technique for measurement of changes of psychological boundaries while using technical devices (TPB-TD). Psikhologicheskie Issledovaniya **2**(22), 5 (2012). http://psystudy.ru. (in Russian, abstr. in English)
14. Esen, B.K., Aktas, E., Tuncer, I.: An analysis of university students' internet use in relation to loneliness and social self-efficacy. Procedia Soc. Behav. Sci. **84**, 1504–1508 (2013). https://doi.org/10.1016/j.sbspro.2013.06.780
15. Fajola, E.H., Vojskunskij, A.E., Bogacheva, N.V.: Chelovek dopolnennyj: stanovlenie kibersoznanija. Voprosy filosofii **3**, 147–162 (2016). (in Russian)
16. Grieve, R., et al.: Development and validation of a measure of cognitive and behavioral social self-efficacy. Pers. Individ. Differ. **59**, 71–76 (2014). https://doi.org/10.1016/j.paid.2013.11.008
17. Gross, E.F., Juvonen, J., Gable, S.: Internet use and well-being in adolescence. J. Soc. Issues **58**(1), 75–90 (2002). https://doi.org/10.1111/1540-4560.00249
18. Heo, J., Oh, J., Subramanian, S.V., Kim, Y., Kawachi, I.: Addictive internet use among korean adolescents: a national survey. PLoS ONE **9**(2), e87819 (2014). https://doi.org/10.1371/journal.pone.0087819

19. How technology changes everything (and nothing) in psychology. 2008 annual report of the APA Policy and Planning Board. Am. Psychol. **64**(5), 454–463 (2009). https://doi.org/10.1037/a0015888

20. Hussain, Z., Pontes, H.: Personality, internet addiction, and other technological addictions: a psychological examination of personality traits and technological addictions. In: Psychological, Social, and Cultural Aspects of Internet Addiction, pp. 45–71. IGI Global (2018). https://doi.org/10.4018/978-1-5225-3477-8.ch003

21. İskender, M., Akin, A.: Social self-efficacy, academic locus of control, and internet addiction. Comput. Educ. **54**, 1101–1106 (2010). https://doi.org/10.1016/j.compedu.2009.10.014

22. Jeong, J.Y., Hyeonyee, K., Ian, H.: Understanding adolescents' problematic Internet use from a social/cognitive and addiction research framework. Comput. Hum. Behav. **29**(6), 2682–2689 (2013). https://doi.org/10.1016/j.chb.2013.06.045

23. Kim, E.J., Namkoong, K., Ku, T., Kim, S.J.: The relationship between online game addiction and aggression, self-control and narcissistic personality traits. Eur. Psychiatry **23**, 212–218 (2008). https://doi.org/10.1016/j.eurpsy.2007.10.010

24. Lai, F.T.T., Kwan, J.L.Y. Socioeconomic determinants of internet addiction in adolescents: a scoping review. In: Bozoglan, B. (ed.) Psychological, Social, and Cultural Aspects of Internet Addiction, Hershey, PA, pp. 221–235. IGI Global (2018). https://doi.org/10.4018/978-1-5225-3477-8.ch012

25. LaRose, R., Eastin, M.S.: A social cognitive theory of Internet uses and gratifications: toward a new model of media attendance. J. Broadcast. Electron. Media **48**, 358–377 (2004). https://doi.org/10.1207/s15506878jobem4803_2

26. LaRose, R., Kim, J., Peng, W.: Social networking: addictive, compulsive, problematic, or just another media habit? In: Papacharissi, Z. (ed.) A Networked Self: Identity, Community, and Culture on Social Network Sites, pp. 59–81. Routledge, New York (2011)

27. Leontiev, D.A.: Personal meaning: a challenge for psychology. J. Posit. Psychol. **8**(6), 459–470 (2013). https://doi.org/10.1080/17439760.2013.830767

28. Maksimova, O.A.: «Cifrovoe» pokolenie: stil' zhizni i konstruirovanie identichnosti v virtual'nom prostranstve. Vestnik CHelGU **22**(313), 6–10 (2013). (in Russian)

29. Malygin, V.L.: i dr: Internet-zavisimoe povedenie. Kriterii i metody diagnostiki. MGMSU, Moskva (2011). (in Russian)

30. Miroshkina, M.R.: Cifrovoe pokolenie. Portret v kontekste obrazovaniya. Mezhdisciplinarnoe issledovanie. Obrazovanie lichnosti **2**, 16–21 (2015). (in Russian)

31. Morosanova, V.I.: Oprosnik «Stil' samoregulyacii povedeniya» (SSPM). Kogito-Centr, Moskva (2004). (in Russian)

32. Mottram, A.J., Fleming, M.J.: Extraversion, impulsivity, and online group membership as predictors of problematic Internet use. Cyber Psychol. Behav. **12**, 319–321 (2009). https://doi.org/10.1089/cpb.2007.0170

33. Regush, L.A., Alekseeva, E.V., Orlova, A.V., Pezhemskaya, Y.S.: Psihologicheskie problemy podrostkov: standartizirovannaya metodika. Nauchno-metodicheskie materialy. In: Gercena, A.I. (ed.) Izd-vo RGPU im, Sankt_Peterburg (2012). (in Russian)

34. Soldatova, G.U., Rasskazova, E.I.: Chrezmernoe ispol'zovanie interneta: faktory i priznaki. Psihologicheskij zhurnal **34**(4), 105–114 (2013). (in Russian)

35. Soldatova, G.U., Nestik, T.A., Rasskazova, E.I., Zotova, E.Y.: Cifrovaya kompetentnost' podrostkov i roditelej. Rezul'taty vserossijskogo issledovaniya. Fond Razvitiya Internet, Moskva (2013). (in Russian)

36. Solpiter D.: Deti i komp'yutery: nastol'naya kniga roditelej Binom, Moskva (1996). (in Russian)

37. Tokunaga, R.S.: Engagement with novel virtual environments: The roles of perceived novelty and flow in the development of the deficient self-regulation of Internet use and media habits. Hum. Commun. Res. **39**, 365–393 (2013). https://doi.org/10.1111/hcre.12008

38. Tokunaga, R.S.: Perspectives on internet addiction, problematic internet use, and deficient self-regulation: contributions of communication research. Ann. Int. Commun. Assoc. **39**(1), 131–161 (2015). https://doi.org/10.1080/23808985.2015.11679174

39. Ushakova, E.S.: Internet-zavisimost' kak problema sovremennogo obshchestva. Lichnost' v menyayushchemsya mire: zdorov'e, adaptaciya, razvitie **1**(4), 44–51 (2014). (in Russian)

40. Vygotskij, L.S.: Sobranie sochinenij. V 6 t. T. 5. Pedagogika, Moskva (1982). (in Russian)

41. Weinstein, A., Lejoyeux, M.: Internet addiction or excessive internet use. Am I Drug Alcohol Abuse **36**(5), 277–283 (2010). https://doi.org/10.3109/00952990.2010.491880

42. Weinstein, A.: Comorbidity of Internet addiction with other psychiatric conditions. J. Behav. Addictions. **4**(Suppl. 1), 43 (2015)

43. Weinstein, A.: Internet addiction: diagnosis, comorbidity and treatment. Med. Psihol. Ross. **4**(33), 3 (2015). http://mprj.ru. https://doi.org/10.1016/j.addbeh.2014.07.031. [in English, in Russian]

44. Winkler, A., Dörsing, B., Rief, W., et al.: Treatment of internet addiction: a meta-analysis. Clin. Psychol. Rev. **33**(2), 317–329 (2013). https://doi.org/10.1016/j.cpr.2012.12.005

Neural Network-Based Exploration of Construct Validity for Russian Version of the 10-Item Big Five Inventory

Anastasia Sergeeva[1][(✉)], Bogdan Kirillov[2], and Alyona Dzhumagulova[1]

[1] ITMO University, Saint Petersburg, Russia
an.se.sergeeva@gmail.com, aledjuna@gmail.com
[2] Skolkovo Institute of Science and Technology, Moscow, Russia
Bogdan.Kirillov@skoltech.ru

Abstract. This study aims to present a new method of exploring construct validity of questionnaires based on neural network. Using this test we further explore convergent validity for Russian adaptation of TIPI (Ten-Item Personality Inventory by Gosling, Rentfrow, and Swann). Due to small number of questions TIPI-RU can be used as an express-method for surveying large number of people, especially in the Internet-studies. It can be also used with other translations of the same questionnaire in the intercultural studies. The neural network test for construct validity can be used as more convenient substitute for path model.

Keywords: Five-factor model · Ten-Item Personality Inventory
Short measures · Personality assessment · Neural networks
Machine learning · Test construction

1 Introduction

Questionnaires are a viable tool in modern psychological research but to be useful they have to be validated in a number of ways. Usual questionnaire validation pipeline involves a test for internal consistency, construct validity and reliability [8]. Our research is centered around second issue, construct validity. Construct validity shows how well the questionnaire can generalize, to what degree the measures that the questionnaire provides can be applied to estimate the characteristics of psychological model.

Construct validity is categorized into convergent and discriminant validity [2]:

1. Convergent validity - "Are two theoretically related ways of estimating the same characteristic really related?";
2. Discriminant validity - "Are two theoretically unrelated ways of estimating the same characteristic really unrelated?";

© Springer Nature Switzerland AG 2018
D. A. Alexandrov et al. (Eds.): DTGS 2018, CCIS 859, pp. 239–249, 2018.
https://doi.org/10.1007/978-3-030-02846-6_19

Both types of validity are important to investigate the construct validity of a questionnaire. Convergent and discriminant validation ensure that the questionnaire does precisely what it is made to do - provides a way to test a psychological model experimentally.

In this study we provide a novel approach to construct validity evaluation. It is based on usage of neural network to predict characteristics of well-established questionnaire from items of the questionnaire under investigation. Using direct prediction we can evaluate convergent validity and we can evaluate discriminant validity using interpretation of trained weights.

The Five Factor Model of personality traits (the Big Five, also known as the OCEAN or CANOE model) is currently among the most used personality traits models. Questionnaires based on its factors (Neuroticism, Extraversion, Openness to experience, Agreeableness and Conscientiousness) are widely used in scientific and industrial applications that require personality diagnosis. The major drawback of existing questionnaires is their size - it ranges from 44 items (BFI [15]) to 240 items [5]. Size makes research more difficult especially in Internet-based cases where lack of outside control (usually provided by researchers in offline cases) and participant's preference to skip tedious tasks condone random response or quitting.

A number of studies [6,11] suggest 10-item personality questionnaires as brief diagnostic tools for they have satisfying psychometric performance. In this study we choose to use a Russian adaptation of TIPI [10] questionnaire due to the TIPI's cross-cultural generalizability (which is shown by a set of TIPI international adaptations [3,12,13,18,20–23] etc.

Currently there are two competing Russian adaptations for TIPI questionnaire, KOBT [17] and TIPI-RU [24]. Their performance in convergent validity are close (TIPI-RU performs slightly better in Extraversion, Agreeableness and Emotional stability, KOBT - in Openness and Conscientiousness). For the current study TIPI-RU is considered a better alternative. TIPI-RU data are freely and openly available at github [25] so the data can be used as an addition to our own sample. We use TIPI-RU as an example for application of our novel convergent validity evaluation method.

2   Materials and Methods

2.1   TIPI-RU Questionnaire

The TIPI-RU is translated and validated version of TIPI questionnaire [10]. The translation can be found in appendix 1 of [24]. Questionnaire consists of 10 questions (below denoted as $TIPI_n$ where n ranges from 1 to 10). The actual big five characteristics are computed from answers according to the following formulae:

$$E = 0.5(TIPI_1 + reverse(TIPI_6)) \tag{1}$$

$$A = 0.5(TIPI_7 + reverse(TIPI_2)) \tag{2}$$

$$C = 0.5(TIPI_3 + reverse(TIPI_8)) \tag{3}$$

$$ES = 0.5(TIPI_9 + reverse(TIPI_4) \tag{4}$$

$$O = 0.5(TIPI_5 + reverse(TIPI_{10})) \tag{5}$$

where *reverse* means taking the opposite value on Likert scale (7 becomes 1, 6 becomes 2 etc.) and big five characteristics are denoted (here and in all plots) by first letters of their corresponding names (Extraversion, Agreeableness, Consciousness, Emotional stability and Openness).

2.2 Dataset

We use the extended dataset of 457 observations that include the one composed by Sergeeva, Kirillov and Dzhumagulova (218 observations). 218 old observations are freely available at [25] and we have collected other 239 by surveying Russian students who did not participate in the research of Sergeeva's group.

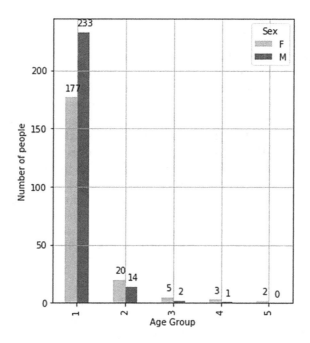

Fig. 1. Gender-Age distribution.

The age groups on the Fig. 1 are: 1 (10−19 years), 2 (20−29 years), 3 (30−39 years), 4 (40−49 years) and 5 (50−59 years).

2.3   Convergent Validity Test via Neural Network

To check convergent validity of TIPI-RU via neural network we use the following scheme:

1. We use 5PFQ as template and we assume that its characteristics (extraversion, emotional stability etc.) can be measured simpler via TIPI-RU questionnaire;
2. If the latter is correct, we can fit a neural network to predict 5PFQ characteristics from answers to TIPI-RU questions;
3. For our approach to be successful we have to address the following issues:
    - Ensure that the network is learning something, preferably a certain mapping from TIPI-RU answers to 5PFQ characteristics;
    - Ensure that even the small network does overfit - it shows that there is a really strong connection between inputs and outputs;
    - Ensure that the result of trained network is different from results of network trained on random permutations of labels.
4. Evaluate the model's performance via quality measures.

Sergeeva, Kirillov and Dzhumagulova among other methods used path model to confirm convergent validity. The neural network approach is conceptually much simpler: path model is based on fitting five different linear regression models to predict a 5PFQ item given the answers to questions that construct its TIPI-RU counterpart, but with neural network we can predict all five 5PFQ characteristics using all 10 TIPI-RU questions with single model. If the network can do it, then TIPI-RU and 5PFQ converge - it proves the convergent validity of TIPI-RU. The process of TIPI-RU computation can be viewed as application of a single hidden layer neural network with the following structure shown on Fig. 2 (for simplicity, we assume that Likert reverses of answers to 2, 4, 6, 8, 10 questions are already taken before the application of the network):

SUM here denotes output of a summarizing neuron with linear activation function and, for simplicity, this toy network has no bias, so the following is correct:

$$SUM = W\dot{X} \tag{6}$$

where W is vector of weights, and X is the vector of inputs to this particular neuron. Same operation is performed at the output layer neurons. The network is fully connected but all edges that don't add up to TIPI-RU scales are set to zero. They are not shown on Fig. 2.

This particular configuration is very hard to reach by gradient descent. It is not impossible, but very improbable for a network to converge there. But the network can converge to a different weight set that allows for non-zero edges that connect output characteristics and questions that aren't included in them. We don't really need to follow the template set by Fig. 2 and can use an arbitrary neural network. Any network that is reasonably small should be enough.

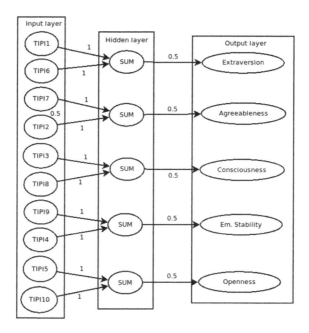

Fig. 2. TIPI computation as neural network. Only non-zero weights are shown.

To ensure that overfitting during the training may be attributed to strong connections between inputs and outputs rather than to model's power we should keep the model as small as possible. There is a trade-off between susceptibility to overfitting and ability to fit anything useful: number of parameters should be large enough to recover a dependency between inputs and outputs but at the same time small enough but it should be still pretty small because the network should find it hard to remember every observation training and validation set. Usually the penalty on model size is considered an auxilliary way of regularization and primarily other ways like L1/L2 regularization are used. For this case we find convenient to use the size as regularizer only.

In the current study we use the following configuration of network (shown on Fig. 3 below):

It is a very small convolutional network that consist of one 2D convolutional layer and two reshaping operations. The first reshape is needed to reshape the incoming TIPI items into "pictures" that are 1 in height, 2 in width and have five channels. The second reshape turns (1,1,5) output of convolutional layer into just 5 answers.

The network has 50 parameters - exactly two thirds of presumed TIPI network shown at Fig. 2 that has 75. A pretty simple way to see whether the network is learning something correct and it is not by happenstance is to destroy any real structure that is present in data, then try to fit the network from such

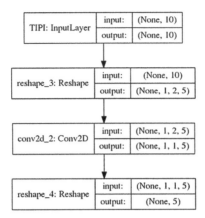

Fig. 3. Actual network that learns TIPI-5PFQ connection.

damaged set and compare the distribution of results with ones of networks trained on unharmed data. To do so, we perform an investigation obeying the following scheme:

1. Train 100 networks on the TIPI-RU dataset - it will provide distribution of MSEs on unharmed data;
2. Shuffle 5PFQ characteristics corresponding to TIPI-RU answers at random, then train a network to predict shuffled 5PFQs from same TIPI-RUs.
3. Do step 2 one hundred times;
4. Check whether two samples of error measures come from the same distribution or not by two-sample Kolmogorov-Smirnov test.

Difference between two distributions shows that there is a true dependency between answers to TIPI questions and 5PFQ characteristics that we got completely destroyed while shuffling the labels. The network is implemented using Keras [4] with Tensorflow [1] as backend. All plots are made with Matplotlib [14].

We make here a reasonable assumption that the best predictions of 5PFQ that a generalizable (not overfitted) model can make from the TIPI-RU data are actually the TIPI-RU values themselves. So we can find a best MSE possible by computing MSE between 5PFQ characteristics and TIPI characteristics both divided by corresponding maximas. A network that gets below that threshold is overfitting.

2.4   Neural Network Performance Measures

This work uses classical loss function for regression: mean squared error (MSE).

$$MSE(\hat{y}, y) = \frac{1}{n} \sum_{i=0}^{n} (y - \hat{y})^2 \qquad (7)$$

where y - real value, $\hat{y}$ - value predicted by model. We choose MSE as loss function instead of mean absolute error because it punishes large deviations from the real value more than small ones. But we use MAE as an additional performance measure:

$$MAE(\hat{y}, y) = \frac{1}{n} \sum_{i=0}^{n} |y - \hat{y}| \tag{8}$$

2.5 Discriminant Validity via Interpretation of Trained Weights

Neural Network Interpretation is a complex task and currently a lot of approaches of solving it exists. For a thorough review one can take a look at [19]. But in this particular case the interpretation becomes very simple. According to definition of convolution [9], the output of single convolution is as follows:

$$Z = \sum_{i=1,j=1}^{n,m} X_{i,j} \times W_{i,j}, \tag{9}$$

where W - weights of single convolutional filter of interest, X - single window from the input image. The resulting output of a convolution filter (without activation function) is constructed by applying this operation to whole input example via a sliding window. The important implication of the convolution is that weights of the convolutional layer after the end of training will mimic structure of its input. A set of CNN interpretation methods is based on this property but for our simple case we need only to visualize the weights for each neuron and it will be enough to observe captured structure. Structure similar to Fig. 2 and Eqs. 1- 5 shows high level of discriminant validity.

3   Results

The network reaches minimal MSE possible somewhere around 60-th epoch of training. Minimal MSE possible is equal 0.05. Validation set is randomly chosen 40% of the whole. Figure 4 shows the history of training:

After reaching the minimal possible MSE the network overfits and drops MSE to 0. During training the distribution of MSEs and MAEs from models trained on reshuffled labels diverges from one of the models trained on correct labels as shown on Fig. 5:

Initially both kinds of models are indistinguishable. But as the training goes, the correct models drive towards zero while permuted ones do not and at the end two distributions are no more the same. The full animation is freely available on YouTube [7]. It is the visual, qualitative way to check whether there is a strong connection between inputs and outputs. The quantitative one that we use is to apply two-sample Kolmogorov-Smirnov test. For computation we use Scipy KS-test implementation [16] and we plot results on Fig. 6:

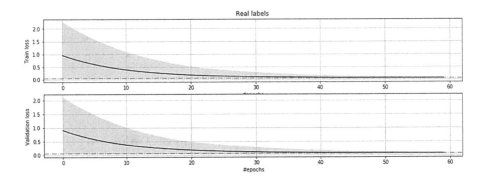

Fig. 4. First sixty epochs of training on real labels. Horizontal line denotes minimal MSE possible. Black line is mean of 100 repetitions, grey area is error region.

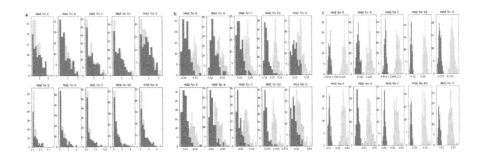

Fig. 5. Divergence of correct and permuted label distributions: **a.** start of training, **b.** 125th epoch, **c.** end of training. Dark grey - correct models, light grey - models trained on permuted labels.

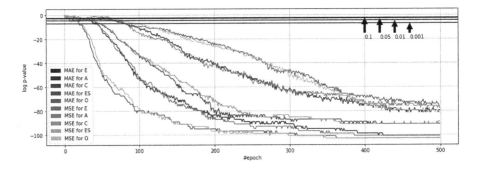

Fig. 6. Two-sample KS-test p-values (natural logarithms of them) for MAE and MSE as a function of training time. Black arrows denote different thresholds.

The question KS-test answers is "Were these two samples drawn from the same distribution?". Null hypothesis is that they are, so for our case p-values should be below the threshold. The differences, as shown on Fig. 6, grow enough to pass even the most strict threshold quite fast. Weight visualization, as present at Fig. 7, shows the structure similar to Fig. 2 and Eqs. 1–5:

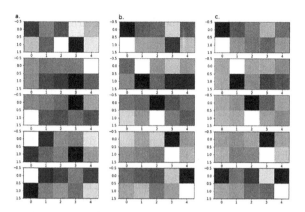

Fig. 7. Evolution of weights during training (average over 100 runs): **a.** start of training, **b.** 70th epoch, **c.** end of training. Single image represents a single convolutional neuron. Black - large positive weight, white - large (in absolute value) negative weight, shades of grey - weights close to 0.

It even captures the sign reversal in Agreableness (second row). Also if one looks closely at the Openness neuron (fifth row), it will be obvious that Openness lacks discriminant validity - there are a lot of other items that are highlighted as strong as the valid column. This inconsistency of Openness was already described by Sergeeva et al. and by other researchers. Neural Network method for validity testing converges with more conventional approaches despite being mostly qualitative.

4   Conclusion

In this study we have introduced a novel qualitative method for evaluation of construct validity of personality questionnaire using a neural network-based approach. The method is easy to implement with modern Deep Learning frameworks and is more interpretable than traditional methods like path models or correlation matrices since it answers much simpler question: "Is there a learnable connection between inputs (answers to questionnaire) and outputs (characteristics of questionnaire made to measure the same constructs)?". An obvious drawback of our method is that no simple way of comparing its performance to path model exists.

Based on our core findings we can recommend using TIPI-RU as a brief method for measuring personality in non-clinical settings like the internet assesment of personality measures as it passes the neural network test. Future studies involve using neural network test for evaluation of other questionnaires and exploring its limits of applicability.

All procedures involving human participants performed in this study went in accordance with the ethical standards of the institutional and/or national research committee and with the 1964 Helsinki declaration and its later amendments or comparable ethical standards.

References

1. Abadi, M.: TensorFlow: large-scale machine learning on heterogeneous systems (2015). http://tensorow.org/
2. Campbell, D.T., Fiske, D.W.: Convergent and discriminant validation by the multitrait-multimethod matrix. Psychol. Bull. **56**(2), 81 (1959). https://doi.org/10.1037/h0046016
3. Chiorri, C., Bracco, F., Piccinno, T., Modafferi, C., Battini, V.: Psychometric properties of a revised version of the Ten-item Personality Inventory. Eur. J. Psychol. Assess. **31**(2), 109–119 (2015). https://doi.org/10.1027/1015-5759/a000215
4. Chollet, F.: Keras (2015). https://github.com/fchollet/keras
5. Costa, P.T., McCrae, R.R.: Revised NEO Personality Inventory (NEO-PI-R) and NEO Five-Factor Inventory (NEO-FFI) professional manual. Psychological Assessment Resources, Inc., Odessa (1992)
6. Crede, M., Harms, P., Niehorster, S., Gaye-Valentine, A.: An evaluation of the consequences of using short measures of the Big Five personality traits. J. Pers. Soc. Psychol. **102**(4), 874 (2012)
7. Divergence video (2018). https://youtu.be/kRuk0Pechxg
8. Furr, R.M.: Psychometrics: An Introduction. Sage Publications, Thousand Oaks (2017)
9. Goodfellow, I., Bengio, Y., Courville, A., Bengio, Y.: Deep Learning. MIT press, Cambridge (2016)
10. Gosling, S.D., Rentfrow, P.J., Swann, W.B.J.: A very brief measure of the Big-Five personality domains. J. Res. Pers. **37**, 504–528 (2003)
11. Gunnarsson, M., Gustavsson, P., Holmberg, S., Weibull, L.: Statistical evaluation of six short Five Factor Model personality measures aiming for potential inclusion in the SOM Institute's national surveys (2015)
12. Herzberg, P.Y., Brähler, E.: Assessing the Big-Five personality domains via short forms. A cautionary note and a proposal. Eur. J. Psychol. Assess. **22**, 139–148 (2006)
13. Hofmans, J., Kuppens, P., Allik, J.: Is short in length short in content? An examination of the domain representation of the Ten Item Personality Inventory scales in Dutch language. Pers. Individ. Differ. **45**(8), 750–755 (2008)
14. Hunter, J.D.: Matplotlib: a 2D graphics environment. Comput. Sci. Eng. **9**(3), 90–95 (2007)
15. John, O.P., Srivastava, S.: The Big-Five trait taxonomy: history, measurement, and theoretical perspectives. In: Pervin, L., John, O. (eds.) Handbook of personality: Theory and research, pp. 102–138. Guilford Press, New York (1999)

16. Jones, E., Oliphant, T., Peterson, P.: SciPy: open source scientific tools for Python (2014)
17. Kornilova, T., Chumakova, M.: Development of the Russian version of the brief Big Five questionnaire (TIPI). Psikhologicheskie Issledovaniya **9**(46), 5 (2016)
18. Laguna, M., Bak, W., Purc, E., Mielniczuk, E., Oles, P.K.: Short measure of personality TIPI-P in a Polish sample. Roczniki Psychologiczne **17**(2), 421–437 (2014)
19. Montavon, G., Samek, W., Müller, K.-R.: Methods for interpreting and understanding deep neural networks. Digital Signal Processing (2017). https://doi.org/10.1016/j.dsp.2017.10.011
20. Muck, P.M., Hell, B., Gosling, S.D.: Construct validation of a short five-factor model instrument: a self-peer study on the German adaptation o the Ten-Item Personality Inventory (TIPI-G). Eur. J. Psychol. Assess. **23**, 166–175 (2007)
21. Oshio, A., Abe, S., Cutrone, P.: Development, reliability, and validity of the Japanese version of Ten Item Personality Inventory (TIPI-J). Jpn. J. Pers. **21**, 40–52 (2012)
22. Renau, V., Oberst, U., Gosling, S.D., Rusifffdfffdol, J., Chamarro, A.: Translation and validation of the Ten-Item-Personality Inventory into Spanish and Catalan. Aloma. Revista de Psicologia, Cincies de l'Educació i de l'Esport **31**, 85–97(2013)
23. Romero, E., Villar, P., Gfffdfffdmez-Fraquela, J., Lfffdfffdpez-Romero, L.: Measuring personality traits with ultra-short scales: a study of the Ten Item- Personality Inventory (TIPI) in a Spanish sample. Pers. Individ. Differ. **3**, 289–293 (2012)
24. Sergeeva, A., Kirillov, B., Dzhumagulova, A.: Translation and adaptation of short five factor personality questionnaire (TIPI-RU): convergent validity, internal consistency and test-retest reliability evaluation. Eksperimentalnaya psikhologiya **9**(3), 138–154 (2016). https://doi.org/10.17759/exppsy.2016090311
25. TIPI-RU data (2017). https://github.com/bakirillov/tipiru

Impulsivity and Risk-Taking in Adult Video Gamers

Nataliya Bogacheva[1]([⊠]) and Alexander Voiskounsky[2]

[1] Sechenov University, 8-2 Trubetskaya Str., Moscow 119991, Russia
bogacheva.nataly@gmail.com
[2] Moscow State University, 11/9 Mokhovaya Str., Moscow 125009, Russia
vae-msu@mail.ru

Abstract. Video games are often seen as a reason for numerous psychological changes, both positive and negative, in players. For instance, many authors believe that video games push children and adults towards risky behaviors and impulsivity. The study aimed to analyze both theoretical and empirical evidences of that sort, as well as to investigate parameters of personal and cognitive impulsivity and risk-readiness in adult video gamers. The sample of gamers included 223 participants, all from Russia. Impulsivity and related personal traits were measured with Eysencks' Impulsiveness Scale (I-7) and Kornilova's Personal Risk Factors Questionnaire. Impulsivity as cognitive style was measured by Kagan's MFFT. No evidence of high impulsivity was found, though video game players, who played more than 12 h per week turned out to be more venturesome, compared to less active gamers. Sex-related differences were investigated: female gamers scored lower in empathy, while male gamers showed higher venturesomeness. In a cognitive style study, video gamers were more accurate compared to non-gamers, and thus showed no tendency for impulsivity. The results are contrasted to the published data, when applicable.

Keywords: Video games · Impulsivity · Risk-Readiness · Adult video gamers

1 The Relationship Between Video Games, Risk, and Impulsivity: An Overview

Video games have become an important part of everyday life for many people, especially children, teenagers and young adults. Some authors even suggest that those who were born after the 1980s belong to "the Gamer Generation" (the so-called "gamers" are people, who actively play video games and see them as a hobby) [1]. The authors claim that video games along with modern technologies brought more changes into everyday life, people's behavior and attitudes than anything else that happened throughout the last decades. Although this conclusion seems slightly exaggerated, we can agree that a significant part of the population of many countries grew with some sort of video games experience. In both USA and Russia, an average "gamer" is in in his/her mid-thirties and belongs to an economically and socially active part of the society [1–3]. Thus, psychological specifics of active video game players become an important research area, but also a biased and controversial one. While the main positive and negative consequences

D. A. Alexandrov et al. (Eds.): DTGS 2018, CCIS 859, pp. 250–263, 2018.
https://doi.org/10.1007/978-3-030-02846-6_20

of video gaming, such as addiction, attention, spatial and logical thinking specifics, emotional changes were pointed out as early as in the mid-80 s (e.g. see [4, 5]), many questions are still unanswered. For example, most contemporary researchers agree that video games can enhance visual-spatial skills in children, adults and elders [6, 7] or that video game addiction, or problematic behavior actually exists, at least in some way, although its criteria are not fully developed yet [8, 9], yet the linkage between violent video games and aggressive behavior still causes a lot of arguments [10–13]. Some commonly shared theories pretend to explain what we actually learn (or unlearn) by playing video games, including problem-solving, critical thinking, imagination [5, 14]. Many authors suggest that video games promote impulsive and risky behavior, especially among youth, that can lead to real life dangers or psychological disorders, such as attention deficit and hyperactivity disorder (ADHD) [10]. In this article, we aim to review such theories and corresponding research in this field as well as to introduce our own empirical study.

1.1   Are Gamers Risky, Impulsive or Both, and Why Should They Be?

The suggestion that video games can promote impulsive behavior is based on several general assumptions. First, to succeed in many video games one needs to be extremely quick. The player does not often have enough time to analyze the situation and while playing it is often more beneficial to be active than to hesitate. Even if the performed action turns wrong most of the mistakes in a video game could be replayed easily, thus the so-called "trial-and-error" behavior is promoted. While the "trial-and-error" strategy is sometimes a good way to learn new things, it can be ineffective or even dangerous in many real-life situations that require critical thinking and reflection [1, 14]. The second assumption, which also underlies the discussion on video game violence states that video gamers transfer their in-game learned behavior into real life.

The resolution of the American Psychological Association (APA), published in 2005 and republished in 2015 despite many critical views, supports this notion by claiming that violent video game exposure can increase "aggressive behavior, aggressive affect, aggressive cognitions" and decrease "prosocial behavior, empathy, and moral engagement" [15]. This statement is based on the works of C. Anderson and his colleagues, whose general aggression model (GAM) framework is based on A. Bandura's social learning theory, L.R. Huesmann's script theory and several other psychological theories of learning, based on the stimuli-reaction paradigm which leaves little freedom to human conscience and will. Briefly, Anderson and his group promote the idea that children, teenagers, and young adults learn behavioral patterns from the mass media they observe and violent video games they play and copy them in their real-life behavior [16]. While GAM predominantly describes aggressive behavior, the same argumentation is often used to explain video gamers' predicted impulsivity.

Anderson's theory strongly implies that video gamers either knowingly or unconsciously fail to distinguish fake reality of video games from the real world. While this notion is often used in mass media to explain mass shootings, suicides, crimes and other unwanted or dangerous behaviors, often blamed to be connected with violent video games, it is also flawed. A thorough research, performed by Sh. Olson and L. Kutner, found that even 10–11 years old teenagers are well aware of the differences between video games and real life. The teenaged boys also pointed out that in a video game they often do things that they could not or would not do in real life because of obvious danger or moral unacceptability [13]. This could mean that the transfer of the in-game behavioral patterns into the real life is actually limited, especially if seen as ineffective or dangerous. On the other hand, discussing cognitive development through video games, for example, when looking for a way to prevent cognitive decline in elders [e.g. 17], the researchers relate to repeated training and subconscious transfer, and not to the deliberate choice of new strategies. Therefore, the question is, whether impulsivity is chosen or unwillingly developed. To answer this question we need to define impulsivity as well as some other related terms.

In psychology, impulsivity (or impulsiveness) is seen as a tendency to act on a whim, with no reflection or consequences consideration, under the spur of the moment, to make random decisions with little or no forethought. This means that impulsive actions are often risky or inappropriate and can lead to unexpected or undesired consequences [18]. In clinical psychology and psychiatry, impulsivity is seen as a symptom of several clinical conditions, including ADHD, kleptomania, gambling, etc. and is often the prime target of therapy [19]. In this meaning, impulsivity is the lack of some cognitive control functions, such as action inhibition, and is not a personality trait and is caused by different reasons [18, 20].

Outside the clinic, impulsivity can be referred to either in cognitive or in personality dimension. For J. Kagan impulsivity is a part of an impulsive-reflective cognitive style. Cognitive styles are stable individual ways to perceive, process and store information [22]. J. Kagan's impulsive-reflective style describes people with different cognitive tempo and accuracy, where impulsivity means quick and inaccurate answers in the situation of uncertainty. Impulsiveness is opposed to reflexivity, which is characterized by slow and accurate answers [21]. Although cognitive styles are stable and highly automatic ways to process information, they can be altered to a certain degree with the use of higher cognitive processes, if proven useless in a current situation.

Impulsivity as a personal trait is rarely connected with its cognitive dimension. In the works of H. Eysenck and S. Eysenck, impulsivity was introduced as a component of extraversion and then turned into a separate measure. Currently, impulsivity scale includes basic impulsivity as the lack of momentary impulse control, venturesome-ness as the willingness to take risks and try out new exciting things and empathy as the readiness to share other people's feelings [23]. S. Dickman pointed that impulsivity can be both "good" and "bad". He describes functional impulsivity as the tendency to act with relatively little forethought for optimal results and the dysfunctional impulsivity that leads to more challenges and difficulties instead. Those tendencies, in fact, indicate separate factors intercorrelated rather moderately and variating in relations to other personality traits, such as venturesomeness or orderliness [24].

While risky behavior is often mentioned in the connection with the impulsivity trait, it is important to notice that they are not synonyms. We have already mentioned venturesomeness as an aspect of Eysencks' impulsiveness scale. However, it is obvious that skydiving, mountain climbing and other activities preferred by venturesome people require something more than the lack of impulse control. While in everyday life, we are more likely to describe risk in terms of loss or harm, it also refers to any situation when the outcomes are uncertain. In this case, a personal ability to accept risks or willingness to averse them is seen as a separate dimension, not necessarily related to the impulsivity. In many cases, risky behavior is the result of a deliberate personal choice, not thoughtlessness [25].

Since in psychological research of video games and gamers, both empirical and theoretical, different terms and definitions are often used to discuss risky or impulsive behavior, we tried to gather the most common definitions and terms, related to this problem, to make the further discussion more accurate and concrete.

1.2 Empirical Studies Analysis

While most theories agree that video games can make people impulsive, the empirical research results are rather diverse.

D. Gentile, a member of C. Anderson's group, presented a 3-year longitude study, in which he not only linked adolescents' violent video game playing with impulsivity and attention deficit but also claimed to show bi-direct causalities [10]. According to this study, violent video games are harmful as they prevent normal cognitive control function in children, which among all can lead to ADHD. Another research supports the linkage between video game use and ADHD, as well as other psychiatric disorders, but only in the context of problematic or addictive gaming behavior. Furthermore, the disorders are viewed as risk factors for gaming addiction development, not vice versa [26]. Finally, there are special video games developed to treat children with diagnosed ADHD [27]. While such products require more controlled trials and independent research, the general idea to influence cognitive control (and cognitive impulsivity) through playing video games seems to be proven worthy [17, 28]. To summarize, cognitive aspects of impulsivity, including ADHD symptoms, show some linkage with regular video game playing, especially in younger gamers. However, it is important to notice that the linkage is mediated by a video game type and by the presence of video game addiction. Controlled use of specially designed video games can be beneficial for cognitive control training and prevention of impulsivity.

There is other evidences that pathological gaming and certain video game genres are related towards higher personality impulsivity and cognitive control deficit. People who play first-person shooter video games (a genre that implies running through an artificial environment and shooting computer or player operated enemies) are more likely to be impulsive and to make risky decisions in simulated gambling situation [29, 30]. The video game addiction also contributes to high personal impulsivity, unlike a different game genre such as strategy games, which demand a slower pace and more thoughtful gaming experience. It is also worth noticing that both shooters and strategies can be designed with more or less graphical violence involved, though this aspect was not a variable. Another popular game type, such as massive multiplayer online role-playing games (MMORPG), known to be an addictive video game genre [31], is on the opposite related to low impulsivity [8]. While non-addicted gamers showed no significant personality specifics, addicted gamers were found less impulsive, both in functional and dysfunctional impulsivity scales. They also received lower scores in self-regulation, which means that both high and low impulsivity can be the result of regulation problems. The study supports the idea that video games of different genres either require different personality characteristics to be played successfully or develop different psychological specifics in their loyal fans, or both.

Numerous studies suggest that video games of certain genres promote risky behavior and even glorify it. A recent longitudinal study claims that teens who excessively play mature-rated video games are more likely, than other teens, to get involved in all types of risky behaviors including smoking, alcohol drinking, fighting or having unprotected sex. The gamers who play mature video games "sometimes" or "once in a while" do not show any difference from those who do not play video games at all [32]. While the strong linkage was found, it is important to notice that mature video games are marked as suitable for ages 17 and up. Playing such games at the age of 14 is rebellious from the beginning, which probably indicates a predisposition for risky behavior or venturesomeness. Nevertheless, this research supports the necessity to follow recommended age restrictions when introducing children to video games and other media. Research also shows that racing simulators can lead to risky driving, but only if such behavior is rewarded throughout the game [33, 34]. As for positive aspects of risk-taking, they are rarely discussed in cyberpsychology. However, J. Beck and M. Wade [1] argue that risk-readiness and risk-taking are the key features of adult gamers, which prove them successful in business. According to them, gamers are not only risky – they also are ready to take responsibilities for those risks. In Russian research of adult online gamers, conducted by A. Avetisova [35], gamers scored higher in both rationality (searching for more information before making a decision) and risk-readiness, which supports this notion. Cognitive studies done by D. Bavelier's research group [36] show that in visual demanding tasks video gamers perform with an increased speed. The gamers do not make more mistakes in stimuli recognition than the control group, which means that the increased performance is risk-taking rather than simply impulsive.

To summarize, there are strong but somewhat controversial empirical evidence that video game experience is linked to (possibly – can alter) impulsive and risky tendencies in children and adults. Negative outcomes are perceived for people with preexisting psychological disorders, including issues with control and self-regulation. People with video game addiction are also in the risk group, and the fans of violent video games, especially shooters require the most attention. On the other hand, participating in other types of video games and for a rational amount of time can be beneficial, as faster decision-making and enhanced readiness for risky decisions are useful for the everyday life.

Research Questions. Most of the video game studies target children, adolescents, or young adults (college students), but only rarely consider older people. Thus, the mean age of a video gamer in Russia (as well as in many other countries) exceeds 30 years old. In our study, we aimed to fill this age gap and to investigate both cognitive and personal impulsivity, as well as different aspects of risk-taking attitudes including venturesomeness and risk-readiness as the ability to make decisions and act in ambiguous or uncertain situations. The main research question was whether the adult gamers in Russia share the specific patterns of impulsivity, which were shown in previous studies? The second question was whether the decision-making specifics of video gamers refer to cognitive or personal impulsivity or risk-readiness, or all of them, as the same behavior can actually root from different personal and cognitive specifics.

2 Empirical Study of Adult Video Gamers' Impulsivity and Risk

2.1 Methods

Participants. Totally 223 participants, 91 males, and 132 females, aged from 18 to 35 years old from Moscow, Russia were recruited via video games forums, social networks, advertising among students and through snowball sampling method. The participation was voluntary and was not rewarded in any material way, although the participants were able to ask any questions about current research or video games studies in psychology in general after they completed all the tasks and thereby received brief lectures in this field if they wanted to.

All the participants were interviewed with demographics (age, educational level, current occupation, marital status) and video game related questions, such as "Do you play video games?", "How much time a week do you spend playing video games?", "What video games genres do you prefer and why?", "Why do you think you keep playing video games?", etc.

All the participants took the same test battery despite their answers, but later they were divided into groups and subgroups accordingly

The control group ("The Non-Video Game Players", the nVGP group) consisted of the participants who self-reported to have little to no video games experience in general, and not interested in playing video games. Those of them who used to play video games earlier in their lives did not play for at least 3 years or longer hitherto.

The comparison group ("The Video Game Players", the VGP group) consisted of those who reported to have significant video game experience (several years of more or less active gaming hitherto) and to be interested in video games playing. They were also playing video games regularly for at least 1 h a week at the time the research was conducted (and most of them played for 3–5 h a week or more).

The whole sample characteristics are presented in Table 1.

Table 1. Sample groups and subgroups characteristics (Study 1)

VGP Group 133 participants: 63 men, 70 women Age: M = 23.6, SD = 4.5		nVGP Group 90 participants: 28 men, 62 women Age: M = 23.7, SD = 4.9
High-VGP subgroup 78 participants: 40 men, 38 women Age: M = 23.6, SD = 4.4	**Low-VGP subgroup** 55 participants: 23 men, 32 women Age: M = 23.7, SD = 4.6	

Table 1 shows that the participants in both groups were approximately of the same age. As for the sex ratio in both groups, the differences were obvious, but justified from the general population perspective. According to the official statistics [2], 52% of modern Russian gamers are males while 48% are females. However, females prevail among those who have no or very little video game experience while the most males show at least some interest in this field. Female gamers all over the world are also slightly older than male gamers and thus could have fallen out our sample [2, 3].

Additionally, the participants in the VGP group were subdivided into subgroups with a different regular intensity of video game playing. Gamers who regularly played for more than 12 h per week formed the high-VGP sub-group, while those who played less than 12 h per week formed the low-VGP sub-group.

All the participants took part in the Study 1 (personality questionnaires). From the initial sample, only 150 participants volunteered to take part in the Study 2 (cognitive style task). The characteristics of that second sample are presented in the Table 2.

Table 2. Sample groups and subgroups characteristics (Study 2)

VGP Group 90 participants: 45 men, 45 women Age: M = 24.0, SD = 4.6		nVGP Group 60 participants: 20 men, 40 women Age: M = 24.5, SD = 4.6
High-VGP subgroup 60 participants: 33 men, 27 women Age: M = 23.9, SD = 4.6	**Low-VGP subgroup** 30 participants: 12 men, 18 women Age: M = 24.3, SD = 4.7	

All the participants completed all the tests tasks separately, in classical paper-and-pencil variants.

Procedure

Study 1. In this study, the participants filled two personality questionnaires, aimed to measure personal impulsivity and attitudes towards risk in the following order:

1. *I-7 Impulsiveness and Venturesomeness Questionnaire.* The original questionnaire was developed by Hans and Sibylla Eysenck and had a total of 54 items, subdivided into three subscales: *Impulsiveness*, *Venturesomeness*, and *Empathy* [23]. The scales were discussed earlier in this paper, so we won't repeat their definition. T. Kornilova and A. Dolnykova [37] introduced the Russian shortened adaptation of the I-7 scale, consisting of 28 questions but following the same structure, and including the same scales as the original.

2. *Personal Risk Factors Questionnaire-21 (PRF-21).* The original scale was developed by T.V. Kornilova, loosely based on the EQS Questionnaire by H. Wolfram (in German) [25]. The PRF-21 questionnaire includes 21 item, subdivided into two scales: *Risk-readiness* and *Rationality*. Risk-readiness describes personal ability to take risks when deciding how to proceed in situations with uncertain outcomes. Rationality indicates personal preferences to look for more information to decrease uncertainty before making the decision. The subscales correlate negatively but do not directly oppose each other, as one can prefer informational seeking, but be ready to take risks after all.

Study 2. The participants (see Table 2) completed **Kagan's Matching Familiar Figures Test (MFFT),** measuring cognitive style "impulsivity – reflexivity". In this test, the participant is shown a picture of a familiar object ("a standard") and eight similar variants with one and only one of them being the same as the standard. The participant is required to find the identical to the standard variant as fast and accurate as possible [21]. The number of mistakes and the cognitive tempo (time needed to give the first answer) were used to determine *impulsive* (many mistakes, fast answers) or *reflexive* (few mistakes, slow answers) cognitive style.

2.2 Results

Note: The results of **I-7 Questionnaire** and **MFFT** showed abnormal distribution, so methods of non-parametric statistics were used. At the same time, we used parametric statistics for **PRF-21** questionnaire results, as the scores were distributed normally.

VGP and nVGP Comparison

I-7 Questionnaire. Impulsivity, venturesomeness and empathy as personal traits are known to have strong and consistent sex-based specifics. To compensate our sample's inequality, we compared the results with the results in the whole sample and found that females in general scored higher in impulsivity (Mann-Whitney U = 2554.0; p = 0.000; M = 3.1, SD = 4.8 (women); M = −2.1, SD = 4.4 (men)) and empathy (U = 4057.5; p = 0.000; M = 5.2, SD = 3.4 (women); M = 3.4, SD = 3.5 (men)), but lower in venturesomeness (U = 4248.0; p = 0.000; M = 0.5, SD = 4.3 (women); M = 2.6, SD = 3.5 (men)). Thus, separate comparisons were made for female and male gamers as well. The results of the comparison for all the scales are shown in Tables 3, 4 and 5.

Table 3. I-7. Impulsivity

	Mann-Whitney U	p	Mean score	SD
VGP Group (all)	5938.5	0.921	1.0	5.3
nVGP Group (all)			1.0	5.4
Female VGP	1872.0	0.170	3.7	4.3
Female nVGP			2.4	5.3
Male VGP	867.0	0.896	−2.0	4.5
Male nVGP			−2.2	4.2

Table 4. I-7. Venturesomeness

	Mann-Whitney U	p	Mean score	SD
VGP Group (all)	5108.5	0.061	1.8	4.2
nVGP Group (all)			0.7	4.1
Female VGP	2120.5	0.820	0.5	4.4
Female nVGP			0.4	4.3
Male VGP	647.0	0.04	3.1	3.5
Male nVGP			1.5	3.5

Table 5. I-7. Empathy

	Mann-Whitney U	p	Mean score	SD
VGP Group (all)	4242.0	0.000	3.8	3.6
nVGP Group (all)			5.5	3.2
Female VGP	1498.0	0.002	4.4	3.6
Female nVGP			6.2	2.7
Male VGP	747.5	0.239	3.1	3.5
Male nVGP			3.9	3.5

The results showed that there are no differences in impulsivity in both VGP and nVGP groups. While men and women have different levels of personal impulsivity, there are also differences between female VGPs and female nVGPs, as well as between their male counterparts. As for venturesomeness, we did not receive differences between VGP and nVGP groups in general, but further comparison suggests that there are actually significant differences between male VGPs and male nVGPs, and male gamers were found more venturesome then male non-gamers. The empathy scale, on the contrary, showed significant differences in the female subgroup only, with female VGPs less empathetic than female nVGPs are.

PRF-21 Questionnaire. We did not find any sex-related specifics for risk-readiness (Student's t = t = −1.184; p = 0.238), while female participants did show significantly lower results in the rationality subscale (t = −3.846; p = 0.000; M = 2.1, SD = 4.0 (women); M = 4.1, SD = 3.6 (men)). There were also no significant differences between all the groups: VGPs were not different from nVGPs (t = 0.284; p = 0.777 for risk-readiness; t = −0.312; p = 0.755 for rationality); female VGPs were not different from female nVGPs (t = 0.627; p = 0.532 for risk-readiness; t = −0.678; p = 0.499 for rationality) and finally, male VGPs were not different from male nVGPs (t = −0.681; p = 0.498 for risk-readiness; t = −0.696; p = 0.488 for rationality). This means that we failed to show any differences between adult video game players and non-video game players at all.

Kagan's MFFT. No significant differences between female and male participants were found in both cognitive tempo (U = 2655.5; p = 0.685) and amount of the mistakes (U = 2263.5; p = 0.057), so we did not compare females and males subgroups further. The whole VGP groups performed the test slightly slower than the nVGP group (M = 60.9 s, SD = 28.3 s versus M = 54.6 s, SD = 27.9 s respectively), but the Mann–Whitney U test showed that the difference is insignificant (U = 2332.5, p = 0.159). Although, the VGP group made significantly fewer mistakes than the nVGP group (U = 2155.0, p = 0.036; M = 5.3, SD = 4.7 for the VGPs; M = 7.5, SD = 6.0 for the nVGPs).

According to these results, we cannot conclude that any of the groups score higher in cognitive impulsivity, since impulsivity as a cognitive style is described by both quick answers and many mistakes. But the results we got mean that VGPs are more accurate than nVGPs and thus cannot be described as "more impulsive" in J. Kagan's terms.

High-VGPs, Low-VGPs and Non-VGPs Comparison

I-7 Questionnaire. The Kruskal-Wallis one-way analysis of variance showed no significant differences between the groups in *impulsivity* (Chi-square = 0.236; p = 0.889), but did found differences in both *venturesomeness* (Chi-square = 7.485; p = 0.024) and *empathy* (Chi-square = 20.029; p = 0.000). Table 6 include mean scores and standard deviations for all the groups.

Table 6. High, low and nonVGPs mean scores in I-7 questionnaire

	Impulsivity	Venturesomeness	Empathy
High-VGP group	M = 1.2, SD = 5.2	M = 2.5, SD = 4.0	M = 3.2, SD = 3.7
Low-VGP group	M = 0.8, SD = 5.4	M = 0.7, SD = 4.2	M = 4.6, SD = 3.4
NVGP group	M = 1.0, SD = 5.4	M = 0.7, SD = 4.1	M = 5.5, SD = 3.2

Pairwise comparison of the subgroups showed the following results: high-VGPs scored significantly higher in venturesomeness then both low-VGPs (U = 1705.0; p = 0.042) and nVGPs (U = 2714.5; p = 0.011), while the latter showed almost equal scores (U = 2394.0; p = 0.738). High-VGPs showed the lowest empathy scores, which differed significantly from both low-VGP (U = 1638.0; p = 0.019) and nVGP groups (U = 2117.5; p = 0.000). Those groups also had no significant differences from each other (U = 2124.5; p = 0.144). Thus, according to I-7 questionnaire, high-VGPs showed the most distinct personality specifics, namely – higher venturesomeness and lower empathy, while low-VGPs did not differ from the nVGPs.

PRF-21 Questionnaire. One-way ANOVA showed no significant differences between the groups in both risk-readiness (F = 0.196; p = 0.822) and rationality scale (F = 2.806; p = 0.063). Pairwise post hoc comparison with the Bonferroni method though showed that low-VGPs were slightly more rational then high-VGPs (p = 0.05; M = 3.8, SD = 3.2 (low-VGPs) and M = 2.2, SD = 4.1 (high-VGPs)). NVGPs scored somewhere in between those two groups (M = 3.1, SD = 4.2) and showed no significant differences with both of the VGPs groups.

Kagan's MFFT. The Kruskal-Wallis one-way analysis of variance showed significant differences between the groups in cognitive tempo (Chi-square = 7.060; p = 0.029), but not in the accuracy (Chi-square = 3.184; p = 0.204). Low-VGPs were the least impulsive among all the groups. They have the longest first response mean time (thus, the difference is statistically insignificant with both the high-VGP group (U = 785.5; p = 0.327) and the nVGP group (U = 679.5; p = 0.059) and the least amount of mistakes (significantly less than in the nVGP group (U = 618.0; p = 0.015), but insignificant in compare with the high-VGP group (U = 1537.0; p = 0.166). See Table 7 for the mean scores.

Table 7. High, low and nVGPs mean scores in MFFT

	Mean time of the first response (cognitive tempo), sec	Mistakes
High-VGP group	M = 59.4, SD = 30.5	M = 5.7, SD = 4.3
Low-VGP group	M = 63.9, SD = 23.5	M = 4.5, SD = 5.4
NVGP group	M = 54.6, SD = 27.9	M = 7.5, SD = 6.0

2.3   Discussion

While not many differences were found in the current research, the zero-results are meaningful as well: they show that despite our hypothesis, adult video gamers are not very different in their personal trait, related to impulsivity and risk, from those who have no video game experience.

While we failed to establish a connection of video game experience with personal impulsivity, we found evidence that video gamers are more accurate and thus less cognitively impulsive compared to non-video gamers, though this difference weakens in the subgroup of high VGPs. We suggest, that in general video game playing experience can raise accuracy in certain visual and visual-spatial tasks (like those shown in the D. Bavelier's research [36]) but with higher involvement in video games the risk of video game addiction rises as well. So, while we did not measure video game addiction in our sample (mostly because there are no reliable video game addiction inventory in Russia, and Internet addiction inventories basically do not distinguish gaming from other types of online activities), we can assume that some of our high-VGP participants were video game addicts and their results altered the general performance of this subgroup. This assumption can also be used to explain the greater personal specifics shown by the high VGPs. On the other hand, lower empathy and higher venturesomeness can also be the reason for extensive video game playing, as video games represent a rather cheap and safe way to get some fresh experience, and in the simplified virtual world with a brief and slang-filled communication, the lack of empathy is less important than in face-to-face communication.

While video games are known to reduce gender differences in spatial performance [38], they do not work the same for the personality traits and attitudes. Most inter-sex differences in our research reproduce those from other studies with no video gamers involved. Thus, lower empathy in female gamers is interesting, as it might indicate which women prefer to play computer games. Or else it might show that video games, after all, alter some personality traits, and probably in a negative way as higher empathy is usually seen as an important part of communication.

2.4   Limitations

One of the main problems of any non-experimental study is the problem of causality. While we do not think it is possible to receive significant changes in personality of adults in an experimental setting (such changes probably require a lot of gaming experience and develop for a long time, if develop at all), the future research might involve different age groups as well as some type of a longitudinal study. We also think that though the mix-gender sample of participants in our study is an advantage (as many video game studies involve single-sex samples), stable gender differences were difficult to control, because we could not recruit enough non-gaming males. Different video game genres references can also alter the results, though most of our participants were unable to choose only one preferable genre.

3 Conclusion

Despite the so-called "common knowledge" marking all gamers to be risky and impulsive, as well as several psychological theories based on different learning models, adult video gamers in Russia, in general, show no significant personal specifics in impulsivity or risk-taking. In the current study the most active (i.e., those who play over 12 h per week) video gamers, especially males, express high interest in new experience and sensation seeking, while active female gamers show significantly low capability of understanding other people's motions; these two findings can be either a predictor or the result of excessive gaming experience.

Acknowledgment. The study was supported by the Russian Science Foundation, project #18-18-00365.

References

1. Beck, J.C., Wade, M.: The Kids are Alright: How the Gamer Generation is Changing the Workplace. Harvard Business Review Press, Harvard (2006)
2. Mail.Ru Group. Profil Rossijskogo Gejmera – 2015. [Russian Gamer Profile – 2015]. https://gamestats.mail.ru/article/profil_rossijskogo_geymera. Last Accessed 30 Jan 2018. (in Russian)
3. ESA. Essential Facts about the Computer and Video Game Industry (2017). http://www.theesa.com/wp-content/uploads/2017/09/EF2017_Design_FinalDigital.pdf. Last Accessed 30 Jan 2018
4. Turkle, S.: The Second Self: Computers and the Human Spirit. Simon & Schuster, New York (1985)
5. Tikhomirov, O.K., Lysenko, E.E. Psihologija Komp'juternoj Igry [Psychology of Computer Games] In Novye Metody I Sredstva Obuchenija [New Methods and Means of Teaching], vol. 1, pp. 30–66. Znanie, Moscow (1988)
6. Green, S., Bavelier, D.: Action-video-game experience alters the spatial resolution of vision. Psychol. Sci. **18**(1), 88–94 (2007). https://doi.org/10.1111/j.1467-9280.2007.01853.x
7. Subrahmanyam, K., Greenfield, P.M.: Effect of video game practice on spatial skills in girls and boys. J. Appl. Dev. Psychol. **15**, 13–32 (1994). https://doi.org/10.1016/0193-3973(94)90004-3
8. Collins, E., Freeman, J., Chamarro-Premuzic, T.: Personality traits associated with problematic and non-problematic massively multiplayer online role playing game use. Pers. Individ. Differ. **52**, 133–138 (2012). https://doi.org/10.1016/j.paid.2011.09.015
9. Griffiths, M.: Online computer gaming: advice for parents and teachers. Educ. Health **27**(1), 3–6 (2009)

10. Gentile, D.A., Swing, E.L., Choon, G.L., Khoo, A.: Video game playing, attention problems, and impulsiveness: evidence of bidirectional causality. Psychol. Popul. Media Cult. **1**, 62–70 (2012). https://doi.org/10.1037/a0026969

11. Ferguson, C.J., Garza, A., Jerabeck, J., Ramos, R., Galindo, M.: Not worth the fuss after all? cross-sectional and prospective data on violent video game influences on aggression, visuospatial cognition and mathematics ability in a sample of youth. J. Youth Adolesc. **42** (1), 109–122 (2013). https://doi.org/10.1007/s10964-012-9803-6

12. Anderson, C.A., Ihori, N., Bushman, B.J., Rothstein, H.R., Shibuya, A., Swing, E.L., Sakamoto, A., Saleem, M.: Violent video game effects on aggression, empathy, and prosocial behavior in Eastern and Western Countries: a meta-analytic review. Psychol. Bull. **136**(2), 151–173 (2010). https://doi.org/10.1037/a0018251

13. Olson, C., Kutner, L. Viewpoints and flashpoints in the study of video game violence and aggression. psychology. J. High. Sch. Econ. **12**(1), 13–28 (2015)

14. Greenfield, P.M.: Technology and informal education: what is taught. What Is Learned. Sci. **323**, 69–71 (2009). https://doi.org/10.1126/science.1167190

15. American Psychological Association. Resolution on Violent Video Games (2015). http://www.apa.org/about/policy/violent-video-games.aspx. Last Accessed 30 Jan 2018

16. Anderson, C.A., Bushman, B.J.: Human aggression. Annu. Rev. Psychol. **53**, 27–51 (2002). https://doi.org/10.1146/annurev.psych.53.100901.135231

17. Van Muijden, J., Band, G.P.H., Hommel, B.: Online games training aging brains: limited transfer of cognitive control functions. Front. Hum. Neurosci. **6**, 221 (2012). http://www.frontiersin.org/Human_Neuroscience/10.3389/fnhum.2012.00221/full. Last Accessed 30 Jan 2018. https://doi.org/10.3389/fnhum.2012.00221

18. Evenden, J.: Varieties of impulsivity. Psychopharmacol. (Berl) **146**(4), 348–361 (1999)

19. Webster, C.D., Jackson, M.A. (eds.): Impulsivity: Theory, Assessment, and Treatment. The Guilford Press, New York (1997)

20. Arce, E., Santisteban, C.: Impulsivity: a review. Psicothema **18**(2), 213–220 (2006)

21. Kagan, J.: Reflection-impulsivity the generality and dynamics of conceptual tempo. J. Abnorm. Psychol. **71**, 17–24 (1966)

22. Ausburn, L., Ausburn, F.: Cognitive styles: some information and implications for instructional design. ECTJ **26**(4), 337–354 (1978)

23. Eysenck, S.B.G., Eysenck, H.J.: Impulsiveness and venturesomeness: their position in a dimensional system of personality description. Psychol. Rep. **43**(3), 1247–1255 (1978). https://doi.org/10.2466/pr0.1978.43.3f.1247

24. Dickman, S.J.: Functional and dysfunctional impulsivity: personality and cognitive correlates. J. Pers. Soc. Psychol. **58**(1), 95–102 (1990)

25. Kornilova, T.V.: Psihologija Riska I Prinjatija Reshenij [Psychology of Risk and Decision Making]. Aspekt Press, Moscow (2003). (in Russian)

26. Andreassen, C.S., et al.: The relationship between addictive use of social media and video games and symptoms of psychiatric disorders: a large-scale cross-sectional study. Psychol. Addict. Behav. **30**(2), 252–262 (2016). https://doi.org/10.1037/adb0000160

27. Akili Interactive Labs. Akili Achieves Primary Efficacy Endpoint in Pediatric ADHD Pivotal Trial. https://www.businesswire.com/news/home/20171203005099/en/Akili-Achieves-Primary-Efficacy-Endpoint-Pediatric-ADHD. Last Accessed 30 Jan 2018

28. Anguera, J.A., et al.: Video game training enhances cognitive control in older adults. Nature **501**, 97–101 (2013). https://doi.org/10.1038/nature12486

29. Metcalf, O., Pammer, K.: Impulsivity and related neuropsychological features in regular and addictive first person shooter gaming. Cyberpsychol. Behav. Soc. Netw. **17**(3), 147–152 (2014). https://doi.org/10.1089/cyber.2013.0024

30. Bailey, K., West, R., Kuffel, J.: What would my avatar do? Gaming, pathology, and risky decision making. Front. Psychol. **4**, 609 (2013). http://www.frontiersin.org/Journal/10.3389/fpsyg.2013.00609/full. Last Accessed 30 Jan 2018. https://doi.org/10.3389/fpsyg.2013.00609

31. Sellers, M.: Designing the experience of interactive play. In: Vorderer, P., Bryant, J. (eds.) Playing Video Games. Motives, Responses and Consequences, pp. 9–22. Lawrence Erbaum, Mahwah, New Jersey (2006)

32. Hull, J.G., Brunelle, T.J., Prescott, A.T., Sargent, J.D.: A longitudinal study of risk-glorifying video games and behavioral deviance. J. Pers. Soc. Psychol. **107**(2), 300–325 (2014). https://doi.org/10.1037/a0036058

33. Beullens, K., Roe, K., Van der Bulck, J.: Video games and adolescents' intentions to take risks in traffic. J. Adolesc. Health **43**(1), 87–90 (2008). https://doi.org/10.1016/j.jadohealth.2007.12.002

34. Deng, M., Chan, A.H.S., Wu, F., Wang, J.: Effects of racing games on risky driving behaviour, and the significance of personality and physiological data. Inj. Prev. **21**(4), 238–244 (2015). https://doi.org/10.1136/injuryprev-2014-041328

35. Avetisova, A.A. Psihologicheskie Osobennosti Igrokov v Komp'juternye Igry [Psychological Specifics of Video Gamers]. Psychol. J. High. Sch. Econ. **8**(4), 35–58 (2011). (in Russian)

36. Dye, M.W.G., Green, C., Bavelier, D.: Increasing speed of processing with action video games. Curr. Dir. Psychol. Sci. **18**, 321–326 (2009). https://doi.org/10.1111/j.1467-8721.2009.01660.x

37. Kornilova, T.V., Dolnykova, A.A.: Diagnostika Impul'sivnosti i Sklonnosti k Risku [The Diagnostics of Impulsivity and Venturesomeness]. Vestnik Moskovskogo universiteta. Serija 14. Psihologija, vol. 3, pp. 46–56 (1995). (In Russian)

38. Feng, J., Spence, I., Pratt, J.: Playing an action video game reduces gender difference in spatial cognition. Psychol. Sci. **18**(10), 850–855 (2007). https://doi.org/10.1111/j.1467-9280.2007.01990.x

The Impact of Smartphone Use on the Psychosocial Wellness of College Students

Anthony Faiola[1(✉)], Haleh Vatani[1], and Preethi Srinivas[2]

[1] University of Illinois at Chicago, Chicago, IL 60612, USA
faiola@iupui.edu, hvatan2@uic.edu
[2] Regenstrief Institute, Indianapolis, IN 46202, USA
presrini@umail.iu.edu

Abstract. Researchers suggest that excessive smartphone use is correlated with negative psychosocial effects, particularly among younger adults—causing feelings of isolation, depression/anxiety, and restlessness. This pilot study on psychosocial wellness, of 22 college students—measured the impact of smartphone use on emotion/mood, dependency, addiction, purpose of life, social communications, and self-consciousness. For our data analysis, we measured frequency with conversion percentages (of 35 questions) using a seven-point Likert-scale of strongly disagree-to-strongly agree, while averaging the scores of each question group pertaining to each hypothesis. While only 22% agreed they were addicted to smartphone use, 68% reported constantly checking their smartphone, with 57% agreeing that they were smartphone dependent. The majority agreed that smartphone use increased anxiety, stress, and feelings of impatience, if their phone was not with them. While the majority agreed that the smartphone is their primary means of communication, 90% agreed that nothing is more fun than using their smartphone.

Keywords: Psychosocial · Self-consciousness · Mood · Emotion
Smartphones

1 Introduction

For centuries, human artifacts of cultural mediation evolved from clubs to hammers to machines of the industrial age [1] During the past few decades, we observed the emergence in development of information technology and electronic tools. However, technology has never become so indistinguishably interwoven within the development of human consciousness until recently. In the last two decades the use of information technology—particularly, the use of smartphones in the recent years has been observed to alter a range of psychosocial conditions. New smartphone applications (apps) continue to arrive on the market, offering a variety of useful tools for staying connected, exploring social media, doing business, playing games, and listening or watching audio/videos. As such, the intertwining nature of technology is having a profound effect on our interpersonal relationships and activities in the world—particularly among the college-age population [2].

Along with the television penetration rate (98%), from 1950 to 2000, by 2000 there was an 80% penetration of computers and cell phones in US—and by 2003, computers

© Springer Nature Switzerland AG 2018
D. A. Alexandrov et al. (Eds.): DTGS 2018, CCIS 859, pp. 264–276, 2018.
https://doi.org/10.1007/978-3-030-02846-6_21

had penetrated 75% of US homes with children, with 63% having access to the Internet [3]. By 2010, Americans spent 1.3 trillion hours on seeking for information, an average of over 12 h per day. [4] This includes the consumption of over 10,845 trillion words translating to 100,500 words per person daily, with the majority of the time being spent seeking information on the Internet [5]. Consequently, the statistics illustrate an extreme change in the way humans interact with and process information.

The exponential use of smartphones in 2007, showed the next stage in excessive and dependent use of technology. Currently, 90% of North Americans possess mobile phones, where about 70% of those are smartphones [6]. As part of everyday life, smartphones provide direct access to people through voice calls, text messages, and support social interaction with a range of networks, such as Facebook and Twitter. As a psychological tool extending the way we distribute information and reinforce personal relationships [7], researchers have defined the smartphone as a type of *"affective technology,"* [8, 9]—linking it to the "emotional" human condition [10, 11] and the means to connect to those around us.

Currently, 98% of college students have a mobile phone, and evidence suggest that daily usage exceeds four hours per day—occurring in a diversity of settings [12, 13]. For example, researchers have demonstrated that texting among college students is so frequent (due to its convenience, speed, and facelessness), that participants exhibited a significant degree of disinhibition and inattention to their immediate social environment or interpersonal relationship [14]. Consequently, considering the diverse use and impact that smartphones have on contemporary human life, particularly young adults, it is expected that an array of behaviors might be observed among this cohort.

In this paper, we introduce our findings of a pilot study on the impact of smartphones on the psychosocial wellness of college students. As such, we measured the impact of smartphone use on emotion and mood, dependency, addiction, purpose of life, social communications, and self-consciousness of college students.

2 Excessive Use and Psychosocial Development

With the recent increase in smartphone adoption [15] and dependence [16] studies continue to identify related neuro-psychological and neuro-social effects of using smartphones. A range of psychiatric disorders leading to abnormal or anti-socio- psychology including: Attention Deficit Hyperactivity Disorder (ADHD), weakening of cognitive focus and shallower thinking skills, reduction of creativity and problem solving skills, a lowered ability to filter out extraneous information, adverse effects upon psychosocial development, hyperactivity and behavioral problems, feelings of isolation, depression, anxiety and restlessness, and an inability to form meaningful and long-lasting relationships, are now confirmed to be correlated with excessive use of smartphones [17].

Research suggests that the brain's contact with external neurological disruptions can transform the course of its development, causing disastrous results, specifically for children and young adults [18, 19]. Recent studies confirm psychiatric disorders in children are correlated with excessive use of and addiction to computer games, as well as the general use of the Internet [20]. Negative outcomes for young adults elicits from an overuse of smartphones, such as sleep deprivation and obesity [21]. Additionally, studies convey excessive use of gaming technologies led to a form of neural rewiring [22], particularly, structural deviations due to exposure to divergent sensory experiences in ways that weaken cognitive focus, resulting in shallower thinking skills. Studies have also shown that the extreme use of the Internet instigated potential adverse effects upon the psychosocial development of adolescents [23]. Such affects have been adversely associated with notable behavioral and social maladjustment, with other outcomes showing hyperactivity and conduct problems. Additional studies show that the forming of meaningful relationships via social media was difficult to establish compared to those in the real world [24]. Online activity did not create lasting friendships, but rather resulted in long-lasting non-casual social connections, while at the same time created weakening real-world relationships.

In the context of Internet use, researchers also characterized extreme or "excessive use" as "poorly controlled preoccupations, urges, or behaviors regarding computer use and Internet access that lead to impairment or distress" (p. 117) [25]. Related to these effects are neural changes in the brain, particularly, structural deviations due to exposure to divergent sensory experiences—particularly on children and adolescents [26]. More significant are recent studies using MRI technology that show the effects of brain activation patterns on middle age adults during Internet searching. These findings indicate that online searching appears much more stimulating than reading from traditional substrates—demonstrating that sensory-rich experiences increase activity of the visual cortices [27]. This suggests that research participants have a significant sensory-rich experience, while searching online that increases activity in the visual brain regions. Outcomes have suggested the potential for negative brain and behavioral effects, including impaired attention and addiction. All in all, these studies provide support for the continued neural alternation that is impacting individuals on many levels, which is most strenuous.

Finally, the use of smartphones is a crucial part of our daily living, however, studies show that the younger population, especially college students, have been using smartphones excessively more than other populations [28–30]. Such use includes a variety of different activities such as texting, emailing, and engaging social media networks [31]. Such findings suggest that the increased use of interactive media, may cause feelings of isolation, depression, anxiety and restlessness—including feeling of anxiety when people are separated from their smartphones [32].

3 Flow Theory and Consciousness

Researchers argue that the experience of "flow" is an emotional condition related to excessive behavior [33]. In such a state, an emotional bond is formed between the psychological and chemical mechanisms of the brain, which occurs through the repeated experience of flow [34, 35]. For example, smartphone users may produce the same self-centric experiences as online gamers—in which their "awareness" of the surroundings disappear from consciousness [36]. Strikingly, Csikszentmihalyi observed that flow was correlated to the loss of self-consciousness. Here, consciousness is a state of psychological immersion often accompanied by positive emotions, where time disappears and the sense of self is lost [37].

Related to "intrinsic motivation", Csikszentmihalyi states that "flow" is a technical term connected to mood and consciousness, a concept related to daily activates and their direct experiences of positive feelings [38, 39]. Through studies that observed technological usage in the 1990s, flow theory was directly applied to the analysis of user experience when seeking information from online sources, with an additional focus on understanding the relation between skill levels and the tendency of experiencing flow [39].

The self-motivating features of flow, exemplifies the appealing nature of smartphones and the users' captivation with such an artifact. The attentive focus allows for a deeper understanding between consciousness and flow, which gives way to a heightened user experience of absorption, enjoyment, and interest [40]. In addition to several studies that have examined emotion and cognitive changes related to mood management, human-to-human and social communication [41–43], the matter of excessive use of smartphones on consciousness or awareness of the world have been increasingly recognized [44, 45]. As such, we argue that researchers should continue to explore the psychological dimensions of what makes social media usage so motivating [46, 47], specifically the impact of the excessive use of smartphones of social engagement.

3.1 Research Question

While psychologists propose possible reasons for smartphone addiction owing to its ability to alter mood and trigger enjoyable feelings [48], it is not fully known if repetition of use is correlated to other negative effects that impact psychosocial development. For this reason, we were compelled to ask what the effects of smartphone use are (among college students), on a variety of psychological domains—such as the (1) emotion/mood, (2) dependency, (3) addiction, (4) purpose of life, (5) social communication, and (6) self-consciousness. From this question, focused on six areas of inquiry, we arrived at seven hypotheses. Table 1 aligns the six topics with the seven hypotheses, with the number of questions asked in each category. The questions can be found in Appendix A.

Table 1. Research topics areas with hypotheses and number of questions.

Topic		Hypotheses	Ques.
		Among college students:	
1	Emotion/Mood	**H1:** Smartphone use significantly impacts stress and anxiety.	1–8
		H2: Smartphone use significantly impacts mood or feelings.	
2	Dependency	**H3:** There is a strong dependency upon smartphones.	9–15
3	Addiction	**H4:** There is a strong degree of addiction to smartphones.	16–22
4	Purpose of Life	**H5:** Smartphone use significantly impacts their personal life.	23–25
5	Social Communications	**H6:** Smartphones are the main form of communication and information gathering.	26–28
6	Self- Consciousness	**H7:** Smartphone usage increases self- consciousness.	29–35

Note: *See Appendix A for list of 35 questions*

4   Methods

4.1   Participants

A convenience sample of 22 full-time and part-time graduate students (63/36% male/female) from the Indiana University School of Informatics (mean age = 26), were recruited for this research study during Spring 2014. All students were part of a graduate level introductory course in informatics. The student participants formed a cohort with diverse ethnicity, age and gender.

4.2   Data Collection

To provide additional insight into the findings of the author's past study on the impact of smartphone use [49], they performed an online post-study pilot questionnaire. (Findings from this study have not been previously reported.) Participants completed the online survey based on their daily experiences using smartphones. The questionnaire consisted of 72 questions, divided into three sections:

1. Smartphone usage—questions regarding quantities of smartphone use and usage time of texting, phone calls, game-play, social media use, etc.
2. Smartphone daily experiences—the largest group of questions, with six subsections, each using a seven-point Likert-scale.
3. Demographics—questions related to Gender, Nationality, Age, and Years of smartphone use, cellphone use, and PC/Laptop use.

Figure 1 shows the online survey interface with the seven-point Likert scale—from: Strongly Disagree (1) to Strongly Agree (7).

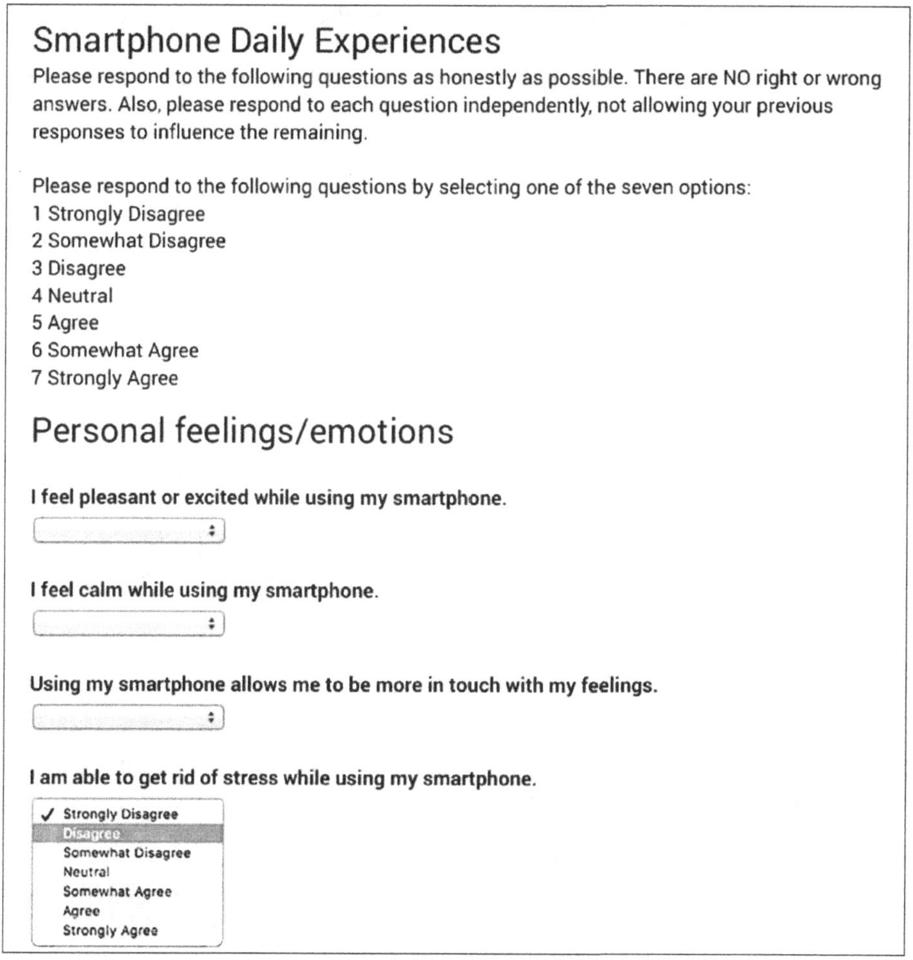

Smartphone Daily Experiences

Please respond to the following questions as honestly as possible. There are NO right or wrong answers. Also, please respond to each question independently, not allowing your previous responses to influence the remaining.

Please respond to the following questions by selecting one of the seven options:
1 Strongly Disagree
2 Somewhat Disagree
3 Disagree
4 Neutral
5 Agree
6 Somewhat Agree
7 Strongly Agree

Personal feelings/emotions

I feel pleasant or excited while using my smartphone.

I feel calm while using my smartphone.

Using my smartphone allows me to be more in touch with my feelings.

I am able to get rid of stress while using my smartphone.

- ✓ Strongly Disagree
- Disagree
- Somewhat Disagree
- Neutral
- Somewhat Agree
- Agree
- Strongly Agree

Fig. 1. Depicts online survey interface.

4.3 Data Analysis

For the reporting of our findings (for this paper), our analysis of the data was executed in three phases. In the first phase we focused on section two (as noted) but narrowed our analysis to only 35 of the 60 questions related to the participants smartphone daily experiences. After our preliminary analysis of the data, we determined that cleaning the data was necessary, i.e., those data points that might be disconnected with the effect that we were trying to isolate—thus allowing us to maintain our focus on those selected topics of interest. We did not observe any obviously erroneous data due to a mistake during data collection or reporting. Also, for this reporting, we are not reporting on participant smartphone usage or demographics—section one and three. After phase one, phase two of our analysis consisted of appropriately aligning the remaining questions under those subsections with their related hypotheses. Phase three consisted of a revised analysis of the 35 questions.

For our data analysis, we measured frequency (within the seven Likert degrees), with conversion percentages—from the scale of strongly disagree to strongly agree. We first averaged the scores of each question, according to the responses from the 22 participants. This was followed by averaging those frequency scores that pertained specifically to each of the hypotheses. As such, we could determine to what degree or percentage the overall response was—within the range of responses from strongly disagree to strongly agree. For example, under hypothesis one, "Smartphone use significantly impacts stress and anxiety," there are four questions that provided frequencies, with a total overall score of 52.27%. This frequency score conveys the degree to which the participants agreed with the stated hypothesis—that the use of their smartphone causes stress and anxiety in their daily life.

5   Results

As noted, the 35 questions were grouped according to their respective hypotheses. As such, we re-state the hypothesis, along with the frequency percentage, and any relevant findings.

H1: Smartphone Use Significantly Impacts Stress and Anxiety
Analyzing the responses of four questions revealed that 52.27% of participants agreed that their anxiety level increases with smartphone use. Added to this percentage is the fact that 72.73% of the participants stated that they feel impatient and fretful when they are not holding their smartphone in hand.

H2: Smartphone Use Significantly Impacts Mood or Feelings
Analyzing the responses of four questions revealed that the negative impact of smartphone use on mood was reported by 60.23% of college students. Impacting this overall score is the fact that 77.27% of participants stated that when they use their smartphone they are enabled to be more in touch with their feelings—while only 45.45% said that feel calm while using their smartphone.

H3: There is a Strong Dependency Upon Smartphones
Analyzing the responses of seven questions, revealed that 57.27% of college students consider themselves dependent on their smartphone. The most influential factors contributing to their dependency were that 72.73% of the students felt safe and secure when they had their smartphones in their procession, while 63.64% had difficulty with going through their daily life without their smartphone in their procession.

H4: There is a Strong Degree of Addiction to Smartphones
Although 68% of participants reported constantly checking their smartphone for different purposes and 63% agreed it was very difficult to live their daily life without their cellphone (note above), only 30.52% of college students agreed (overall of seven questions) that they were addicted to their cellphones. In this section of seven question,

the question that received the lowest score (at 22.73%) was: "I believe I am addicted or have an abnormal dependency on my smartphone."

H5: Smartphone Use Significantly Impacts Their Personal Life. (Positively)

Analyzing the responses of three questions, our findings revealed that smartphone use improves purpose of life among college students at 59.09%–including its positive impact on their personal life by giving them greater values and helping them to be successful. Among this group of three questions, "believing that nothing is more fun than using my smartphone," received a score of 90.91%, the highest percentage of any of the 35 questions.

H6: Smartphones are the Main Form of Communication and Information Gathering

Analyzing the responses of three questions, we observed that smartphones are considered as the main form of communication by 66.67% of the participants and the main tool for finding information by 72.73%, rather than ask others to help them.

H7: Smartphone Usage Increases Self-consciousness. (Positively)

Analyzing the responses of seven questions, our findings uncovered that 54.55% of participants (overall) indicated that they have a better level of self-consciousness when they use their smartphone. Interestingly, 72.73% agreed that when using their smartphone makes them less aware of the close surroundings—while 63.64% agreed that it is difficult to talk on their phone if they think others are watching them.

6   Discussion

In this study we examined the effects of smartphone use on the psychosocial wellness among college students, with a focus on six psychological domains: emotion/mood, dependency, addiction, purpose of life, social communications, and self-consciousness. Our findings were independent of sex, age group, or level of smartphone usage among our participants. Our study suggests addictive usage behaviors, 68% reporting constantly checking their smartphone, and smartphone dependency, 57% reporting smartphone dependency as well. However, despite these findings, only 22% of participants in our study agreed to have smartphone addiction and dependency.

According to analysis of our survey data, smartphone use increases anxiety and stress level among college students, and our participants agreed they feel impatient if they do not have their phone in their possession. Although we cannot deduce causality in this study, these findings are aligned with the previous research findings that correlate excessive use of smartphones with increased level of anxiety and restlessness.

Another component of our findings suggest smartphones are the main tool for communication and information seeking among college students. As noted, these findings are in agreement with past studies investigating smartphone usage among college students. Taken together, a conclusion that the two most significant smartphone usage patterns among college students are: (1) maintaining social relationships and (2) accessing the online digital information.

The most outstanding finding emerging from our data analysis is the impact of smartphones on the personal life of college students—by more than 90% agreeing on nothing is more fun than using their smartphones. This discovery provides new perceptions into the notion that excessive smartphone use has an individuals' social life. This behavior, we believe, may also suggest that excessive and/or repeated use may promote the experience of pleasure and improvement of mood, but may also lead to an increase in a lack of awareness of the close environment. In the former case, the risk of forming habitual usage and addiction are present.

Regarding limitations to the study—first, the sample size of the study limited our ability to extend our findings inferentially to the greater population of college students. For this reason, we identified it as exploratory in nature, and thus, a pilot study. Other contributing limitations to a small sample might include the narrow characteristics of our participants—drawing upon one class of college students, from one university. We suggest that future studies extend the investigation to be more inclusive of socio- demographics, ethnicity and cultural backgrounds, devices, location, and type of activities.

7   Conclusion

Excessive smartphone use, of any degree, can have physical, psychological and social implications for any individual. However, our findings specifically suggest that college students depend heavily on their smartphones—to a level that if they were not in possess of them, they would not feel secure or in control. While smartphones are their primary mode of communication and information seeking, excessive use may have a significant influence on mood, while increasing the level of stress and anxiety.

In sum, although smartphones can translate into tools that empower and expand one's communication capacity, they may adversely affect psychosocial wellness, as well as disrupt one's accurate sense of consciousness—of both persons and objects around them. Our challenge remains therefore, to not only understand the influence of these mediational technologies, but increasingly to identify those explicit threats to mental and social wellness—leading to the shaping of lives that are positive, productive, and socially engaged.

Acknowledgement. The authors would like to acknowledge the student work of Alexandra Dirico, University of Illinois at Chicago, for her contribution in the development of this paper.

A Appendix

List of research questions provided participants according to topics, with responses according to a seven-point Likert scale.

Emotion and Mood

1. I am able to get rid of stress while using my smartphone.
2. I feel impatient and fretful when I am not holding my smartphone.
3. I sometimes become irritated while using my smartphone.

4. Not having my phone in my possession (at home/outside home) makes me feel nervous/anxious.
5. I feel pleasant or excited while using my smartphone.
6. I feel calm while using my smartphone.
7. Using my smartphone allows me to be more in touch with my feelings.
8. I become depressed or sad if I am not able to use my smartphone.

Dependency

9. Having my smartphone with me at all times gives me the feeling of safety and security.
10. I feel more comfortable using my smartphone for communication as opposed to other forms of communication.
11. It is very difficult to consider my daily life without having my smartphone.
12. I use my smartphone each day longer than I had intended.
13. I often consider that I should shorten my smartphone use.
14. I feel I waste time when I am on my smartphone more than necessary.
15. I feel I am more in control when using my smartphone.

Addiction

16. I believe I am addicted or have an abnormal dependency on my smartphone.
17. I feel the urge to use my smartphone again immediately after I stopped using it.
18. I cannot bear the thought of not having my smartphone with me at all times.
19. Regardless of the circumstances around me, I would never give up the use of my smartphone.
20. Not being able to use my smartphone would be as painful as losing a friend.
21. I miss planned or anticipated work or responsibilities due to smartphone use.
22. I have my smartphone (and using it) on my mind even when I'm not using it.

Purpose of Life

23. I feel that the use of my smartphone gives greater value to my life.
24. I believe there is nothing more fun to do than use my smartphone.
25. Without my smartphone, I believe I could not be successful in the world.

Social Communications

26. I constantly check my smartphone so I will not miss communication between other people via email, text, Twitter or Facebook, etc.
27. I check social networking sites like Twitter or Facebook right after waking up in the morning or right before going to bed at night.
28. I prefer searching for information that I need by using my smartphone rather than asking people in real time.

Self-consciousness

29. It is hard for me to talk on my smartphone when I think others are watching me.
30. I feel nervous when I am talking on my smartphone in public.
31. I pay attention to my inner feelings when I am using my smartphone.

32. Using my smartphone makes me less aware of my close surroundings.
33. I am often unaware if I am speaking too loudly on my smartphone when I am in public.
34. I often lose track of time when I am using my smartphone.
35. When I am on my smartphone, my attention is only focused on it.

References

1. Ingold, T.: Eight themes in the anthropology of technology. Soc. Anal. **41**, 106–138 (1997)
2. Yang, C.C., Brown, B.B., Braun, M.T.: From Facebook to cell calls: layers of electronic intimacy in college students' interpersonal relationships. New Media Soc. **16**, 5–23 (2014). https://doi.org/10.1177/1461444812472486
3. Basics, T.V.: A Report on the Growth and Scope of Television. Television Bureau of Advertising (2012). http://www.tvb.org/
4. Bohn, R.E., Short, J.E.: How Much Information? 2009 Report on American Consumers. University of California, San Diego, Global Information Industry Center (2009). http://hmi.ucsd.edu/pdf/HMI_2009_ConsumerReport_Dec9_2009.pdf
5. Roberts, D.F., Foehr, U.G.: Trends in media use. Futur. Child. **18**, 11–37 (2008)
6. Pew Research Center Internet & Technology. Mobile Fact Sheet (2018). http://www.pewinternet.org/fact-sheet/mobile/
7. Katz, J.E.: Mainstreamed mobiles in daily life: perspectives and prospects. In: Handbook of Mobile Communication Studies, pp. 433–446 (2008). https://doi.org/10.7551/mitpress/9780262113120.001.0001
8. Fortunati, L.: Electronic Emotion: The Mediation of Emotion via Information and Communication Technologies, pp. 1–31. Peter Lang, Oxford (2009)
9. Silva, S.R.: On emotion and memories: the consumption of mobile phones as 'affective technology'. Int. Rev. Soc. Res. **2**, 157–172 (2012). https://doi.org/10.1515/irsr-2012-0011
10. Gross, J.J., Thompson, R.A.: Emotion Regulation: Conceptual Foundations, pp. 3–24. Guilford Press, New York (2006)
11. Lazarus, R.S.: Emotion and Adaptation. Oxford University Press, New York (1991)
12. Smith, A., Rainie, L., Zickuhr, K.: College students and technology. Pew Research Center Internet & Technology (2011)
13. Thomée, S., Härenstam, A., Hagberg, M.: Mobile phone use and stress, sleep disturbances, and symptoms of depression among young adults-a prospective cohort study. BMC Public Health **11**, 66 (2011). https://doi.org/10.1186/1471-2458-11-66
14. Harrison, M.A., Gilmore, A.L.: U txt WHEN? College students' social contexts of text messaging. Soc. Sci. J. **49**, 513–518 (2012). https://doi.org/10.1016/j.soscij.2012.05.003
15. The Mobile Consumer: A Global Snapshot. Nielsen Holdings, New York (2013)
16. Lee, U., Lee, J., Ko, M., Lee, C. et al.: Hooked on smartphones: an exploratory study on smartphone overuse among college students. In: Proceedings of the 32nd Annual ACM Conference on Human Factors in Computing Systems, pp. 2327–2336 (2014)
17. Weiss, M.D., Baer, S., Allan, B.A., Saran, K., Schibuk, H.: The screens culture: impact on ADHD. Atten. Deficit Hyperact. Disord. **3**, 327–334 (2011)
18. Healy, J.M.: Endangered Minds: Why Children Dont Think And What We Can Do About I. Simon and Schuster, New York (2011)
19. Zhou, Y., Lin, F.C., Du, Y.S., Zhao, Z.M., Xu, J.R., Lei, H.: Gray matter abnormalities in Internet addiction: a voxel-based morphometry study. Eur. J. Radiol. **79**, 92–95 (2011)

20. Perlow, L.A.: Sleeping with Your Smartphone: How to Break the 24/7 Habit and Change the Way you Work. Harvard Business Review Press, Brighton (2012)

21. Whitbourne, S.K.: Your smartphone may be making you... not smart (2011). https://www.psychologytoday.com/blog/fulfillment-any-age/201110/your-smartphone-may-be-making-you-not-smart

22. Chirico, D.M.: Building on shifting sand the impact of computer use on neural & cognitive development. Wald. Educ. Res. Inst. Bull. **2**, 13 (1998)

23. Kormas, G., Critselis, E., Janikian, M., Kafetzis, D., Tsitsika, A.: Risk factors and psychosocial characteristics of potential problematic and problematic internet use among adolescents: a cross-sectional study. BMC Public Health **11**, 595 (2011). https://doi.org/10.1186/1471-2458-11-595

24. Lenhart, A., Madden, M.: Teens, Privacy and Online Social Networks. How teens manage their online identities in the age of MySpace. Pew Internet & American Life Project Report (2007). http://www.pewinternet.org/~/media//Files/Reports/2007/PIP_Teens_Privacy_SNS_Report_Final.pdf.pdf

25. Weinstein, A., Lejoyeux, M.: New developments on the neurobiological and pharmaco-genetic mechanisms underlying internet and videogame addiction. Am. J. Addict. **24**, 117–125 (2015). https://doi.org/10.1111/ajad.12110

26. O'Keeffe, G.S., Clarke-Pearson, K.: The impact of social media on children, adolescents, and families. Pediatrics **127**, 800–804 (2011). https://doi.org/10.1542/peds.2011-0054

27. Small, G.W., Moody, T.D., Siddarth, P., Bookheimer, S.Y.: Your brain on Google: patterns of cerebral activation during internet searching. Am. J. Geriatr. Psychiatry **17**, 116–126 (2009). https://doi.org/10.1097/JGP.0b013e3181953a02

28. Contractor, A.A., Weiss, N.H., Tull, M.T., Elhai, J.D.: PTSD's relation with problematic smartphone use: mediating role of impulsivity. Comput. Hum. Behav. **75**, 177–183 (2017). https://doi.org/10.1016/j.chb.2017.05.018

29. Kim, Y., Jeong, J., Cho, H., Jung, D. et al.: Personality factors predicting smartphone addiction predisposition: behavioral inhibition and activation systems, impulsivity, and self-control. PloS One **11**, e0159788 (2016). https://doi.org/10.1371/journal.pone.0159788

30. Roberts, J., Yaya, L., Manolis, C.: The invisible addiction: Cell-phone activities and addiction among male and female college students. J. Behav. Addict. **3**, 254–265 (2014). https://doi.org/10.1556/JBA.3.2014.015

31. Lin, L., Sidani, J.E., Shensa, A., Radovic, A., et al.: Association between social media use and depression among US young adults. Depress. Anxiety **33**, 323–331 (2016). https://doi.org/10.1002/da.22466

32. Hartanto, A., Yang, H.: Is the smartphone a smart choice? The effect of smartphone separation on executive functions. Comput. Hum. Behav. **64**, 329–336 (2016). https://doi.org/10.1016/j.chb.2016.07.002

33. Csikszentmihalyi, M.: Introduction in Optimal Experience Psychological Studies of Flow Consciousness. Cambridge University Press, New York (1988)

34. Moneta, G.B., Csikszentmihalyi, M.: The effect of perceived challenges and skills on the quality of subjective experience. J. Pers. **64**, 275–310 (1996). https://doi.org/10.1111/j.1467-6494.1996.tb00512.x

35. Peifer, C.: Psychophysiological correlates of flow-experience. Adv. Flow Res., 139–164 (2012)

36. Pace, S.: A grounded theory of the flow experiences of Web users. Int. J. Hum Comput Stud. **60**, 327–363 (2004). https://doi.org/10.1016/j.ijhcs.2003.08.005

37. Csikszentmihalyi, M.: Flow: The Psychology of Optimal Experience. Harper & Row, New York (1990)

38. Csikszentmihalyi, M.: Finding Flow: The Psychology of Engagement with Everyday Life. Basic Books, New York (1997)
39. Ryan, R.M., Deci, E.L.: Self-determination theory and the facilitation of intrinsic motivation, social development, and well-being. Am. Psychol. **55**, 68 (2000)
40. Rodriguez-Sanchez, A.M., Schaufeli, W.B., Salanova, M., Cifre, E.: Flow experience among information and communication technology users. Psychol. Rep. **102**, 29–39 (2008). https://doi.org/10.2466/pr0.102.1.29-39
41. Zillmann, D.: Mood management in the context of selective exposure theory. Ann. Int. Commun. Assoc. **23**, 103–123 (2000). https://doi.org/10.1080/23808985.2000.11678971
42. Brandon, J.: Are we losing the emotion from communication. In: Depth Does Relying on Technology Harm Our Senses and Emotions (2013)
43. Bandura, A.: Social cognitive theory of mass communication. Media Psychol. **3**, 265–299 (2001)
44. Turkle, S.: Alone Together. Basic Books, New York (2011)
45. Bianchi, A., Phillips, J.G.: Psychological predictors of problem mobile phone use. CyberPsychology Behav. **8**, 39–51 (2005). https://doi.org/10.1089/cpb.2005.8.39
46. Finneran, C.M., Zhang, P.: Flow in computer-mediated environments: promises and challenges. Commun. Assoc. Inf. Syst. **15**, 82–101 (2005)
47. Ghani, J.A., Supnick, R., Rooney, P.: The experience of flow in computer-mediated and in face-to-face groups. ICIS **91**, 229–237 (1991)
48. Faiola, A., Newlon, C., Pfaff, M., Smyslova, O.: Correlating the effects of flow and telepresence in virtual worlds: Enhancing our understanding of user behavior in game-based learning. Comput. Hum. Behav. **29**, 1113–1121 (2013). https://doi.org/10.1016/j.chb.2012.10.003
49. Faiola, A., Srinivas, P., Duffecy, J.: The effects of excessive smartphone use on mood, consciousness, and psychosocial wellness: the sometimes negative side of flow experience, slated. J. Happiness Stud. (In Review)

Detecting and Interfering in Cyberbullying Among Young People (Foundations and Results of German Case-Study)

Sebastian Wachs[1], Wilfried Schubarth[1], Andreas Seidel[1], and Elena Piskunova[2(✉)]

[1] Potsdam University, K-Liebknecht-Str. 24/25, 14476 Potsdam, Germany
wachs@uni-potsdam.de
[2] Herzen State Pedagogical University, n.r. Moika, 48, 191123 St.-Petersburg, Russia
L_piskunova@mail.ru

Abstract. Information and communication technologies (ICT) play more and more significant role in the lives of children and young people. Adolescents use ICT to communicate with others via chat, instant messenger, online communities, etc. They take online offers and services concerning music, pictures or videos to entertain themselves; to search for new knowledge and information, and to acquire and to use the online game offers. The majority of ICT is used in a constructive and peaceful manner, bringing no negative online experiences, which can be perceived as stressful, but this is not the case for all adolescents. The online world similarly as the physical world can bring danger, and the possible danger that adolescents encounter today in the online world is cyber-bullying.

In the present contribution, we address to the online risk of cyberbullying among adolescents. The article deals with the following questions - how cybermobing can be defined from the research perspective, how many adolescents are affected by it in Germany, what makes cyberbullying specific, what do we know about the victims and the perpetrators, what are the possible consequences, and what are the recommendations for adolescents and adults (parents, teachers and educators) who are dealing with cyberbullying problems.

Keywords: Bulling · Cyberbullying · Adolescents

1 Introduction: Detecting Cyberbullying Among Young People

Information and communication technologies (ICT) such as Internet, PCs, mobile phones/smartphones, tablet computers and social networks become increasingly important and evident in the lives of children and adolescents. Young people use ICT to communicate with other people through chat, messenger, online community, etc., to use online offerings and services related to music, photos or videos, to look for new knowledge and information and to adapt the findings for their own purposes and to play online games. Even if most people use ICT constructively and peacefully and do not acquire any negative experience on the Internet, which can be perceived as stressful, this, is not the case by far for all adolescents. The online world is as dangerous as the physical world. Online risk exactly as risks in the physical world are very multi-faceted and range

© Springer Nature Switzerland AG 2018
D. A. Alexandrov et al. (Eds.): DTGS 2018, CCIS 859, pp. 277–285, 2018.
https://doi.org/10.1007/978-3-030-02846-6_22

from spam, (concealed) sponsorship, phishing or games of chance (with commercial interests) through pornographic or problematic sexual content, unwanted sexual messages, sexual harassment, assault or abuse (sexuality); racist, biased, misleading, extremist information, advice, guidance (as for example, right-wing extremist, or salafist on-line groups) (values/ideologies) to violent, hate-filled websites/posts, comments in social networks, calls, applications and instructions on self - or other-damaging actions (aggression) [5: 13 et seq.].

In this article, we focus on the online risk of cyberbullying among adolescents. We study how cyberbullying can be defined from a scientific perspective, how many adolescents are affected by it in Germany, how cyberbullying can be specified, what do we know about the victim and the perpetrator, what are the possible consequences, and what are the recommendations for adolescents and adults (parents, teachers and educators) in dealing with cyberbullying. As a kind of introduction, a case example will give a true-to-life insight into the phenomenon. The following case goes back to the description of a 16-year-old cyberbullying perpetrator, who reported her experience in the youth magazine "Spiesser".

> "Why did I insult, humiliate and threaten others? I can hardly explain that. [...]

It was so easy to write a common comment. To post insults under images. Or just the nasty things of others to 'like'. Today I know it was because I did not look the victims into the eyes. I faced nobody who could give me immediately a spell like a counter-attack. I received confirmation from others, because often my comments have been 'liked'. Outside of the World Wide Web nothing happened, it had no consequences for me. In the schoolyard, a teacher might come and interfere. If I harassed somebody in the team, the trainer prevented me. But if I insult and annoy someone online for hours, no one comes. From some of the chats the administrator threw me out, from SchülerVZ I was deleted more than once. However, what does it matter? Then I logged on under a different name. My family knows nothing of it all.

When I was 16, my friend did not know my "other side." Until he became my ex-boyfriend. After that I was completely crazy. I terrorized him, threatened him. And when he had a new girlfriend, she became my enemy, too. I did not know her at the time. [...]

The occasions could also be null and void. As in the case of the girl from the parallel class, which had the mega hot boots that I wanted, but they were too expensive. Immediately, she was a snob, an arrogant bitch. Online, I spread rumors that she bought the marks, and so on. To other girl I put the stamp of 'school slut' online, because she had spoken with my friend. [...]

Today, I fear. Once to become a victim of someone like me. And also that my family knows what I've done" [3].

Cyberbullying is a new form of traditional bullying or bullying without the use of ICT. Cyberbullying can be understood as an aggressive pattern of behavior in which one person or group uses ICT to cause harm to a weaker person or a group with full intent and repeatedly [9: 376]. In the case of partially inflationary use of the word in the media as well as in everyday life, a distinction from isolated online conflicts makes sense. As in the physical world, one cannot call any quarrel or dispute bullying, nor can one do so in the virtual world. We speak about cyberbullying when in addition to the use of ICTs, there must be a deliberate intention to harm a person, and abuses must occur repeatedly over a long period, and there must be an imbalance of power between the offender and the victim [8: 3].

In the study, at least four forms of cyberbullying were defined [7]:

- Harassment: Sending offensive, insulting, hurtful and threatening text messages, picture or video messages to the victim, to frighten and intimidate.
- Denigration: The spreading of text messages, audio or video material with the aim of destruction of social relations or the reputation of the victim.
- Betrayal: The publication and dissemination of intimate information about the victim via ICT to harm the victim.
- Exclusion: The exclusion out of online communities and online groups (e.g. Facebook groups, WhatsApp groups) in order to isolate the victim.

Now that those four forms are combined with the use of various ICT, it quickly becomes clear that cyberbullying is a very multifaceted phenomenon. Table 1 provides an overview, but not exhaustive, of possible examples of cyberbullying [13: 85].

Table 1. Examples of cyberbullying

Medium	Use
Mobile/Smartphone	Abusive and annoying phone calls (anonymous) threats and intimidation by telephone, threats of physical violence, etc. Creating, manipulating and sharing of victim images and videos and posting them online to harm the victim Spreading rumors, slander and libel
Instant Messenger (WhatsApp, ICQ, etc.)	Sending offensive messages, pictures or videos Blocking the victim, deleting friend lists and excluding group chats using a different account by offenders to send malicious messages or to threaten the victim with false identification Posting scornful, offensive comments, photos or videos about the victim
Social networks (Facebook, Twitter, YouTube, Google+, etc.) Chat rooms (YouNow, Knuddels, Habbo Hotel, etc.)	Threatening the victim, pursue him or her Creating groups of hatred Excluding the victim from joint groups Identifying the victim by creating a second fake victim profile and give as a victim

A frequently discussed question is whether cyberbullying is actually old wine in new bottles or not. In fact, cyberbullying and traditional mobbing have some common features, as repetition of the assault, violation intent of the offender, as well as demonstration the power imbalance to the victim. Nevertheless, cyberbullying is characterized by certain peculiarities, which suggest that the use of ICT is not the only significant distinction [13: 82 et seq.]:

- By cyberbullying, medial skills (e.g., using Photoshop) have an advantage over physical traits (e.g., physical superiority).
- Cyberbullying can be performed in any place and at any time. The victim may be attacked after school, even, for example, in their own room. Due to this independence of time and place of cyberbullying, the victim loses all areas of retreat and rest.
- By cyberbullying perpetrators can easily hide their identity through the use of ICT, which can lead to increased social insecurity among victims.
- Cyberbullying is often characterized by particularly untestrained and violent attacks, which can be explained by the absence of social control by adults in the virtual worlds and the absence of a personal (face-to-face) situation between the victim and the offender. Due to the absence of direct contact, the perpetrator does not receive tangible feedback on the emotional consequences of attacks on the victim. This, in turn, impedes feelings such as remorse, compassion and guilt on the part of the offender.
- Publicly carried out the attack on the victim can often be viewed online by many others over long periods of time. This may be an additional humiliation for the victim and thus an additional burden.

2   Cyberbullying as not an Everyday Occurrence or a Rarity

Various studies of cyberbullying show that the spread of intimidation is neither a commonplace phenomenon, nor a rare one. In a representative survey of 1,734 German adolescents between the ages of 14 and 20, 6% reported that they were victims of cyberbullying. 7.5% said that they had acted as perpetrators, and 1.2% reported that they were involved in cyberbullying as perpetrators and victims [6, p. 13 et seq.]. Cyberbullying is by no means a phenomenon that is limited to German teenagers, it is spread throughout Europe: in the representative survey EU Kids Online in 25 European countries, covering approximately 25,000 adolescents aged from 9 to 16 years, 6% of respondents reported victims, and 3% were perpetrators [4: 22].

We cannot say that cyberbullying is more common among boys or girls. Different studies come to different conclusions where either differences do not exist or are likely to be related to boys or girls. Based on a review fulfilled by Tokunaga [10], it can be concluded that cyberbullying is a phenomenon that is almost evenly distributed between the two genders [10, p. 280]. However, some differences in cyberbullying of girls and boys occur: boys will likely participate in direct cyberbullying (e.g., harassment), while girls often use indirect forms (such as the secret betrayal or insult) [12: 742 et sec.].

By age, young people in 7th and 8th grades represent the greatest risk of becoming victims of cyberbullying [10].

The next step is to clarify which signs of increased risk of cyberbullying from the perspective of victims and offenders could be identified in the study.

Possible risk factors and consequences of cyberbullying

In general, any person who uses ICT, can be victims of cyberbullying. However, some risk-enhancing factors could be identified in the research [1, 2, 10, 11, 14, 15]. The increased risk of becoming a victim of cyberbullying can be identified by teenagers, who are introverted above average, fearful, sensitive and insecure in dealing with others, who demonstrate the excessive use of ICT and deal permissively with private information and contact details; their parents maintain authoritarian parenting style; who have few or no friends and no connection to any fixed group or community; usually at schools where the victims of bullying study there are no prevention measures or programs against cyberbullying. The enhanced risk to become the perpetrators of cyberbullying have the teenagers, who are impulsive, have a low ability to empathy, low levels of social competence and moral sense of justice; they use ICT excessively; who frequently drink alcohol; whose parents maintain an authoritarian parenting style, in whose families violence occurs and a harsh tone prevails; the perpetrators most study at schools with no measures or prevention programs against cyberbullying.

When the possible causes of cyberbullying have been described, we refer to the possible consequences of cyberbullying for both victims and perpetrators in short. It should be noted that the above mentioned factors can increase the risk of becoming a victim or a perpetrator, but not necessarily lead to that. Similarly, these factors apply to many, but not all of the victims and perpetrators.

In addition to the identification of risk-enhancing factors the central concern of the research was to identify the possible consequences of cyberbullying. The experience of cyberbullying for the victim leads to serious consequences in completely different areas. The emergence and intensity of impact is influenced by the personal perception of the victim as well as by the duration and intensity of the attacks. On the basis of the literature review the next consequences may be summarized [1, 2, 10, 12, 15]:

- physical complaints such as headaches or abdominal pain, loss of appetite and sleep problems;
- strengthening of ongoing negative emotions such as anger, shame, sadness, despair, loneliness, helplessness, nervousness, irritability and anxiety;
- social problems, such as leaving social work with peers and joint family activities;
- increasing psychological problems such as negative self-esteem and, correspondingly, depression, eating disorders, Internet addiction;
- abuse of licit and illicit drugs;
- school problems, such as acute episodes of absenteeism, concentration difficulties and unexpectedly falling grades.

It seems appropriate to note that not only victims but also perpetrators of cyberbullying through their actions can have such consequences as long-term damage to their positive development. Possible consequences can be summarized as follows:

- frequent involvement in other forms of aggressive and criminal behaviour (including vandalism, shoplifting, drug abuse);
- increased negative feelings (such as hate, stress or anger) and increased readiness for aggression;
- increased experience of violence with friends, family members and love partners;
- fewer opportunities to learn pro-social behavior;
- psychological problems (e.g. depression, self-esteem, low self-esteem);

- school problems (such as drop in performance, frequent conflicts with teachers, exclusion from school);
- social issues (e.g. peer rejection, fewer friends and social contacts).

The description of possible consequences of cyberbullying has made it clear that cyberbullying influences both parties involved with serious and lasting adverse effects which can go hand-in-hand. Cyberbullying is thus a serious development risk.

In the following part recommendations will be given, what adolescents and close persons can do to prevent cyberbullying, and what they can do if they are faced with cyberbullying.

3   Recommendations for the Prevention and Control of Cyberbullying

To protect yourself from victims of cyberbullying, teens needs first to reflect critically on their own attitude to the media. At the same time, it is important to find a relative interaction with ICT, which is a useful addition to offline activities, but does not completely replace it, as intensive ICT activities lead to narrowing of the space of behavior. A critical reflection of the media's own behavior includes not only the question of how much time you spend online, but much more, how to behave during online time. Victims of cyberbullying often talk about themselves on the Internet too much, so it is important to ask oneself with whom to share certain information and how to use optimally the privacy settings for oneself. One should also be aware of general guidelines for handling sensitive data (not share passwords with other people and change them regularly).

Issues of media behavior, especially the ethical issues – what a person is allowed to use ICT for, but also what a person is not allowed to do – particularly in the prevention of cyberbullying in relation to the offender are of most importance. Adolescents should be made aware of, on the one hand, how difficult is to delete the digital material from the Internet, on the other hand, the lack of suitability of ICT to the conflict or the potential effects of ICT on human communication and interaction behavior (e.g., disinhibition effect). Young people should also be aware that if, for example, the perpetrators put defamatory images of their victims in social networks, and they share or positively evaluate (like) – they are the accomplices. In the case of the mediation of critical media behavior parents, but even more schools play a central role by negotiating with adolescents about their media behavior to be open and unbiased, by discussing with them certain standards of behavior such as rules for fair dealing with each other not only at school but also in the online context. The rules should be formulated in a mandatory way (e.g., "I never use ICT to harm others") and due to this assumed by the teenagers. This joint negotiation may also be a part of a firmly implemented cyberbullying prevention program at schools.

A further preventive measure against cyberbullying aims to social relationships with peers. In this case, it may be particularly useful to encourage and to support adolescents to build social relationships with peers. For adolescents, the shortcomings in the social sphere (i.a., less positive peer relations, less in-depth friendships, more often, exclusion from peer activities, lower levels of popularity), show a higher risk of becoming a victim of cyberbullying. Regular and intensive contact with peers is not only a social protection, but also a great factor of the personal development, recognition and social participation. The promotion of social relations and social skills can also help adolescents to be involved in it. Parents and schools can help by providing growing space and opportunities to develop social relationships.

Further preventive measures take into account the psychological characteristics. As an example can be frequent low self-confidence of victims referred to, which can be strengthened by approaches of empowerment. Thus, adolescents may be able to act confidently in conflict situations and to protect themselves for longer periods of assaults, and enduring silently the offender's attacks. Adolescents can also learn not to look for reasons for cyberbullying in themselves. But even if empowerment is very important, it cannot be the only approach, and other copping strategies (e.g., appealing for help, ignoring, blocking, sharing) in dealing with Cyberbullying are discussed and tested.

Definition of perpetrators' psychological manifestations can also help to determine preventive measures. Perpetrators of cyberbullying are characterized by a lower ability to feel empathy, and the use of ICT can make it still more difficult for them to feel and demonstrate empathy. Therefore, it is important to make it clear for adolescents that even if the attacks are virtually carried out, it makes real harm to the victim.

Adolescents should also learn to control their impulses. Due to the lack of direct interaction in the on-line context attacks are carried out faster and carelessly, without considering about the consequences. Let us equate, for example, an insult online with an insult offline. This association does not degenerate tents conflicts and disputes and escalate to cyber-bullying, but it is necessary to provide young people with rules of peaceful conflict resolution (for example, mutual respect, non-violence). In the peaceful solution to the conflict, parents can provide a role and behavior model, based on their own educational and cultural foundations of behavior. The school offers in addition, the appropriate framework for the testing and further development of these skills.

If adolescents are already victims of cyberbullying, the top priority is to stop immediately cyberbullying, demonstrate readiness to provide full assistance to victims and work together with victims on a solution. Previous studies have shown that this helps victims to handle better the consequences of cyberbullying when they experience peer support, and when their family environment is supportive and characterized by trust and emotional warmth.

If a person is bullied online, he or she should show the negative attitude towards the attacks. No comments can be interpreted by the offender as a tacit consent and encourage, in turn, to continue attacks.

The first assistance to the victim may be willingness to help, demonstration of the concern by listening to attentively and taking the deal seriously. The next step can be to encourage the victim to seek help from adults, or to apply himself or herself for help from a trusted adult.

Cyberbullying incidents should be clearly documented. This can be done by using screenshots and saving e-mails, SMS, posts, pictures, and information about the perpetrators (including nickname, real name). Adults can help those affected to document the attacks, and to pass the information at school forward, because very often the victim and the perpetrator come from the same environment. To delete, for example, offensive or infringing image or video files, can the operator of the relevant service. In the case of particularly serious cases of harassment, slurs and gross violations of personality rights, the police may be informed. One may also determine the identity of the perpetrators, if they acts anonymously, via a mobile network operator and Internet service provider.

As victims of cyberbullying trust adults – parents, teachers, educators, adults should talk regularly and openly about the online activities of children and youth and associated risks. The online events should be paid the same attention as what is happening in everyday life in the physical world. In this way can both shame and trust be built up. Beneficial is, when adults take on not only controlling attitude, but interested and unprejudiced position, give support, particularly when children and adolescents are in situations which they cannot solve alone. It may also be useful to identify assistance opportunities on the net such as common utility services, providing anonymous help – u.a. to learn from trained peers which, for example, www.juuuport.de provides.

If mobbed young persons turn for assistance to the adult, it is important to refrain from blame and to show understanding. By the time the victims turn to the help of adults, they usually have suffered from the attacks of the perpetrators over a longer period of time. Therefore, it is important to involve the victims in the active solution process, to help them to leave the state of powerlessness and loss of control. A ban from the ICT use, is not advisable; it is the fear of a ban on the use of ICT which does makes many victims not to talk with adults about what is happening.

Many of the recommendations for dealing with cyber-harassment are focused on the victim. However, the offenders also need appropriate solutions, concerning how they can cease their destructive behavior and get parents' support if the situation gets out of control. Parents of offenders should speak to the children in detail and impartially about all the people concerned (victims, friends), to understand the situation from different perspectives. The offender must then be led to understanding of the damage he or she was doing with this behavior on the victim. Likewise, strategies of prevention of such conduct in the future may be taught, for example, through role play, and tested.

ICT have changed children's maturation. They offer new opportunities but also bring new risks. It is important that adolescents are as early as possible prepared to these changing conditions. The prevention of violence and, therefore, cyberbullying needs to be understood as a social task which can't be solved by individual actors as young people, parents or teachers, or individual institutes of socialisation like family, school or day-care institutions.

References

1. Baldry, A.C., Farrington, D., Sorrentino, A.: "Am I at risk of cyberbullying"? A narrative review and conceptual framework for research on risk of cyberbullying and cybervictimization: the risk and needs assessment approach. J. Aggress. Violent Behav. **23**, 36–51 (2015)

2. Kowalski, R.M., Giumetti, G.W., Schroeder, A.N., Lattanner, M.R.: Bullying in the digital age: a critical review and meta-analysis of cyberbullying research among youth. J. Psychol. Bull. **140**(4), 1073 (2014)
3. Leutner, P.: Mobbing unter Jugendlichen: "Die Schule wurde meine persönliche Hölle" 4. Teil: Jana, 20: "Sofort war sie ein arrogantes Miststück". J. Spiegel (2012). http://www.spiegel.de/ schulspiegel/leben/mobbing-im-internet-und-in-der-schule-taeter-und-opfer-erzaehlen-a-873794-4.html
4. Livingstone, S., Haddon, L., Görzig, A., Ólafsson, K.: EU kids online. Final report, September 2011 (2011). http://www.lse.ac.uk/media%40lse/research/EUKidsOnline/EU%20Kids%20II %20(2009-11)/EUKidsOnlineIIReports/Final%20report.pdf
5. Paus-Hasebrink, I., Ortner, C.: Online-Risiken und-Chancen für Kinder und Jugendliche: Österreich im europäischen Vergleich. In: Bericht zum österreichischen EU Kids Online-Projekt (2008). http://www.kinderundmedien.at/fileadmin/user_upload/Bibliothek/ eukidsonlineabschlussbericht.pdf
6. Porsch, T., Pieschl, S.: Cyberbullying unter deutschen Schülerinnen und Schülern: Eine repräsentative Studie zu Prävalenz, Folgen und Risikofaktoren. J. Diskurs Kindheits-und Jugendforschung **9**(1), 7–22 (2014)
7. Riebel, J., Jaeger, R.S., Fischer, U.C.: Cyberbullying in Germany – an exploration of prevalence, overlapping with real life bullying and coping strategies. J. Psychol. Sci. Q. **51**(3), 298–314 (2009)
8. Smith, P.K.: Cyberbullying: Challenges and opportunities for a research program. J. Olweus Eur. J. Dev. Psychol. **9**(5), 553–558 (2012)
9. Smith, P.K., Mahdavi, J., Carvalho, M., Fisher, S., Russell, S., Tippett, N.: Cyberbullying: Its nature and impact in secondary school pupils. J. Child Psychol. Psychiatry **49**(4), 376–385 (2008)
10. Tokunaga, R.S.: Following you home from school: a critical review and synthesis of research on cyberbullying victimization. J. Comput. Hum. Behav. **26**(3), 277–287 (2010)
11. Wachs, S.: Moral disengagement and emotional and social difficulties in bullying and cyberbullying: Differences by participant role. J. Emot. Behav. Difficulties **17**(3–4), 347–360 (2012)
12. Wachs, S., Wolf, K.D.: Zusammenhänge zwischen Cyberbullying und Bullying – Erste Ergebnisse aus einer Selbstberichtsstudie. J. Praxis der Kinderpsychologie und Kinderpsychiatrie **60**(9), 735–744 (2011)
13. Wachs, S., Hess, M., Scheithauer, H., Schubarth, W.: Mobbing an Schulen: Erkennen, Handeln. Vorbeugen. Kohlhammer, Stuttgart (2016)
14. Wachs, S., Junger, M., Sittichai, R.: Traditional, cyber and combined bullying roles: differences in risky online and offline activities. J. Soc. **5**(1), 109–135 (2015)
15. Zych, I., Ortega-Ruiz, R., Del Rey, R.: Systematic review of theoretical studies on bullying and cyberbullying: facts, knowledge, prevention, and intervention. J. Aggress. Violent Behav. **23**, 1–21 (2015)

International Workshop on
Computational Linguistics

Anomaly Detection for Short Texts: Identifying Whether Your Chatbot Should Switch from Goal-Oriented Conversation to Chit-Chatting

Amir Bakarov[1,2(✉)], Vasiliy Yadrintsev[2,4], and Ilya Sochenkov[2,3]

[1] The National Research University Higher School of Economics, Moscow, Russia
amirbakarov@gmail.com
[2] Federal Research Center 'Computer Science and Control' of Russian Academy of Sciences, Moscow, Russia
vvyadrincev@gmail.com
[3] Skolkovo Institute of Science and Technology, Moscow, Russia
i.sochenkov@skoltech.ru
[4] Peoples' Friendship University of Russia (RUDN University), Moscow, Russia

Abstract. Goal-oriented conversational agents are systems able converse with humans using natural language to help them reach a certain goal. The number of goals (or domains) about which an agent could converse is limited, and one of the issues is to identify whether a user talks about the unknown domain (in order to report a misunderstanding or switch to chit-chatting mode). We argue that this issue could be resolved if we consider it as an anomaly detection task which is in a field of machine learning. The scientific community developed a broad range of methods for resolving this task, and their applicability to the short text data was never investigated before. The aim of this work is to compare performance of 6 different anomaly detection methods on Russian and English short texts modeling conversational utterances, proposing the first evaluation framework for this task. As a result of the study, we find out that a simple threshold for cosine similarity works better than other methods for both of the considered languages.

Keywords: Anomaly detection · Novelty detection
Conversational agent · Chatbot · Distributional semantics
Word embeddings

1 Introduction

The task of *anomaly detection* (also called *outlier detection*) is to find in a given set objects highly deviating from others. Such deviating objects are called *anomalies*, or *outliers* [1]. The task of anomaly detection is considered as a supervised machine learning task, and it is actually similar to the classification

D. A. Alexandrov et al. (Eds.): DTGS 2018, CCIS 859, pp. 289–298, 2018.
https://doi.org/10.1007/978-3-030-02846-6_23

task, but the primary difference is that in the former the number of positive (non-deviating) samples in the training set is dominant, while the number of negative samples (deviating samples, anomalies) is low (e.g. there could be 1% of anomalous examples in the training set). The anomaly detection task is usually confused with a *novelty detection task*. Actually, goal of that task is basically the same (to find the outlying objects), but the point is that there no anomalous objects in the training set, so the model is being trained only on one class of objects and learns to find objects highly deviating from the ones participating in the training [2]. This is why the task of novelty detection is also called *one-class classification task*.

Neither anomaly detection task, nor novelty detection task are not actually widespread in the natural language processing. We are aware of a certain amount of cases whether anomaly detection systems work under the hood of recommender systems or document classification models dealing with large documents [3]. However, in this work we propose another application of this task to NLP: we rely on it for textual data consisting of very short texts (1 or 2 sentences), considering the problem of automated intent classification in conversational agents.

Conversational agents (also called *dialog systems*) are systems that are able to converse with a human on a natural language, imitating dialogue with a real human being [4]. Usually taxonomy of conversational agents proposes two distinctive axis. On the first axis agents are distinct by amount of their word knowledge: there are the *open-domain bots* which could converse in an unlimited number of domains of human knowledge (sports, science, literature), and the *closed-domain bots*, which could support only one or two topics of conversion [5]. On the other axis, conversational agents are distinct by the purpose of their use: there are so-called *general conversation agents*, or *chatbots* which do not consider a certain goal of dialogue and can just chat about everything, and there are goal-oriented conversational agents that should help a user to reach a goal through a short conversation (for example, to order a pizza).

The goal-oriented agents are the main interest of business and industry nowadays since they help to automatize some human work (of a call center, for instance) or to propose a much more friendlier interface for certain complicated systems (for example, they can help searching though FAQ) [6]. Usually such agents are not limited with a single possible goal, so they support different goals in the same domain (or even in different ones): to order pizza, to reserve table in a pizza restaurant, and so on. However, extending an agent to multiple domains or multiple goals usually is not a hard task: it could be considered as a classification problem which is widely known in NLP and machine learning community and has been successfully resolved from different perspectives [7].

However, the main issue comes when one wants to extend the conversational agents to a chatbot, i.e. to implement both of the behavior models (a goal-oriented talk and a general talk). In this case, the agent should recognize whether a user wants to reach a certain goal or whether she wants just to talk about something, switching between these modes [8]. Reducing this task to the aforementioned classification problem possibly should not work since utterances used in a general talk could highly differ from each other (by domain, by style, by other things), and if we consider all general utterances as a single class, that will be a highly heterogeneous class.

We argue that in this case the issue could be considered as an anomaly detection (or even novelty detection) problem that we described in the start of this section. Actually, we have a number of homogenous objects, for which we can generate a train set (in our case homogenous objects are examples of utterances of a certain domain or goal), and we have objects which could have unlimited number of possible domains (utterances for a general talk), for which we are not able to generate a train set. Then, the first ones could be considered as normal objects, and the second ones could be considered as anomalous objects. We can try to train a model to distinguish the first ones from the second ones (inducing the anomaly detection task), or to try to find how much a new object deviates from the known ones (inducing the novelty detection task). We can even create a system able to work in a number of multiple domains or goals by using a separate model. The anomaly detection model would distinct general utterances from goal-oriented ones, and the second one (a simple multi-class classifier) could perform a classification to distinguish goal in a more narrow domain.

The idea is that such systems should help a user reach a certain goal if the utterance belongs to one of known domains (defined by the developers), and enable a chit-chatting mode or report misunderstanding (saying something like *I'm not able to help you with this question*) in the other case. Conversational agents usually recognize domain by comparing semantics of a new utterance with an already known semantics of each of the domains (that could be defined with keywords, for example). Semantics processing is usually performed with different semantic modeling approaches like distributional semantic models. Such models had a recent success in a broad range of various natural language processing tasks, and in this work we will also rely on distributional semantics to model the meaning of utterances and find the degree of similarity between pairs of them.

So, the main aim of our work is to try to apply different anomaly detection algorithm to the problem of detection the unknown (off-topic) conversational utterances. We argue that the task of anomaly detection was never considered before in a natural language community before from the perspective of conversational agents and short text data, so we consider our work as a first towards exploration of application of anomaly detection methods for short texts. So, our main contribution consists in comparing and evaluating different anomaly detection methods on a benchmark of short texts.

Another of our major contribution is in creation of a cross-lingual evaluation benchmark for this task. We propose it for two different languages, Russian and English. We crawl data from Web forums and manually annotate it to create the first datasets for off-topic anomaly detection for short texts. Moreover, in the Russian natural language processing community the task of anomaly detection (as well as the task of novelty detection) was never considered before, so we are first to introduce and investigate this task for Russian.

All in all, this study is organized as follows. In Sect. 2 we put our paper in the context of previous works. In Sect. 3 we extensively describe our dataset, while in Sect. 4 we describe the setup of our experiments. In Sect. 5 we propose the obtained results and a discussion on them. Section 6 concludes the paper.

2   Related Work

The roots of the task of anomaly detection goes back to the 19th century [9], when this task was firstly formulated. Nowadays the scientific community is aware of a broad range of methods for anomaly detection, like *One Class SVM* or *Isolation Forest*; a survey of existing anomaly detection methods goes out of scope of this work, and an interested reader could see a survey of modern methods of anomaly detection by Chandola [1] or a survey of novelty detection techniques [10].

Being a very mainstream problem for different fields of machine learning, the anomaly detection task is rarely has been applied to different natural language processing issue. We are aware only of certain works that rely on detection of anomalies for textual data. Baker et al. was first to propose such task, considering novelty detection from the perspective of topic detection and tracking [11]. The first extensive work on anomaly detection for textual data considered document classification through One Class SVM [12], and then it was extended by Guthrie, who deal with an issue of detection of documents with unusual genre or sentiment in a document collection [13]. Later, Kumaraswamy et al. explored importance of domain knowledge provided in first-order logic in the task of anomaly detection for textual data [14]. In 2016, Camacho-Collados proposed an outlier detection in word sets as an evaluation benchmark for word embeddings [15], while Pande and Prohuja were first to investigate application of word embeddings to the task of anomaly detection [16].

So, mostly works on anomaly detection for textual data were performed as a part of document classification task, considering processing of linguistic or stylometric features (like average word length) and sparse vectors. The main difference of our work is that we process short texts and sentences, relying on compositional textual representations obtained as a function of dense vectors produced by distributional semantic models. Additionally, all studies that we mentioned considered only English data, but we are not aware of any work related to anomaly detection for Russian.

In other words, our work could be considered the first towards multiple different scopes.

3   Short Text Anomaly Detection Dataset

We are not aware of any suitable dataset for the anomaly detection task for short texts, so this study presents such a dataset. We suggest that it will help other researchers to evaluate different techniques of anomaly detection for a similar task. Our dataset consists of English and Russian parts, and could be called cross-lingual.

Since we resolve the task of anomaly detection from the scope of a conversational agent, we made our dataset consist of real conversational utterances. We considered Web forums as the best source of data because usually user messages there have a conversational style and presented as short texts of 1–2 sentences. Moreover, posts on Web forums are taxonomically separated for different domains (in other words, the range of covered topics is wide), so we could propose a multi-domain analysis with such data.

To make this dataset, we crawled two collections of posts from the most popular Web forums in each language: Dvach in Russian (https://2ch.hk) and Reddit in English (https://www.reddit.com)[1].

Each part consisted of 11 domains with different topics consisting of 100 posts each in which 10 were homogeneous (so each posts in each domain was topically related to all other objects in the same domain) and 1 was heterogeneous (the posts were not necessary to be topically related to other objects in this domain). In other words, 11 homogeneous domains modeled the known class and heterogeneous domain modeled the anomalous (or novel) class.

We used already defined Web forum taxonomy of domains to assess posts with domains (each post could have only 1 domain). To create assessments for English, we used a subreddit structure of Reddit which is presented as a pool of topical sub-forums dedicated to discussions on a certain topic. We sampled 10 different subreddits trying to pick the most diverse domain as possible, and then sampled 100 posts from each one:

- **r/science** (discussions on news in science and technology)
- **r/politics** (discussions on political news and events)
- **r/askhistorians** (discussions on a historical science)
- **r/space** (discussions on news related to space and astronomy)
- **r/minecraft** (discussions on a Minecraft video game)
- **r/sex** (discussions on sexual activities)
- **r/guns** (discussions on guns and pistols)
- **r/food** (discussions on food and cooking)
- **r/music** (discussions on different artists, genres and news of music industry)
- **r/motorcycles** (discussions on motorcycles)

To create the anomalous class we used the subreddit with jokes (**r/jokes**) which pretend to contain short texts not limited with a single domain (so there could be jokes about politics, about school, and so on).

As for Russian part, we used a hierarchical taxonomy of Dvach, which proposes different subforums (dubbed "boards") split into threads of discussion of more narrow topics. According to the aforementioned methodology, we picked threads that have the most diverse domains by our opinion:

- **Greek Literature** (discussions on literature of ancient Greece)

---

[1] Actually, the collection of Reddit posts is based on an already crawled corpus available at https://github.com/linanqiu/reddit-dataset.

- **Borussia Dortmund** (discussions on Borussia Dortmund football club)
- **Coffee** (discussions on coffee)
- **Java** (discussions on Java programming language)
- **Fountain Pens** (discussions on fountain pens)
- **Bread Bakery** (discussions on bread bakery)
- **Hairstyles** (discussions on hairstyles)
- **Keyboards** (discussions on computer keyboards)
- **Higher School of Economics** (discussions on the National Research University Higher School of Economics)
- **macOS** (discussions on Macintosh Operating System)

To create the anomalous class we used randomly sampled posts from /b board which does not limit to a single topic of discussion and allows conversations on every possible topic.

We also asked three bilingual Russian-English volunteers to check the accuracy of the automated assessments considering that each of the posts actually belongs to the proposed domain. If the assessor marked certain domain assessment as incorrect, we re-sampled the post from this domain, and re-checked it with all three assessors. We have done this iteratively until all 100 posts in each 10 domains were considered as correct by all assessors. All in all, each part of our cross-lingual datasets consisted of 1100 posts, so the whole amount of posts in our dataset was 2200.

4   Experimental Setting

In this work we testes applicability of the following techniques of anomaly detection (they were mentioned but not compared by performance at [10]):

1. **One-Class Support Vector Machine**. Draws a soft boundary on training objects, considering all objects falling outside the boundary as anomalous [12].
2. **Isolation Forest**. Isolates objects of the dataset by sampling features for which a threshold value would be randomly selected between the maximum and minimum values of that feature. Objects falling out of the threshold would considered as anomalies [17].
3. **Local Outlier Factor**. Computes the local density deviation on every object of the dataset, considering as anomalies samples that have a substantially lower density than their neighbors [18].
4. **Threshold for Standard Deviation of Classifier's Predictions**. The idea is to train a multi-class classifier (we used a logistic regression in this study) on "normal classes", and then for each new object compute a standard derivation of probabilities of classifier's predictions for classes. If standard derivation will be lower than the threshold (should be defined in advance), the object will be labeled as anomalous.

5. **Threshold for Distance to Topical Keywords**. The idea is very similar to the previous method, but here we creates a set of references (bags of keywords) for each class of training data. Each set of references is generated as a set of keywords reporting the topic of a set of short texts through a topic modeling technique (we used LDA in this study [19]). Then for each new object we should compute distance (we used cosine measure) between new object and every reference; if the distance *to each* reference will be lower than the threshold, the object will be labeled as anomalous.

6. **Threshold for Reconstruction Error**. The idea is to train an autoencoder [20] on training data and for each new object compute a reconstruction error between regression target and actual value. The objects which error will be higher than the threshold would be considered as anomalies.

In our experiments the data was lemmatized and cleared from stop-words using NLTK stopword lists [21]. As a morphological analyzer, we used *pymorphy2* [22] for Russian and UDPipe [23] for English.

To obtain representations of the short texts (conversational utterances) we used the averaged vectors of all the words in the sentence, considering this as the most effective and robust approach for obtaining compositional distributional representations (out-of-the-vocabulary words were dropped) [24]. To obtain the word embeddings we used two Word2Vec models for each language [25]: one trained on the Russian data of Dvach [26] and one trained on the English news corpus. For each method we tuned the best hyperparameters on each dataset which were obtained by grid search; code, datasets and links to models are available at our GitHub[2].

5   Results

Table 1. Performance of each of the compared methods on each dataset measured in accuracy.

	Dvach	Reddit
One Class SVM	0.5	0.53
Isolation Forest	0.47	0.47
Local Outlier Factor	0.47	0.47
Threshold for Standard Deviation	**0.68**	**0.71**
Threshold for Distance to Keywords	0.5	0.5
Threshold for Reconstruction Error	0.5	0.5

For evaluation we used the whole anomalous class and 100 posts from shuffled data of other classes, the remaining amount (900 posts of 10 domains) was used

---

[2] https://github.com/bakarov/conversational-anomaly.

for training the models. Quantitative results of the comparison are presented in Table 1 where we show accuracy of work of each of compared methods on each dataset (we use this measure since we have an equal class balance, and we actually are not interested in a high precision or recall particularly, and just want to obtain a general performance score). It is observable that the most simple technique which is a threshold for semantic distance had the highest score. We could explain this by the fact that other methods are not capable of working with high-dimensional data whether it is unknown which vector components could actually be significant.

To more properly analyze the obtained results, we made a graphical representation through a method of *t-SNE* [27]. We projected the whole dataset in a two-dimensional space, marking anomalies with red-color points and normal objects with black-color points. These visualizations illustrate how well each methods works, comparing the picture with a gold-standard topology propose at the leftmost picture (Fig. 1).

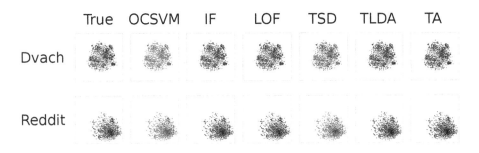

Fig. 1. Topology of anomalous (red) and normal data (black) in a two-dimensional space for both datasets. The leftmost column with pictures is a gold-standard topology that should be obtained, others illustrate topology predicted by an each method. (Color figure online)

The figures suggest that it is hard to visually distinguish separate clusters of classes (may be only just a few ones), and anomalous data is strongly mixed with normal data in both datasets. We think that such fact could be an explanation for low results of all of the employed methods. However, it is visually observable that the autoencoder marked data that is far from the centroid as anomalous, so it can be concluded that it could work better whether the normal classes and anomalous classes would not be so strongly mixed.

6   Conclusions

In this study we investigated the applicability of mainstream methods for anomaly and novelty detection to the data consisting of short texts, and created two publicly available datasets for off-topic anomaly detection based on data crawled from Russian and English web forums. We compared different techniques for anomaly detection on these datasets, concluding that using threshold on SD of predictions of a metric classifier is the most efficient method on our datasets.

The proposed survey and obtained results could help researchers and industrial teams to improve their results in making the most efficient architecture of a conversational agent. Another important contribution is that out work is the first towards the task of anomaly detection applied to textual data in Russian.

We think that obtained results could be called interesting and worth further investigation. In the future we plan to extend this work, considering different approaches to distributional compositionality. We also want to more properly explore impact of different features of used word embeddings model (window size, training algorithm, pick of a corpus, and so on), proposing the task of anomaly detection as an extrinsic evaluation benchmark for word embeddings. Extending the dataset to other languages (especially the low-resource ones, like Chuvash) also goes to our plans.

Acknowledgements. We thank three anonymous reviewers for helpful and attentive reviews. We also thank our colleague, Andrey Kutuzov, for productive discussion on this paper.

This study was supported by the Ministry of Education and Science of the Russian Federation (grant 14.756.31.0001) and partially funded by RFBR according to the research project №15-29-06031.

References

1. Chandola, V., Banerjee, A., Kumar, V.: Anomaly detection: a survey. ACM Comput. Surv. **41**, 15:1–15:58 (2009). https://doi.org/10.1007/978-1-4899-7502-7_912-1
2. Markou, M., Singh, S.: Novelty detection: a review-part 1: statistical approaches. Signal Process. **83**(12), 2481–2497 (2003). https://doi.org/10.1016/j.sigpro.2003.07.018
3. Guthrie, D., Guthrie, L., Allison, B., Wilks, Y.: Unsupervised anomaly detection. In: IJCAI, pp. 1624–1628 (2007)
4. Lester, J., Branting, K., Mott, B.: Conversational agents. In: The Practical Handbook of Internet Computing, pp. 220–240 (2004)
5. Chen, H., Liu, X., Yin, D., Tang, J.: A survey on dialogue systems: recent advances and new frontiers. arXiv preprint arXiv:1711.01731 (2017). https://doi.org/10.1145/3166054.3166058
6. Cui, L., Huang, S., Wei, F., Tan, C., Duan, C., Zhou, M.: Superagent: a customer service chatbot for e-commerce websites. In: Proceedings of ACL 2017, System Demonstrations, pp. 97–102 (2017). https://doi.org/10.18653/v1/P17-4017
7. Venkatesh, A., et al.: On evaluating and comparing conversational agents. arXiv preprint arXiv:1801.03625 (2018)
8. Mathur, V., Singh, A.: The rapidly changing landscape of conversational agents. arXiv preprint arXiv:1803.08419 (2018)
9. Edgeworth, F.: XLI. on discordant observations. Lond. Edinb. Dublin Philos. Mag. J. Sci. **23**(143), 364–375 (1887). https://doi.org/10.1080/14786448708628471

10. Pimentel, M.A., Clifton, D.A., Clifton, L., Tarassenko, L.: A review of novelty detection. Signal Process. **99**, 215–249 (2014). https://doi.org/10.1016/j.sigpro.2013.12.026

11. Baker, L.D., Hofmann, T., McCallum, A., Yang, Y.: A hierarchical probabilistic model for novelty detection in text. In: Proceedings of International Conference on Machine Learning (1999)

12. Manevitz, L.M., Yousef, M.: One-class SVMS for document classification. J. Mach. Learn. Res. **2**, 139–154 (2001)

13. Guthrie, D.: Unsupervised Detection of Anomalous Text. Ph.D. thesis, Citeseer (2008)

14. Kumaraswamy, R., Wazalwar, A., Khot, T., Shavlik, J.W., Natarajan, S.: Anomaly detection in text: the value of domain knowledge. In: FLAIRS Conference, pp. 225–228 (2015)

15. Camacho-Collados, J., Navigli, R.: Find the word that does not belong: a framework for an intrinsic evaluation of word vector representations. In: Proceedings of the 1st Workshop on Evaluating Vector-Space Representations for NLP, pp. 43–50 (2016). https://doi.org/10.18653/v1/W16-2508

16. Pande, A., Ahuja, V.: WEAC: word embeddings for anomaly classification from event logs. In: 2017 IEEE International Conference on Big Data (Big Data), pp. 1095–1100. IEEE (2017). https://doi.org/10.1109/BigData.2017.8258034

17. Liu, F.T., Ting, K.M., Zhou, Z.H.: Isolation forest. In: Eighth IEEE International Conference on Data Mining, 2008. ICDM 2008, pp. 413–422. IEEE (2008). https://doi.org/10.1109/ICDM.2008.17

18. Breunig, M.M., Kriegel, H.P., Ng, R.T., Sander, J.: SIGMOD Conference on New York. LOF: Identifying Density-Based Local Outliers

19. Blei, D.M., Ng, A.Y., Jordan, M.I.: Latent dirichlet allocation. J. Mach. Learn. Res. **3**, 993–1022 (2003)

20. Ng, A.: Sparse autoencoder. CS294A Lect. Notes **72**, 1–19 (2011)

21. Bird, S.: NLTK: the natural language toolkit. In: Proceedings of the COLING/ACL on Interactive Presentation Sessions, pp. 69–72. Association for Computational Linguistics (2006). https://doi.org/10.3115/1225403.1225421

22. Korobov, M.: Morphological analyzer and generator for russian and ukrainian languages. In: Khachay, M.Y., Konstantinova, N., Panchenko, A., Ignatov, D.I., Labunets, V.G. (eds.) AIST 2015. CCIS, vol. 542, pp. 320–332. Springer, Cham (2015). https://doi.org/10.1007/978-3-319-26123-2_31

23. Straka, M., Hajic, J., Straková, J.: Udpipe: trainable pipeline for processing conll-u files performing tokenization, morphological analysis, pos tagging and parsing. In: LREC (2016)

24. Li, B., et al.: Investigating different syntactic context types and context representations for learning word embeddings. In: Proceedings of the 2017 Conference on Empirical Methods in Natural Language Processing, pp. 2421–2431 (2017). https://doi.org/10.18653/v1/D17-1257

25. Mikolov, T., Sutskever, I., Chen, K., Corrado, G.S., Dean, J.: Distributed representations of words and phrases and their compositionality. In: Advances in Neural Information Processing Systems, pp. 3111–3119 (2013)

26. Bakarov, A., Gureenkova, O.: Automated detection of non-relevant posts on the Russian imageboard "2ch": importance of the choice of word representations. In: van der Aalst, W.M.P. (ed.) AIST 2017. LNCS, vol. 10716, pp. 16–21. Springer, Cham (2018). https://doi.org/10.1007/978-3-319-73013-4_2

27. Van der Maaten, L., Hinton, G.: Visualizing data using t-SNE. J. Mach. Learn. Res. **9**(2579–2605), 85 (2008)

Emotional Waves of a Plot in Literary Texts: New Approaches for Investigation of the Dynamics in Digital Culture

Gregory Martynenko[1] and Tatiana Sherstinova[1,2(✉)]

[1] Saint Petersburg State University, Universitetskaya nab. 11, 199034 St. Petersburg, Russia
{g.martynenko,t.sherstinova}@spbu.ru
[2] National Research University Higher School of Economics, Saint-Petersburg, Russia

Abstract. Digital technologies provide new opportunities for the study of objects of cultural heritage. The paper deals with investigation of the dynamics in literary and musical texts. It is hypothesized that, from a linguistic point of view, it is not by accident that the action in text develops from the beginning (the exposure) through the introduction to the climax, and from the climax to the denouement, but it always has a certain tendency, which can be visualized. In the given research three 'small genres' are being investigated: Russian short stories, Russian classical sonnets, and classical Russian romances which belong to a hybrid genre of both musical and verbal nature. Generalized profiles of the plot development were made by means of statistical time series method, but with different parameters for different genres. Thus, literary texts were analysed based on measurement of sentence length, poetry texts were measured by stress index, whereas romances were measured both by poetry stress index and musical pitch/duration index. The other variables related to plot development may be used as well. The dynamics of each genre is visualized by means of curves resembling the 'line of beauty' proposed by William Hogarth. In conclusion, the received results are compared with dynamic contours obtained by applying sentiment analysis to a big data collection of texts belonging to world classical literature. The obtained results testify that there exist some universal regularities in text and plot generation, which may be revealed independently to research methodology.

Keywords: Digital culture · Russian cultural heritage · Narrative · Literature
Short story · Poetry · Romance · Plot · Dynamics · Visualization · Time series
Interdisciplinary research

1 Introduction

Since the 1960-ies, the notion of narrative became the main subject of literary studies, and the theory of narrative (narratology) became an important component of the theory of literature that studies dynamic regularities in literary texts. The Oxford scientist Jonathan Culler believes that people have an inherent fondness to narration and perception of stories [2, pp. 95–97]. Culler connects this human peculiarity with an inherent

D. A. Alexandrov et al. (Eds.): DTGS 2018, CCIS 859, pp. 299–309, 2018.
https://doi.org/10.1007/978-3-030-02846-6_24

predisposition of people to narratives of different kinds (i.e., narrative competence)—by analogy with a linguistic competence.

The way of narration is structured by the plot, storyline and composition—which are the central categories of the theory of literature. No matter how these terms are interpreted in different scientific schools, they share a special kinship with each other as they all are related to the development of action in narrative texts. Every storyline has a beginning, a midpoint and an end, i.e. an exposure (introduction), a development of action, a culmination (climax) and a denouement. Many literary critics consider the plot (narrative arc) as a scheme of events, whereas the storyline is its concrete implementation. As for the composition, it refers to the order of dynamic components of the plot [13, pp. 393–394].

In linguistics, the dynamic aspect of text (discourse) occupies a significantly smaller place than in literary criticism. Most often, it is associated with frequency characteristics of text variables. Here, we will remain within the framework of this linguistic paradigm, but we will connect statistical characteristics with the development of narration in text. Our main task is to reveal statistical regularities in the dynamics of linguistic variables and to visualize these regularities using graphical profiles, representing these patterns in visual form as the combinations of falls and growth.

In this paper, the development of action in texts is viewed through the prism of statistical technique, which uses time series method. It was hypothesized that, from a linguistic point of view, it is not by accident that the action in text develops, but it always has a certain tendency, which can be visualized. We have investigated and compared three different genres of texts—short stories, sonnets and Russian classical romance, which belong to a hybrid genre of both musical and verbal nature. All these texts belong to 'small forms', in which it seems to be easier to follow the trend of their plots.

2 The Dynamics of Plot Development in Short Stories by Anton Chekhov

In the book [8], the development of the action in short stories by four famous Russian writers—Anton Chekhov, Leonid Andreev, Alexander Kuprin and Ivan Bunin—are considered. In this paper, we will show the dynamics of five Chekhov's stories basing on measurement of sentence length.

Let us suppose that some stylistic variables (for example, sentence length), following the development of the plot, behave not accidentally, but obey some hidden patterns that reflect the movement from the beginning to the climax and from the climax to the denouement. For example, as practice shows, short sentences are usually associated with 'the increase of energy in text', i.e. the behavior of characters becomes more emotional and energetic. This is reflected not only in characters' speech, but also in the author's speech. It is no coincidence that, if the writer uses very short sentences, the critics often call such works as 'neurasthenic' prose [14]. This may be due not only to the emotional background of the literary text in concern, but also due to the social upsurge and acceleration of social life. For example, they say about 'revolutionary prose'. Revolutionaries are usually persons of very few words. Rather, they are people of decisive action. As for 'heavy prose' consisting of long sentences, it usually refers to the quiet, unhurried parts of texts. It may also characterize some texts in the whole, which are usually descriptive.

For investigating the text dynamics, we used two variables: sentence length and paragraph size. The experiment consisted of several steps:

1. It was hypothesized that the narrative depends on both sentence length and paragraph length. For example, one can expect that in descriptive parts of narration the length of sentences and paragraphs will be, on average, longer, whereas in action parts—shorter. Moreover, the more emotional the fragment is, the more agitated becomes the narrative, and the more lapidary its structures are. As for the descriptive and 'philosophical' text fragments related with thoughts and reflections, they are typically rather lengthy.
2. In each story the length of each sentence and each paragraph were measured.
3. Further, the linear sequence of sentences and paragraphs of each story, regardless of its general size, was divided into 10 equal parts, and in each of them the average value of the corresponding variable is calculated. This allows us to compare the dynamic structures of stories of different sizes.
4. For each variable, a time series was built, in which the ordinal number of some text segment stands for an argument, and the average length of paragraph or sentence in this segment is a dependent variable.
5. The obtained mean values in each series are smoothed by one of the variants of moving averages method.
6. The resulted sequence of smoothed average values were tested for a certain trend tendency.
7. The dynamic profiles were built for each story, for each author and for the corpus as a whole, reflecting changes in variables as the plot develops from the introduction towards the central conflict (culmination) and then to final denouement.

It is difficult to expect that such dynamic profiles for different short stories will be identical. In Table 1 the correspondent values for five stories by Anton Chekhov are shown [8].

Table 1. Time series of the mean length of sentences in five short stories by Anton Chekhov

	1	2	3	4	5	6	7	8	9	10
'An upheaval'	14.4	13.8	13.3	13.0	13.4	13.9	14.5	14.2	13.5	12.6
'An Actor's End'	19.4	16.8	14.7	13.0	12.7	12.9	14.0	14.6	14.9	14.9
'Panic fears'	30.1	25.4	21.0	17.5	15.3	14.5	13.7	13.2	12.0	11.2
'Overdoing it'	16.3	15.0	14.2	12.1	12.2	12.4	14.0	15.0	13.2	10.9
'Rothschild's fiddle'	19.8	18.7	17.5	16.5	15.4	15.6	16.1	17.3	18.7	19.9
Mean	20.0	17.9	16.1	14.4	13.8	13.9	14.5	14.9	14.5	13.9
Coeff. var.	0.28	0.24	0.19	0.14	0.11	0.08	0.08	0.12	0.18	0.22

Let us discuss the results obtained.

Among the five stories, only two have the same geometry—'*An upheaval*' (rus. '*Perepolokh*') and '*Overdoing it*' (rus. '*Peresolil*'). It is curious that both stories names are 'dynamic', connected with some kind of acute situation, eventual 'over-the-top'. However, if to compare all the graphs, they have one thing in common: they all have a descending geometry in the initial (exposure) section. It means that Chekhov begins his narrative approximately in the same manner. The local minimum point, if to use the synergistic terminology, is a bifurcation point in which there is an alternative to choosing further direction of movement: it may proceed to decrease or change its way for an ascending mode. In four stories, Chekhov chooses an upward movement and only in '*Overdoing it*' we observe a downward movement. In the next step, that is, going on forth to the denouement, we have two more bifurcation points, in which the descending-ascending branching is also observed—three downwards and one upward. The generalized scheme of these Chekhov's stories from the point of view of the dynamics of sentence length is shown in Fig. 1 [8], and that for the intertextual coefficient of variation of the mean sentence length is given in Fig. 2 [8].

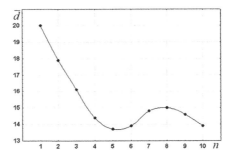

Fig. 1. Generalized dynamic profile of five short stories by Anton Chekhov

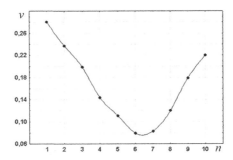

Fig. 2. Generalized time series of the intertextual coefficient of variation of the mean length of sentences in five short stories by Anton Chekhov

It is seen that in the second case even more evident and contrasting dynamics is observed. Here, the dynamic series has an evident U-shape with a local minimum around the central zone of the series, that is, in the area of mirror symmetry. The difference between the two edge values and the minimum value is quite large. This means that the maximum instability of the series is localized in beginnings and denouements of stories, whereas its minimum takes place in the climax. This can be understood in such a way that in the initial and final parts of Chekhov's stories both very short and very long sentences can be used. The first case corresponds to the 'easy' beginnings or endings, and the second case refers to the heavy, massive, and lengthy ones.

As for the shape of the generalized dynamic profile measured for sentence length, its resulting trend has something in common with the ideas of William Hogarth (1697–1764), an English artist and art historian of the 18th century who claimed that he uncovered the secret, with the help of which the Greeks succeeded in surpassing other peoples in arts. This secret he saw in following a wavy, snake-like or spiral line, which he called 'the line of beauty' [4]. This line consists of two bends directed in different directions in the form of the letter '*S*', thus looking like an incomplete clothoid (Cornu spiral).

Later, Hogarth's ideas were continued in the aesthetics of *serpentine line*, called *Figura serpantinata*. This idea was extended to sculptural and painting compositions, in which its author (painter or sculptor) figured the human body and other artistic objects the form of a spiral twist in order to achieve lightness, dynamics, and gracefulness of the image. The classic examples of Hogarth's 'line of beauty' and similar serpentine lines are Leonardo da Vinci's 'Leda and the Swan' and Maurits Escher's 'Whirlpools'.

3 The Dynamics of Russian Classical Sonnet

3.1 Sonnet as a Genre

Sonnet is the most common genre among the rigid forms of poetry, which are the strictly canonized combinations of stanza forms. In Europe, these rigid forms (sonnet, ballad, rondo and many others) appeared in medieval Romance poetry [12].

Sonnet organization being ultimately formalized orders and disciplines the elements of poetic creativity, and creates favorable conditions for crystallization and concentration of poetic thought. In its classical Romance version, sonnet includes two quatrains and two subsequent tercets with a definite rhyme system within each stanza and between stanzas. Such organization of poetic text was designed to ensure the distribution of meaning in the sonnet 'space' on the way from its *central idea* (thesis, beginning) and its *accompanied idea* (antithesis) through their synthesis to the final resolution [3]. Of course, in sonnets, this distribution of meaning is not always respected. However, this regularity does act (sometimes latently), which is manifested by the behavior of sonnet structural characteristics, as it will be discussed below.

In this paper, sonnet dynamics will be shown on two famous collections of sonnets —Konstantin Balmont's collection 'Sonnets of the Sun, Honey, and the Moon' published in 1917 [1] and Igor Severyanin's collection 'Medallions' publishes in 1934 [11].

3.2   The Dynamics of Stress Index

First, we will use the parameter called the *stress index* (*SI*), which is the ratio of stressed syllables to all syllables. To built time series for sonnets, we have to measure it in each line. The number of stress syllables in a line even within a rigid poetry forms varies quite strongly, though theoretically it should be constant. Moreover, the number of syllables in lines slightly varies too, although according to the canon, it should be constant, too.

Empirical data for the dynamics of stress index in sonnets by Balmont and Severyanin are shown in Fig. 3 [7].

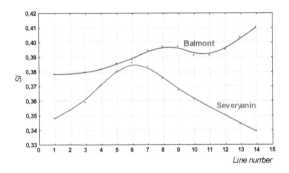

Fig. 3. The dynamics of stress index (*SI*, measured vertically) in the sonnets by Balmont and Severyanin

These curves have a very bizarre shape. Having essentially different starting positions, they increase rather quickly as the line number increases. Reaching the minimum of discrepancies somewhere in the area of the second quatrain, they start to scatter away, reaching the maximum of their differences in the last line of the poem. Both graphics, as well as for story dynamics, resemble the *S*-shaped line of beauty by William Hogarth. In this case, the figure, bounded by two individual lines, represents a semblance of a human figure, being viewed from right to left. If to put this figure 'on legs', i. e. to rotate it into a vertical position, then this analogy will be even more impressive. Two lines of beauty, interacting, are a good visual confirmation of the British art critic's idea.

3.3   The Dynamics of Word Length

Figure 4 presents information about the dynamics of words length. It is seen that in this case two curves form the geometric figure that resembles the one that we have just obtained for stress index profile of sonnets.

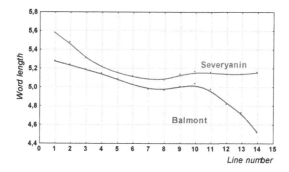

Fig. 4. The dynamics of word length in the sonnets by Balmont and Severyanin

The difference between these two figures consists only in the fact that their bends are oriented in opposite directions, that is, there exists an inversely proportional relationship between the variables under study.

Thus, we can draw the following conclusions [7]:

(1) A sonnet is a dynamic structure, the energy of which is realized mainly in the transition from one line to another (or from one stanza to another stanza).
(2) An evident pattern concerning the inverse proportionality of lexical and tonic structures was revealed.
(3) The proposed approach allows to disclose the latent dynamics in the meaning of sonnets.

4 The Dynamics of Russian Classical Romance

4.1 Parametrization of Verbal and Musical Components

As in the previous sections, here we apply the time series method. For the study of romance, we used two variables: (1) *the stress index* (i. e. the ratio of the number of stressed syllables to all syllables, *SI*), which is used for describing the poetic component of the romance, and (2) *pitch/duration index* (*PDI*) calculated as a multiplication of duration and pitch for each note, which is used for describing its musical component [6].

The choice of these variables is due to the hypothesis that there is a correlation between these indices and their components: in particular, the stress in poetic text is related to the duration of syllable sounding, as well as to the intensity of pronouncing of the correspondent vowel. A certain connection exists between the length and height of sound in vocal music.

When choosing a technique, we took into account the fact that poetry scholars usually study the ratio of meter and rhythm, as well as the type of poetic meter and its deviations. As for musical texts, the length and height for each note are given in explicit form by a musical notation. With this approach, the volume of sound is the variable that remains out of the study. Such deficiency may be partially compensated by appeal to sound recordings which integrally take into account all relevant characteristics, including the individual peculiarities of the performers.

In the obtained dynamic series, we took into account three components: general trend, cyclical fluctuations, and random variations. Trend and cyclical fluctuations were determined by means of the moving average method. We also used the method of least squares and the method of polynomial smoothing. The first is used to 'catch' cyclical fluctuations, while the second allows to draw the trend.

Given the complexity of interaction of musical and poetry components, we tried to find some element, which should lead to a formal description of dynamics in this hybrid text genre.

At the initial stage, we tried to find out whether there is a correlation between the rhythmic structure of the verse, which lies on the basis of the romance, and the corresponding dynamics of its vocal line.

In this research, the piano part was temporarily excluded from consideration, what is probably an excessive reduction. However, at the pilot stage of the study such an approach can be justified.

4.2   The Dynamics of Pitch/Duration Index

Now, let us proceed to analyze the internal temporal dynamics of Sergei Rachmaninoff's romances: *All was Taken from Me*, *The Fountain* (both published in 1906, lyric by Fyodor Tyutchev), and *What Happiness* (1912, lyric by Afanasy Fet) [9]. We will consider these romances in terms of the dependence of *PDI*-index on the ordinal number of measures. The correspondent dynamic profiles of these romances are shown in Figs. 5, 6 and 7 [6].

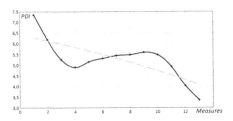

Fig. 5. Dynamic profile of the *PDI*-index in Rachmaninoff's romance '*All was Taken from Me*'

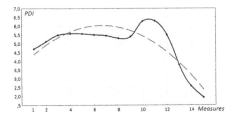

Fig. 6. Dynamic profile of the *PDI*-index in Rachmaninoff's romance '*The Fountain*'

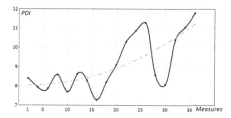

Fig. 7. Dynamic profile of the *PDI*-index in Rachmaninoff's romance '*What Happiness*'

The dynamic curves of the three romances (as well as the others) have a complex character. They combine a dynamic upward trend with cyclical oscillations. Cyclical fluctuations (a wave-like curve) are built by the least squares method, and the trend (a dotted curve) is determined by the exponential smoothing one. In the charts, we find several local extrema, as well as several inflection points.

The curves have the form of continuous chains of *S*-shaped Hogarth's curves forming a serpentine line. More than that, the length of each *S*-shaped segment on the average is about seven to eight musical measures (bars). This leads us to pose the question: is it possible to consider a linear segment of such length to be a kind of minimal prosody unit of a verbal text which is woven into a musical text? If so, it could be called a *musical poetic stanza*. The number of such phrases in romances, naturally, is related to its size. The longer the romance, the more phrases it has.

The mean value of this indicator for Rachmaninoff's songs is 7.52, SD is 2.01. The coefficient of variation (i.e., the ratio of standard deviation to the mean) is 18.9%, which indicates a very high concentration of empirical values near the mean. In the humanities, such stability is a very rare phenomenon. This is supported as well by the value of error, which equals to 0.788 (for 95% confidence level), that is, having even a small sample size, the average number of cycles in the period falls into the interval of (6.73–8.31). With 99% confidence level, this interval is somewhat widened (5.94–9.10). This 'hard' variant is very close to Miller's magic number of 7 ± 2, which allows us to be even more firmly established in our conclusion.

In most cases, the number of complete Hogarth's curves coincides with the number of stanzas in the poem underlying the romance. Thus, in the romance '*All was Taken from Me*' there are one stanza and one complete curve. In the romance '*The Fountain*' there is one stanza of eight lines, but Rachmaninoff actually divided it into two quatrains, apparently guided by the fact that he used only the first part of the poem by Tyutchev. Therefore, it was his intention to divide the remaining eight lines into two parts. As for the song '*What Happiness*', in original Tyutchev's poem there are three stanzas, and four Hogarth's curves. Rachmaninoff allowed himself to divide the first stanza into two 'small' *S*-shaped curves.

As a rule, a quarter of musical stanza corresponds to the verse (poetic line), its half —to a couplet, and the quatrain (stanza) corresponds to a complete musical stanza (or the full Hogarth's curve). In music, structurization is expressed more explicitly. This is not only the height and duration of notes, but also pauses (both inter-phrasal and inter-linear), as well as various labels indicating rhythm and tempo changes, accents, leagues, loudness indicators (forte, piano), etc. All this remains outside current discussion, although in the notation of Rachmaninoff's romances such information is presented in quantity.

The results obtained in this section make it possible to come to the following conclusions:

1. The poetic text, getting into the 'musical environment' undergoes substantial transformation. It is exposed to more relief, evident and clear structuring, and acquires additional energy.
2. The dynamics of romances is characterized by a combination of dynamic trend and cyclical fluctuations. Apparently, the same dynamics will be observed for all narrative structures.

5   Conclusion

As we could see, between the dynamic structures of the narrative, the sonnet and the romance, there is an evident parallelism, supported by a pattern of dynamic contours, generally ascending to Hogarth's line of beauty.

In recent years, in linguistics a new direction has been formed related to the study of the pragmatic aspect of language as a semiotic system, i. e. the shift in the interest of linguists from propositional values (facts, events) towards estimated values, from a proposition to modality, from dictum to modus. On the basis of studying evaluative resources of languages, a scientific discipline has emerged that deals with the study of opinions (*sentiment analysis*), revealing the attitude of communicants to the information circulating in society. Often, this attitude is associated with the emotional evaluation of messages, given that this estimate may change over time [5].

Within this direction, a very intriguing research discipline was born, called *hedonometrics* [10]. Having analysed the large text corpus, consisting of 1700 masterpieces of world literature with the use of multivariate statistical analysis methods, the scientists have found that the emotional trajectories of the entire set of texts may be reduced to 6 typical plots. This method of work allows to distinguish typical narrative arcs of world

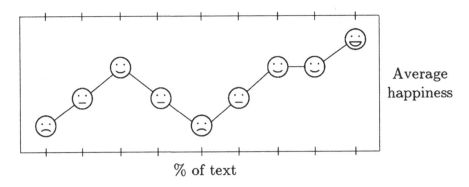

Fig. 8. Hedonometric analysis of literary texts and the plot of average happiness

literature. One of the most typical structures among them is a dynamic contour of the form shown in Fig. 8 [10].

Evidently, this figure resonates with the dynamic contours obtained above in the analysis of the dynamics of short story, sonnet and romance. The results presented in this paper testify that there exist some universal regularities in text and plot generation, which may be revealed independently to research methodology.

Acknowledgements. The research is supported by the Russian Foundation for Basic Research, project #17-29-09173 "The Russian language on the edge of radical historical changes: the study of language and style in prerevolutionary, revolutionary and post-revolutionary artistic prose by the methods of mathematical and computer linguistics (a corpus-based research on Russian short stories)".

References

1. Balmont, K.D.: Sonety Solnca, Meda i Luny [Sonnets of the Sun, Honey, and the Moon], V.V. Pashukanis edit. house, Petrograd (1917)
2. Culler, J.: Literary Theory: A Very Short Introduction. Oxford University Press, New York (1997). https://doi.org/10.1093/actrade/9780199691340.001.0001
3. Gasparov, M.L.: Sonnet. In: Kratkaja Literaturnaja Entsiklopedija [The Brief Literary Encyclopedia], vol. 7, p. 67. Prosveshhenie, Moscow (1972)
4. Hogarth, W.: The Analysis of Beauty: Written with a View of Fixing the Fluctuating Ideas of Taste, London (1753). Printed by J. Reeves for the author
5. Khokhlova, M.V.: Analiz tonal'nosti [Sentiment analysis]. In: Prikladnaja i komp'juternaja lingvistika [Applied and computational linguistics], LENAND, Moscow, pp. 245–258 (2016)
6. Martynenko, G.: Structural interaction of poetry and music components in songs by Sergei Rachmaninoff. In: Eismont, P., Konstantinova, N. (eds.) LMAC 2015. CCIS, vol. 561, pp. 127–139. Springer, Cham (2015). https://doi.org/10.1007/978-3-319-27498-0_11
7. Martynenko, G.Y.: Ritmika russkogo soneta po gorizontali i vertikali: Konstantin Bal'mont i Igor' Severianin [Rhythm of the Russian sonnet in horizontal and vertical dimensions: Konstantin Balmont and Igor Severyanin]. In: Jazyk i Metod: Russkij Jazyk v Lingvisticheskikh Issledovanijakh XXI Veka [Language and Methodology. The Russian Language in Linguistic Studies of the XXI-th Century]. Jagiellonian University, Krakow. Issue 4 (2018, in Print)
8. Martynenko, G.Y.: Vvedenije v chislovuju garmoniju teksta [The Introduction to Numeral Harmony of Text]. St. Petersburg State University, St. Petersburg (2009)
9. Rachmaninoff, S.: Romancy [Songs], vol. I–II. Muzyka [Music], Moscow (1989–1990)
10. Reagan, A.J., Mitchell, L., Kiley, D., Danforth, C.M., Dodds, P.S.: The emotional arcs of stories are dominated by six basic shapes. arXiv:1606.07772 [cs.CL] (2016). https://arxiv.org/pdf/1606.07772.pdf
11. Severyanin, I.: Medal'ony [Medallions]. Belgrade (1934)
12. Shengeli, G.: Tverdye formy [Solid forms]. In: Tehnika stiha [Technique of verse], pp. 288–297. Goslitizdat, Moscow (1960)
13. Slovar' literaturovedcheskikh terminov [The dictionary of literary terms]. Prosveshhenie, Moscow (1974)
14. Vinogradov, V.V.: Stilistika. Teorija pojeticheskoj rechi. Pojetika. [Stylistics. The theory of poetic speech. Poetics]. USSR Academy of Sciences, Moscow (1963)

Application of NLP Algorithms: Automatic Text Classifier Tool

Aleksandr Romanov$^{(\boxtimes)}$ ⓘ, Ekaterina Kozlova ⓘ,
and Konstantin Lomotin ⓘ

National Research University Higher School of Economics, Myasnitskaya Ulitsa,
101000 Moscow, Russian Federation
a.romanov@hse.ru, hse.kozlovaes@gmail.com,
ke.lomotin@gmail.com

Abstract. This research is dedicated to the design of a decision support system for categorization of scientific literature. The purpose of this work is to research possible ways to apply the machine learning algorithms to the automation of manual text categorization. The following stages are considered: preprocessing of raw data, word embedding, model selection, classification model, and software design. At the first stage, in collaboration with VINITI RAS, the training set of 200,000 Russian texts was formed. At the second stage, the word embedding model was justified as Word2 Vec vector representation from text matrix by "sum" convolution with dimensionality 1500. At the third stage, the quality of the classifiers was estimated, and the logistic regression algorithm with the highest F1 score (0.94) was selected. And at the final stage, the ATC (Automatic Text Classifier) application, which embeds the results obtained on the previous stages, was developed. The overall application structure was described. It consists of compact program modules that can be replaced or adapted to the incoming text and gain the most classification score.

Keywords: Decision support system · Supervised learning · Boosting
Support vector machine · Multilayer perceptron · Text analysis
Natural language processing · Decision tree

1 Introduction

In contemporary world conditions, many organizations encounter a problem of unstructured information streams. Electronic document management systems, search engines, and mail servers suffer from data that do not correspond to the subject area of the organizations. High-end IT-companies offer expensive complex solutions [1–3] which might be not accessible for small or low-budget organizations. Usually, these software products provide excess functionality for corporations with highly specialized field of activity. Moreover, once installed software requires qualified staff to be used properly.

This problem can be solved by the specialized software that is able to serve only narrow subject area and needs almost no tuning up. To create such adaptive software, machine learning algorithms usage can be highly effective. The program, designed in

© Springer Nature Switzerland AG 2018
D. A. Alexandrov et al. (Eds.): DTGS 2018, CCIS 859, pp. 310–323, 2018.

such a way, requires a number of training samples referring to the subject area. Narrow specialization allows developers to provide minimal user interface and encapsulate all the algorithms involved.

The first section of this paper contains some information about VINITI RAS, a short overview of used classification methods, as well as solutions for building vector representation of the text. Then, the results of the conducted experiments are presented and analyzed in order to select the model to embed it into the ATC tool. After training, the data, provided by VINITI, is analyzed.

In the second part of the research, the way to build such a system for the subject area, that covers the incoming stream of scientific publications from all over the world, is described. The system, which is based on the model selected in the first section, is able to perform text categorization according to the pre-defined set of topics.

After that, several directions for the future work are expounded, and then the conclusion is drawn.

2 Training Set Generation from VINITI RAS Abstracts Database

2.1 General Information

VINITI RAS (All-Russian Institute for Scientific and Technical Information of Russian Academy of Sciences) [4] is an institution aimed to collect, categorize and store technical and scientific publications from all over the world. Collected information can be sent by request, for example, from any research university or design office to introduce the results of the newest researches. Moreover, the institute itself performs numerous scientific researches. VINITI is a basic organization of the Commonwealth of Independent States for exchange of scientific information.

VINITI intends to involve modern approaches to process such information streams. Most sources of publications, such as eLIBRARY.RU [5], are accessed via the Internet. Vast stacks of papers were substituted with contemporary database servers. Now, the ATC tool is introduced to replace manual categorization of thousands of publications by using the program module that encapsulates the applied machine learning algorithms.

2.2 Objectives

In this task, it was proposed to build up a system which is able to generate a list of topics from two classification structures for the arbitrary text. The list of results has to be ordered by the probability of input text belonging to the corresponding topic. Thus, this task is highly similar to the development of the decision support system.

In this research, categorization was considered according to plain classification system named "SUBJ" or "Section code".

SUBJ includes 16 topics on the different areas of science:

1. e1 – COMPUTERS; ELECTRONICS;
2. e2 – ASTRONOMY;

3. e3 – BIOLOGY; MEDICAL SCIENCES;
4. e4 – GEOGRAPHY; GEOPHYSICS;
5. e5 – GEOLOGY; EARTH SCIENCES; MINES AND MINING INDUSTRY;
6. e7 – LIBRARY AND INFORMATION SCIENCES;
7. e8 – MATHEMATICS;
8. e9 – MACHINERY; ENGINEERING – INDUSTRIAL ENGINEERING;
9. f1 – METALLURGY;
10. f2 – ENGINEERING – MECHANICAL ENGINEERING;
    PHYSICS – MECHANICS;
11. f3 – ENVIRONMENTAL STUDIES; CONSERVATION;
12. f4 – TRANSPORTATION;
13. f5 – PHYSICS;
14. f7 – CHEMISTRY; ENGINEERING – CHEMICAL ENGINEERING;
15. f8 – ECONOMICS;
16. f9 – ENGINEERING – ELECTRICAL ENGINEERING; ENERGY.

Label "e8 – MATHEMATICS" was ignored due to the poor number of texts and complex markup elements used in VINITI to indicate formal expressions (e.g. formulas). In rare cases, one text may belong to several labels.

3   Overview of the Classification Methods

In general, classification task in machine learning was formulated as follows.

Given: training set of samples $\{(x_i, y_i)\}$ of cardinality M and K classes, where $1 \leq i \leq M$, x_i is a vector $[x_1, x_2, \ldots, x_N]$, is a sample vector with N features; y_i is a class label for i–th sample $(y_i \in \{y^1, y^2, \ldots, y^K\}$ – a set of all class labels). Consider h_θ as a parameterized model such that $h_\theta(x_i) = y_i$, and for any new x $h_\theta(x) = y$. Vector θ is a vector of model parameters. Thus, a classification task consists of three challenges:

1. Select a model that is able to fit the data.
2. Choose hyperparameters for the selected model.
3. Derive the parameters θ (using any optimization algorithm).

To obtain a working solution for this task, numerous algorithms were applied. Here is an overview of the methods used in this research.

3.1   Logistic Regression

Logistic Regression [6] is a classification model that multiplies all features of a vector x on parameters θ. Result of multiplication is passed in logistic function that gives value in a continuous range from 0 to 1. The answered class and its probability are based on this number. Class labels in this model can be only 0 or 1. Such algorithm as multinomial logistic regression is also applicable to the task of text categorization; however, in this study it is not considered due to the absence of its implementation in the used libraries.

$$h_\theta(x) = \frac{1}{1 + \exp(-(\theta_0 + \theta_1 * x_1 + \ldots + \theta_N * x_N))} \tag{1}$$

Logistic Regression can work with two classes. For this reason, the data were transformed with One vs All [7] algorithm. One target feature y with K class labels was split to K features $y_1, y_2, \ldots, y_K$ with $y_{ij} = 1$ when original $y_i = y_j$. Then the logistic regression models were trained on each $y_1, \ldots, y_k$ target features to train model h_θ^i to answer, if it is class y_i or not, and with what probability.

3.2 Naive Bayesian Classifier

Naive Bayesian Classifier [8] is a probabilistic classifier based on Bayes' theorem with the assumption of features independence. For this model probabilities of y_i, where $i = 1 \ldots k$ and conditional probabilities of class y_i, where given x_j, where $j = 1 \ldots N$, are computed. Then received probabilities are used to find class with maximum probability.

$$y = \text{argmax}_{i=1\ldots k}\left(p(y_i)\prod_{j=1}^{N}p(x_j \mid y_i)\right) \tag{2}$$

3.3 Support Vector Machine

Support Vector Machine (SVM) [9] is a machine learning algorithm that can be used both to classification and regression tasks. As well as Logistic Regression, it works only with binary classification. For that reason, One vs One [7] algorithm was used for preprocessing the data. It means that $\frac{k(k-1)}{2}$ classifiers were trained on pairs of classes y_i and y_j, where i, j in $1 \ldots k$, $i \neq j$. Class labels for this classifier must be from $\{-1, 1\}$.

$$h_\theta(x_i) = sign(\theta_0 + \theta_1 * x_{i1} + \ldots + \theta_N * x_{iN}) \tag{3}$$

Sometimes kernels $K(x_i, x)$ are used to improve results on linearly inseparable classes.

3.4 Artificial Neural Network

Artificial neural network (ANN) [10] is a computing system that corresponds to mammalian brains. It can perform many different tasks, and the classification is one of them. ANN are divided into different types such as: Recurrent, Feed-forward, Convolutional, Long short-term memory (LSTM), etc. Feed-forward ANN with from one to three hidden layers was explored. Implementations of more complicated structures are experimental and could not be embedded in ATC software.

3.5   Decision Tree

Decision Tree [11] is a decision support model that uses a tree-like graph where each leaf corresponds to an answer. In classification task decision tree leaf relates to a class. Trees have depth, splitting criterion and minimum amount of objects in a leaf as most important hyperparameters. They have tendency to overfitting [12]; for this reason, it is important to limit the depth and/ or a minimum amount of objects in a leaf.

3.6   Random Forest

Random Forest [13] is a machine learning algorithm consisting in the use of an ensemble of decision trees. Each tree casts a unit vote for a class label at input x, and the most popular one becomes the classifier's response. Usually, it has from five to several dozens of trees. Each tree is less deep and complicated, however, it gives a chance to avoid overfitting, and it leads to getting better results. For this classifier amount of trees, their depth and minimum amount of objects in a leaf are most important hyperparameters.

3.7   Boosting Algorithms

AdaBoost [14] is a machine learning meta-algorithm that builds a composition of the basic learning algorithms to improve their effectiveness. In this meta-algorithm, each following classifier is built on objects that are poorly classified by previous classifiers. Trees with maximum depth two and three were considered as basic learning algorithms.

Gradient boosting [15] is a machine learning meta-algorithm that uses particular method to optimize arbitrary differentiable loss function. It can be used for classification and regression tasks. In this research, it was used with same trees structures as for AdaBoost.

4   Text Vectorization with Word2Vec

The classification methods, listed above, work exclusively with the vector data representation. For this reason, it is required to present the contents of the text in natural language in numerical form of fixed dimension. Due to the small average length of the texts (about 80 words), we had to abandon the statistical approaches (TF-IDF, Bag-Of-Words). The most promising tool at the moment is the semantic model Word2Vec.

Word2Vec (W2V) [16–18] is a combination of two-layer neural networks that can analyze linguistic contexts of words in order to create word embedding. Then, each word corresponds to a vector in semantic space. Synonyms vectors are close that allows the classifier model to treat synonyms equally. Each word corresponds to a vector, and hence, the text corresponds to a matrix of dimension amount of words by amount of features in Word2Vec model. But all the classifiers need a fixed dimension vector for work; so, the data needs to be transformed. The approach with summing and searching for a maximum over all the words in the text is used. These approaches were applied for several dimensions of vectors Word2Vec.

5   Training Data Analysis

Due to the collaborative work with VINITI RAS, a set of short abstracts of scientific articles was obtained. The input data contains columns "title", "keywords", "text", and labels of all the classification systems assigned to the text by specialists. The title of the article, the key words, and the article text were combined and submitted for further processing together. The effect of each of these columns on the classification quality separately was not considered. Such an analysis may show interesting results, but is not the subject of this research. VINITI markup language includes a set of special characters for indicating font modifiers and ASCII symbols. It may also contain TeX formulas as well (Fig. 1). The initial task included the ability to process texts in English and in Russian as well. Nevertheless, experiments with English texts are not of interest due to the non-representative dataset.

```
id_publ title   ref_txt kw_list SUBJ    IPV RGNTI   eor
17.01-13Д.336   _ЁСтруктура модели стратегии противодействующей стороны _ЁНечеткое (лингвистическое)
моделирование все шире применяется во многих прикладных областях для решения трудноформализуемых задач.
Преимуществом этого метода является возможность эффективного анализа сложных систем, не поддающихся
точному математическому описанию. В данной работе предлагается использование лингвистического
моделирования для решения задачи оценивания стратегии._ёраг Лингвистическая модель представляет собой
нечеткое отношение следующего вида: $$_ёпо{LIN}=_ёbigcup_ёlimits^{k}_{i=1}(a_{1,j}_ёcap
a_{2,j}_ёcap_ёldots _ёcap a_{m,j})_ёto (b_{1,i}_ёcap b_{2,i}_ёcap_ёldots_ёcap b_{n,i}),$$где
$a_{i,j}_ёin A_{j}$, $j=1,_ёldots,m$; $i=1,_ёldots,k$; $k$~--- число правил; $A_{j}$~--- множество
значений входных лингвистических переменных, $b_{i,j}_ёin B_{1}$, $l=1,_ёldots,n$; $i=1,_ёldots,k$;
$B_{1}$~--- множество значений выходных лингвистических переменных. Значениям лингвистических переменных
$a_{i,j}_ёin A_{j}$ соответствуют нечеткие множества с функцией принадлежности $_ёmu_{a_{ij}}(X_{j})$, а
значениям лингвистических переменных $b_{i,j}_ёin B_{i}$~--- нечеткие подмножества с функцией
принадлежности $_ёmu_{b_{i,j}}(Y_{i})$ _ЁПенза\_Ёмоделирование\_Ёприкладное ПО е8  13Д 50.05.17   ###
```

Fig. 1.   The sample of VINITI markup.

The set of abstracts of Russian articles includes approximately 200,000 texts written by qualified experts in VINITI. Figure 2 depicts the distribution of the texts on the SUBJ topics.

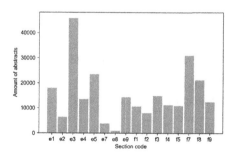

Fig. 2.   Distribution of the Russian abstracts on the SUBJ topics.

Apparently, the samples are unbalanced by categories (from 45934 in "e3" to 3858 in "e7"). On the one hand, it reflects the fact that Russian-speaking scientific community produces more publications on Biology and Chemistry. On the other hand, these topics contain similar lexis, and many abstracts belong to more than one branch of science. Hence, the distribution depends on the labels given to the texts by VINITI experts. Nevertheless, the presence of unbalanced classes obstructs the classification [19, 20]. One of the possible ways to handle it is aligning the amount of the abstracts to the maximum among all the classes. In previous researches, this approach did not lead to quality increase [21]; consequently, the unbalanced training set was used for the experiments. As a result, the ATC tool gained two advantages: firstly, the relatively short model training time; secondly, the model was prepared to the certain stream of publications with described distribution. This approach has also some drawbacks: the system became less robust to the changing diversity of incoming data.

Each topic is characterized by a certain set of keywords in conjunction with their context.

Average amount of words in the texts on each topic, written in Russian, is shown in Fig. 3. Totally, an average abstract has approximately 80 words.

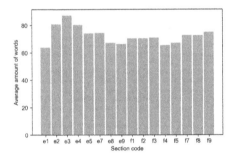

Fig. 3. Average amount of words in the Russian texts on each topic.

Topic "e3 – BIOLOGY" is not only the most numerous one. It contains the texts with the highest amount of words.

6   Data Preprocessing

Preprocessing of the texts includes the following steps

1. Removing of VINITI markup.
2. Removing of stop words and punctuation marks.
3. Normalization.

6.1   Removing of the VINITI Markup Elements

Raw sample texts encapsulate fragments of VINITI intrinsic markup alphabet. It allows VINITI information services to represent formulas, Greek letters and other special symbols as a plain text. Nonetheless, these elements are not valuable for classification.

After primary processing, the text still contains the names of variables in equations, special letters etc. For example, word "model" with context "θ" can indicate that the abstract describes the paper on machine learning, and the most appropriate topic is "e7".

6.2   Removing of Stop Words and Punctuation Marks

This step of preprocessing uses the predefined lists of stop words (pronouns, conjunctions, prepositions, etc.). The list of stop words was formed manually during the research using open dictionaries. Punctuation marks are described using regular expression mechanism. After that, the text contains only words associated with each other only through grammatical forms and context.

6.3   Normalization

The purpose of normalization can be referred to unifying words in different forms (e.g. verb in the Past with ending "–ed"). Usually, to perform this operation, one of two approaches is applied:

1. Stemming [22].
2. Lemmatization [23].

Stemming algorithms are used for cutting words endings and attempt to leave the word root.

Lemmatization is a procedure of actual transformation of a word into its normal form. The most of practical implementations of lemmatization use dictionaries. Some of them are able to predict a normal form for the unknown word.

In this work, PyMystem3 [24] lemmatization tool was used in the preprocessor module.

Eventually, after all permutations, the text represents so-called "Bag-Of-Words" – a set of words associated with each other only with context. If each form of the word is considered as a unique sequence, the overall amount of words in the processed text is much less than in the raw one.

7   Vector Representation Development

As it was justified above, Word2Vec tool was selected for vector representation of the training set generation. It converts a single word into vector with fixed number of elements that reflect the context of the word in the training set. Thus, each text can be represented as a matrix of M rows and N columns, where M is a number of words in the text, and N is a predefined amount of semantic features (elements of the vector) – Word2Vec hyperparameter.

Classifying model is able to process vectors, not matrices. In order to reduce text matrix dimension and form the vector with fixed number of elements that does not depend on the text size, two methods were used:

1. The maximum value from each column of text matrix selection (the "max" method).
2. The sum of all elements for each matrix column computing (the "sum" method).

For these methods, series of experiments with the most appropriate machine learning model were conducted.

8   Results

In order to estimate quality of the classifier, the training set was divided into the actual training set and the testing set for validation; a third part of the texts was used for testing. Both sets were represented as vectors with dimensionality of 50 formed with "sum" method. Results are presented in Table 1. To assess the classifier as a decision support system, the quality metrics were calculated for answers with the highest metrics: for one answer (the "One attempt" column), for two answers (the "Two attempts" column), and for three answers (the "Three attempts" column).

Logistic regression algorithm and artificial neural network showed highly similar F1-score (0.63–0.94). Computational complexity of ANN exceeds the computational complexity of logistic regression, because each formal neuron has a nonlinear transformation at the output, whereas to compute logistic regression, it is necessary to perform only one transformation (in both cases logistic function was used). Therefore, logistic regression is more applicable in the task of creation of the applied tool.

After that, some experiments with vector model parameters were conducted. Testing was performed for the vectors of 50, 100, 500, and 1500 elements, as well as for both considered methods of matrix convolution. The results are depicted in Fig. 4.

Quality of "max" vectors classification depends on the vector dimensionality more strongly than quality of "sum" vector categorization.

Accordingly, it follows that vectors with 500 features, created with "sum" method, are more satisfactory in terms of performance and quality.

9   ATC Tool Design

9.1   Architecture Design

The application is aimed to be used by people without a special education in machine learning, and its key features are:

1. Simplicity of the user interface.
2. Ability to operate VINITI text formats as well as plain text.
3. Easy deployment.

Table 1. Quality metrics for tested models

Model	Accuracy score	Precision score	Recall score	F1– score
One attempts				
Logistic regression	0.78	0.65	0.64	0.63
Artificial neural network	0.78	0.61	0.63	0.61
Decision tree	0.58	0.43	0.5	0.44
Random forest	0.73	0.58	0.58	0.57
Gradient boosting built from decision trees	0.77	0.61	0.65	0.61
AdaBoost built from decision trees	0.62	0.47	0.51	0.46
Naive Bayesian classifier	0.29	0.06	0.08	0.05
Support vector machine	0.25	0.58	0.06	0.03
Two attempts				
Logistic regression	0.91	0.89	0.88	0.88
Artificial neural network	0.92	0.87	0.9	0.88
Decision tree	0.67	0.57	0.62	0.57
Random forest	0.86	0.84	0.81	0.82
Gradient boosting built from decision trees	0.90	0.79	0.82	0.79
AdaBoost built from decision trees	0.81	0.73	0.78	0.74
Naive Bayesian classifier	0.42	0.14	0.13	0.09
Support vector machine	0.42	0.08	0.12	0.09
Three attempts				
Logistic regression	0.95	0.95	0.93	0.94
Artificial neural network	0.96	0.93	0.95	0.94
Decision tree	0.71	0.65	0.67	0.65
Random forest	0.90	0.90	0.88	0.89
Gradient boosting built from decision trees	0.95	0.86	0.87	0.86
AdaBoost built from decision trees	0.89	0.85	0.87	0.85
Naive Bayesian classifier	0.55	0.21	0.20	0.18
Support vector machine	0.53	0.15	0.19	0.16

Also, the application must be prone to wrong input data samples.

Due to the experimental nature of the product, the program architecture should provide an opportunity to introduce changes without modifying of other program components. This principle was applied in the module structure. The user can change the modules in a special setting window. The application dynamically loads the required module – basic functional unit, e.g. preprocessor for English abstracts in plain format, or classifier for Russian texts. The modules can be simply updated. All modules are inherited from "Module" class providing common interface and meta-information for its displaying in the module manager tool.

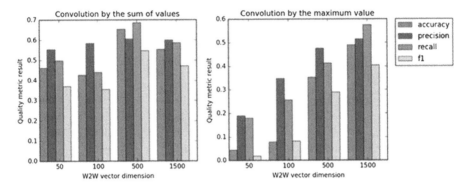

Fig. 4. The results of logistic regression testing with different vector model configurations.

The scheme of the general application structure is presented in the Fig. 5.

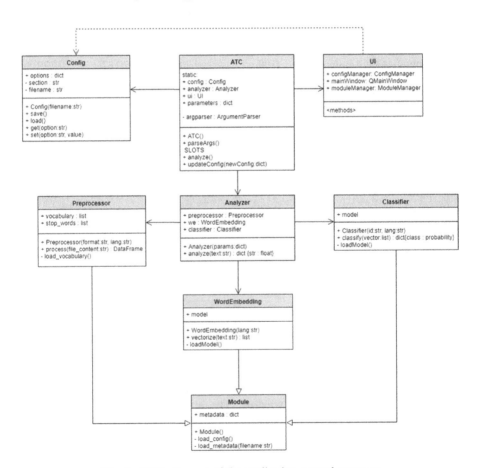

Fig. 5. UML diagram of the application general structure.

The main window in graphical user interface contains input and output fields, a menu for selecting the language and classification system, and a button for invocation of the setting window. The following features were added to simplify the communication with users:

1. Ability to select the font size.
2. Auto-recognizing of the typed text language.
3. Result export in VINITI database compatible format.

9.2   Implementation

To implement the application, Python programming language was used. Python is a high-end language to perform machine learning related researches with vast variety of optimized frameworks and rich built-in modules library.

For reducing the influence on the user's operating system environment, portable Python interpreter was used. The Python programming language is cross-platform; nevertheless, some libraries have their own requirements for the operating system. Non-proprietary WinPython v3.6.2 is delivered with following pre-installed libraries and frameworks:

1. SciKit-Learn (v.0.18.1).
2. Pandas (v.0.20.1).
3. NumPy (v.1.12.1).
4. NLTK (v.3.2.3).
5. PyQt (v.5.6.0).
6. PyMystem3 (v.0.1.5).

The latest version of the ATC tool requires 64-bit MS Windows operating system.

9.3   Performance Analysis

Classifier training takes significantly different amounts of time, but there are no fundamental differences when using trained models. Classification of one text by using a graphical interface takes about 0.5–0.7 s. In the MULTIDOC mode, processing of 1000 documents takes 25–27 s. The difference is due to the need to load the machine learning model which is stored in RAM during the processing.

10   Future Work

Modern natural language processing (NLP) algorithms are usually experimental and abstract. However, they use the newest achievements of machine learning scientific community. The purpose of the future research is to explore how high-end models are applicable in the considered task.

The following approaches and models are supposed to enhance classification quality:

1. LSTM neural networks [25].
2. Convolutional neural networks [26, 27].
3. Semantic networks [28].
4. ANN attention mechanisms [29].

11 Conclusion

In this paper, practical aspects of the applied NLP algorithms are considered. The choice of the word embedding model was justified – the most appropriate vector representation was derived with Word2Vec tool from text matrix by "sum" convolution with dimensionality 1500.

After that, the model selection and testing result were provided. The logistic regression algorithm gained the highest F1 score (0.94). In contradistinction to the artificial neural network with the similar quality, logistic regression requires only one nonlinear transformation; hence, it is more preferable in terms of performance.

The overall application structure was described. It consists of compact program modules that can be replaced or switched between themselves to adapt to the input text.

References

1. ABBYY Smart Classifier. https://www.abbyy.com/en–eu/smartclassifier
2. Nübel, R., et al.: Bilingual indexing for information retrieval with AUTINDEX. In: LREC Proceedings, pp. 1136–1143. ELDA, Las Palmas (2002)
3. Text Analytics. https://www.meaningcloud.com/solutions/text–analytics
4. All-Russian Institute for Scientific and Technical Information. http://www.viniti.ru/
5. eLIBRARY.RU. https://elibrary.ru
6. Hosmer, Jr., D., Lemeshow, S., Sturdivant, R.: Applied Logistic Regression, 2nd edn. Wiley, New York (2013). https://doi.org/10.1002/sim.4780100718
7. Rifkin, R., Klautau, A.: In defense of one-vs-all classification. J. Mach. Learn. Res. 5(1), 101–141 (2004)
8. Murphy, K.: Naive Bayes classifiers. In: Lectures, pp. 1–5. University of British Columbia, Vancouver (2006)
9. Tong, S., Koller, D.: Support vector machine active learning with applications to text classification. J. Mach. Learn. Res. 2(11), 45–66 (2001). https://doi.org/10.1162/153244302760185243
10. Zhang, Z.: Artificial neural network. In: Zhang, Z. (ed.) Multivariate Time Series Analysis in Climate and Environmental Research, pp. 1–35. Springer, Cham (2018). https://doi.org/10.1007/978-3-319-67340-0
11. Safavian, S., Landgrebe, D.: A survey of decision tree classifier methodology. IEEE Trans. Syst. Man. Cybern. 21(3), 660–674 (1991). https://doi.org/10.1109/21.97458
12. Dietterich, T.: Overfitting and undercomputing in machine learning. CSUR ACM Comput. Surv. 27(3), 326–327 (1995). https://doi.org/10.1145/212094.212114
13. Breiman, L.: Random forests. Mach. Learn. 45(1), 5–32 (2001). https://doi.org/10.1023/A:1010933404324

14. Rätsch, G., Onoda, T., Müller, K.: Soft margins for AdaBoost. Mach. Learn. **42**(3), 287–320 (2001). https://doi.org/10.1023/A:1007618119488
15. Friedman, J.: Stochastic gradient boosting. Comput. Stat. Data Anal. **38**(4), 367–378 (2002). https://doi.org/10.1016/S0167-9473(01)00065-2
16. Mikolov, T.: Efficient estimation of word representations in vector space. In: International Conference on Learning Representations, arXiv:1301.3781 (2013)
17. Mikolov, T.: Distributed representations of words and phrases and their compositionality. In: Advances in Neural Information Processing Systems, pp. 3111–3119. NIPS, Montreal (2013)
18. Word2Vec. https://radimrehurek.com/gensim/models/word2vec
19. Akbani, R., Kwek, S., Japkowicz, N.: Applying support vector machines to imbalanced datasets. In: Boulicaut, J.-F., Esposito, F., Giannotti, F., Pedreschi, D. (eds.) ECML 2004. LNCS (LNAI), vol. 3201, pp. 39–50. Springer, Heidelberg (2004). https://doi.org/10.1007/978-3-540-30115-8_7
20. Kubat, M., Holte, R., Matwin, S.: Machine learning for the detection of oil spills in satellite radar images. Mach. Learn. **30**(2), 195–215 (1998). https://doi.org/10.1023/A:1007452223027
21. Lomotin, K., Kozlova, E., Romanov, A.: Comparative analysis of classification methods for text in UDC code generation problem for scientific articles. In: Information Innovative Technologies, pp. 359–363. Association of graduates and employees of AFEA named after prof. Zhukovsky, Moscow (2017)
22. Hull, D.: Stemming algorithms: a case study for detailed evaluation. JASIS **47**(1), 70–84 (1996). https://doi.org/10.1002/(SICI)1097-4571(199601)47:1%3c70:AID-ASI7%3e3.3.CO;2-Q
23. Chrupala, G.: Simple data-driven context-sensitive lemmatization. Nat. Lang. Process. **37**, 121–127 (2006)
24. PyMystem3. https://pypi.org/project/pymystem3/
25. Gers, F., Schmidhuber, J., Cummins, F.: Learning to forget: continual prediction with LSTM. In: Ninth International Conference on Artificial Neural Networks, pp. 850–855. IET, Edinburgh (1999). https://doi.org/10.1162/089976600300015015
26. Simard, P.: Best practices for convolutional neural networks applied to visual document analysis. ICDAR **3**, 958–962 (2003)
27. Kim, Y.: Convolutional neural networks for sentence classification. In: Proceedings of EMNLP 2014, arXiv:1408.5882 (2014)
28. Sowa, J.: Principles of Semantic Networks: Explorations in the Representation of Knowledge. Morgan Kaufmann, San Mateo (2014). https://doi.org/10.1016/c2013-0-08297-7
29. Liu, Y.: Learning natural language inference using bidirectional LSTM model and inner-attention. arXiv:1605.09090 (2016)

Structural Properties of Collocations in Tatar-Russian Socio-Political Dictionary of Collocations

Alfiya Galieva[✉] and Olga Nevzorova

Institute of Applied Semiotics of the Tatarstan Academy of Sciences,
Kazan, Russia
amgalieva@gmail.com, onevzoro@gmail.com

Abstract. This paper discusses some of the issues and challenges encountered during the compilation of the Tatar-Russian Socio-Political Dictionary of collocations, which is based on the data of the available corpora of the Tatar language. The area of collocations within the language system is of particular importance, and the well-known language-specificity of collocations suggests the need for bilingual collocation dictionaries.

The main criteria for selecting linguistic data are those of objective (frequency in the corpus) and subjective evaluation (evaluation of the word from the point of view of its thematic, stylistic and collocational value). The main unit in the Dictionary is the noun or verb phrase formed by filling one of possible semantic-syntactic positions of the word and meeting the criteria of semantic completeness. As an exception, we also included certain combinations of header words with postpositions derived from nouns, as long as the corresponding collocations are typical for socio-political discourse.

As it is a nontrivial task to fix basic forms of word combinations in morphologically rich languages, special attention is paid to the issue of lemmatization of collocations in the Dictionary (representing grammatical voice forms, fixing and translating predicative phrases, lemmatizing items with polyfunctional affixes, etc.).

Keywords: Socio-political vocabulary · Collocation dictionary
The Tatar language · Bilingual dictionary · Corpus

1 Introduction

Development of new lexicographic resources for minority and low-resource languages is a task of current importance that has scientific and practical dimensions. This paper presents the project on developing the Tatar-Russian Socio-Political Dictionary of Collocations (http://spdict.turklang.tatar/). The area of collocations within the language system is of particular importance, and the well-known language and domain specificity of collocations requires developing particular principles and methodology for building such resources.

Compiling the Tatar-Russian Socio-Political Dictionary of Collocations is carried out due to a combination of factors, such as:

© Springer Nature Switzerland AG 2018
D. A. Alexandrov et al. (Eds.): DTGS 2018, CCIS 859, pp. 324–335, 2018.

- socio-political vocabulary is a significant formation of any language in active use, and it is undergoing permanent changes;
- available bilingual Russian-Tatar dictionaries for general purposes and special lexicons contain outdated data and are lacking new words and phrases, thus failing to reflect the current state of the language; besides they contain rather a limited number of collocations;
- Tatar corpora provide reliable information about Tatar socio-political vocabulary and lexical co-occurrences in actual use, so they are to be used in compiling new generation dictionaries.

The Tatar-Russian Socio-Political Dictionary of collocations is based on the data of the available Tatar corpora. The use of corpus-based dictionaries is but a recent trend, especially as far as it concerns minor languages (Jantunen 2016). This article discusses the main design decisions adopted in compiling the Tatar-Russian Socio-Political Dictionary of Collocations. The direction of compiling the dictionary is from Tatar to Russian (and not vice versa) because the dictionary is aimed at detecting and fixing main features of the present-day Tatar socio-political lexicon. The practical interest in collocations is registered due to the fact that they are considered an important source for producing naturally sounding speech, which is one of the primary goals in language teaching. Collocations refer to how words go together or form fixed relationships; they are regarded as essential building blocks of the natural language. The new Dictionary is demanded by linguists, journalists and professional translators as well as in education process both in secondary and high school.

The remainder of this paper is organized as follows. Section 2 gives a brief overview of related work, Sect. 3 outlines main available resources used to compile the Dictionary, Sect. 4 presents methodology of the dictionary development, and considerable attention is paid to the issue of presenting the encountered challenges. Finally, Sect. 5 lists the conclusions that can be derived from this research.

2 Related Work

In recent decades the lexicographic and practical value of collocations has become evident and a large number of linguists and editors were involved in projects related to this issue. In (Stubbs 2002) collocation is defined as "a lexical relation between two or more words which have a tendency to co-occur within a few words of each other in running text" (Stubbs 2002: 24). Collocations in this broad sense may include a wide range of heterogeneous sets of words. So there are some difficulties concerning understanding the nature, structure and definition of collocations. Partington (1998) classifies the definitions of collocations into textual (co-occurrence in a text), statistical (co-occurrence with a greater than random probability) and psychological (co-occurrence due to a psychological link between words). Collocations are highly specific for a particular language; they are conventionalized and may have contextual restrictions. Competently composed collocation dictionaries are a valuable resource for translators and language learners (for beginners as well as advanced students).

Language learners and users draw much of their vocabulary knowledge from context, apart from explicit instruction. Nation (2001: 318) summarizes the discussions about the importance of collocations with the following arguments: (1) language knowledge is collocational knowledge; (2) fluent and appropriate language use requires collocational knowledge; and (3) many words are used in a limited set of collocations and knowing these is a part of what is involved in knowing the words.

Available collocation dictionaries of English (Benson et al. 1986; Hill and Lewis 1999; Mcintosh et al. 2002; Rundell 2010) provide a wide repertoire of collocations and target different audiences. Similar dictionaries were launched for other languages: Spain (Bosque, Bosque 2004, 2006), French (Beauchesne 2001), Russian (Borisova 1995), etc. The fact that collocations are highly language-specific, poses a problem of compiling bilingual collocation dictionaries. Such dictionaries (Benson and Benson 1993; Klégr et al. 2005; Benson et al. 1997; Konecny and Autelli 2014; Orenha-Ottaiano 2016; etc.) provide information about interlingual lexical correspondences and are aimed at encouraging learners and translators to more actively use collocations and incorporate them into their mental lexicon.

Due to the historical destiny of the Tatar people and the influence of geopolitical factors, the main stream of bilingual Tatar lexicography was Tatar-Russian and Russian-Tatar (Ganiev 1997; Agishev 2001; Safiullina 1996, 2002). The brief review of Tatar-Russian lexicography is presented in (Safiullina 2016). The practical need in English-Tatar dictionaries caused the compilation of bilingual English-Tatar dictionaries during the last decades (Arslanova and Safiullina 1999; Garifullin 2007; Safiullina 2014), all of them aimed at being used in education and having a limited volume.

In 90's the rising official status of the Tatar language and the subsequent practical needs engendered the development of special lexicons, including bilingual Russian-Tatar socio-political dictionaries (Nizamov 1995, Amirov 1996, Ganiev 1997). By now these dictionaries contain the lexical data that is obsolete in many respects, and the number of collocations in them is very limited. The available Tatar-Russian combinatory dictionaries are compiled for education purposes (Agishev 2001) or represent mainly phraseological data (Safiullina 2002). Besides, all of them had been compiled before the development of Tatar corpora. So Tatar lexicography needs corpus based dictionaries that would provide relevant, statistically verified information about word meanings, distributions and contextual environments.

3   Available Tatar Corpora

The socio-political domain is a broad sphere of contemporary social relations comprising the following main topics: politics and state administration; international relations; economy and financial issues; industry; army and military sphere; social sphere; culture and art; religion; sports, etc. This domain is one of the most dynamically developing spheres of present-day life, with the socio-political vocabulary rapidly developing and being enriched with new lexical items reflecting the realities of the time. Linguistic data from constantly updated corpora are very important for a comprehensive and objective study of the current Tatar language.

The Tatar-Russian Socio-Political Dictionary of collocations is based on the data of the main available corpora of the Tatar language: the Corpus of Written Tatar (Corpus of Written Tatar 2017) and the Tatar National Corpus (Tatar National Corpus 2017). These corpora include texts of various genres, from official documents and scientific publications to fiction, media texts, and school books. The corpora have comparable volumes and are supplied with morphological description, i.e. information about the part of speech of the word stem and the set of its grammar features. Besides, the data from a special collection – Socio-Political Subcorpus of the Tatar language (Socio-Political Subcorpus 2017) – is employed. This subcorpus is composed of texts of electronic media on social and political topics, as well as texts of legal documents.

Corpus technology greatly facilitates obtaining the empirical data for the Dictionary and their processing; in particular, it allows obtaining objective data about the frequency, distribution and compatibility of lexemes. The search system of the corpora makes it possible to conduct search by lemma (lexeme), by word form, as well as by a set of morphological parameters specified by the user. Linguistic data from constantly updated corpora are of great importance for a comprehensive and objective research into the processes that take place in the modern socio-political discourse, and also for fixing the actual state of the lexicon in lexicographic resources.

4   Methodology of Compiling the Dictionary

In this section, the principles of data collection and analysis are discussed, as well as the main challenges and solutions that were obtained.

4.1   Main Stages of Building the Dictionary

The methodology of compiling the Dictionary included the following main stages:

- selecting header words using corpus data;
- retrieving collocations in Tatar corpora and linguistic dictionaries;
- translating collocations into Russian.

The initial stage implied compiling the frequency list of actual terms (the list of one-word terms as potential header words) using the Socio-Political Subcorpus. First we automatically generated the list of the most frequently used noun word forms, which were lemmatized, and then the list of potential header words was manually compiled. The main criteria for selecting vocabulary are based on objective (frequency in the Socio-Political Subcorpus) and subjective evaluation (considering the words' thematic, stylistic and collocational value and their use in texts of social, political and cultural topics). The current list of potential header words is composed of 1000 items.

Then, using bi-gram models (a sequence of two adjacent elements) obtained from the Corpus of Written Tatar and the Tatar National corpus, we have built a frequency list of collocations for each frequent term.

The limitations for cutting elements from the collocations list were based on the frequency of using linguistic items in the corpora, and these limitations were determined empirically; in the current version of the dictionary the lower threshold for including a collocation is its occurrence in at least 50 corpus contexts (actually for the overwhelming majority of collocations this threshold is significantly higher because corpus collocations are given for word forms, not for lemmas) (Galieva et al. 2017).

The Dictionary of collocations contains meaningful common word combinations of different structures such as *дәүләт сәясәте* 'policy of the state' (N + N, POSS_3), *табигый байлыклар* 'natural resources' (ADJ + N, PL), *музей ачу* 'to open the museum' (N + V, VN).

Collocations in the Dictionary are represented in basic forms (issues concerning lemmatization are discussed in a special section below). In the current version of the Dictionary most of collocations are composed of two notional components. As an exception, we also included certain combinations of header words with postpositions derived from nouns, as long as the corresponding collocations are typical for socio-political discourse, for example:

карар нигезендә 'on the basis of a resolution' (header word *КАРАР* 'resolution, decree');
закон каршында 'facing the law' (header word *ЗАКОН* 'law');

The entries of the new dictionary are currently limited by nouns and relative adjectives arranged alphabetically. the structure of an entry is built as follows:

- header word (capitalized) – Tatar word, frequently used in socio-political discourse;
- Russian translation of the header word (only senses relevant for socio-political discourse are provided);
- lexical compatibility to the right, Russian translation of collocations;
- lexical compatibility to the left, Russian translation of collocations.

When selecting the collocations, we considered the syntactic structure of each of them and the morphological parameters of their constituents. The current version of the Dictionary contains more than 30 basic structural types of collocations. Distribution of main structural types of collocations is represented in Table 1.

Table 1. Distribution of main structural types of collocations

Structural type of collocation	Example	English transla-tion	Number in the Diction-ary
N + N, POSS_3	банк хезмәткәре	'bank employee'	779
ADJ + N	мәдәни тормыш	'cultural life'	534
N + N, PL, POSS_3	дин нигезләре	'foundations of religion'	152
N + V, VN	музей ачу	'opening of the museum'	226
N, ATTR + N	абруйлы шәхес	'authoritative person'	73

4.2   Lemmatization of Collocations

Compiling dictionaries for languages of rich morphology sets the problem of lemmatization of linguistic items – reducing them to the uniform structure to be represented in the Dictionary.

Turkic languages have complicated morphology and syntax, which presents a challenge for language processing. A significant feature of the Tatar language is that there are no inflectional paradigms in traditional sense, as closed sets of word forms (for example a set of all inflected forms based on a single stem). A paradigm in Turkic languages may have fuzzy borders, because the number of potential elements that may be added (inflectional affixes) is indefinite. In phrases – for example, in noun phrases, each component may join its own set of grammatically permitted affixes, and the number of potential grammatical forms of the noun phrase is uncertain, so in texts noun phrases may be used in a huge amount of individual forms. For example, the noun phrase *банк хезмәткәре* 'Bank employee' may have the first or the second component or both of them inflected, depending on grammatical number, case, possessive or other affixes. Table 2 represents some of them. All collocations given here are represented in the initial form *банк хезмәткәре* 'Bank employee'.

Table 2 Some grammatical forms of noun phrase *банк хезмәткәре* 'Bank Employee'

Grammatical form of the noun phrase	Structure	English translation	Number in Corpus
банк хезмәткәре	Bank + employee-POSS_3	'Bank employee'	90
банк хезмәткәрләре	Bank + employee-PL, POSS_3	Bank employees'	129
банкы хезмәткәрләре	Bank-POSS_3 + employee--PL, POSS_3	'employee of the Bank'	20
банклар хезмәткәрләре	Bank-PL + employee--PL, POSS_3	'employees of Banks'	1
банк хезмәткәрләренең	Bank- + employee--PL, POSS_3, GEN	'of Bank employees'	18
банк хезмәткәреме	Bank- + employee-POSS_3, GEN, INT	'is (it a) Bank employee?'	1

Since the number of individual word forms and individual usage forms of phrases in Turkic languages may be large, there is a question of what linguistic items should be lemmatized for a collocation dictionary and how.

When collecting collocations we fix them in their basic form, preserving the basic grammatical structure of the collocation. The total automatic lemmatization of linguistic data is of little avail, because it destructs grammatically conditioned word combinations. The basic form of a collocation is determined according to the initial form of the main component of the phrase. A grammatical form of a word combination (phrase) is determined by grammatical properties of its main component that can join dependent components of certain types and only in grammatically admitted word forms. However in corpus contexts a collocation may be found in a great number of grammatically permitted forms. That is why we built individual lemmatization rules for all basic grammatical types of collocations to unify them. Some of these rules are presented in Table 3.

Table 3. Examples of lemmatization rules for basic grammatical types of collocations

Basic grammatical types of collocations	Example of collocations before lemmatization	Example of collocations after lemmatization	English translation	Lemmatization rule
ADJ + N	тарихи шәхеслэрне	тарихи шәхес	'historical figure'	All inflected affixes of the noun are cut
N + N, POSS,3	авыл клубларында	авыл клубы	'village club'	All inflected affixes of the nouns are cut, except the possessive affix demanded by the noun phrase structure
N + V	шәхесне формалаш-тырганда	шәхес формалаш-тыру	'personality formation'	The verb is given in the form of a verbal noun, and all inflected affixes of the noun are cut, except those that are demanded by the verb government

One of the issues concerning lemmatization is where a word form and a lexeme should be demarcated; for a number of cases it turned out to be nontrivial. Further we discuss particular examples of choosing the basic form and the appropriate way of fixing collocations in the Dictionary.

Representing Grammatical Voice Forms. The Tatar language, like other Turkic languages, has 5 grammatical voices (the Basic, Passive, Reflexive, Reciprocal and Causative voices); voice affixes may be treated as inflectional or constructive; they are added immediately to the verbs' stems and form extended stems (Lewis 2001: 143). The precise meanings of voice derivates cannot always be deduced logically from the basic verb forms and may be individual for each stem and ambiguous in many cases.

This is the reason why such voice derivates are treated dissimilarly in different theoretical frameworks and are fixed in dictionaries unsystematically.

While lemmatizing verb voice forms in the Dictionary of Collocations we treat them as individual items related to the same header word (noun):

> *Ярышлар үтү*
> competition-PL pass-VN
> 'to be held – on competitions' (the header word ЯРЫШ 'competition').
> *Ярышлар үткәрү*
> competition-PL pass-CAUS, VN
> 'to hold competitions' (the header word ЯРЫШ 'competition').

Fixing and Translating Predicative Phrases. In Tatar there are predicative phrases consisting of a syntactic subject and a predicate with a non-finite verb form. The basic form of such constructions is the form of verbal nouns that enables to express subject-predicate coordination and does not express any grammatical tense or mood, so these subject-predicate constructions cannot be treated as grammatically complete sentences. Because of their absence in Russian, as well as in English, such subject-predicate constructions that can be used as constructive elements of sentences, are difficult to translate and their structure is hard to explain. One decision may be to fix verbs in one of the finite forms with succeeding translation as a sentence:

> *Ярышлар узды.*
> competition-PL pass-CAUS, PAST_DEF
> 'Competitions have been held'.
> *Ярышлар узачак.*
> competition-PL pass-CAUS, FUT_DEF
> 'Competitions will be held'.

However such option leads to a preferred presentation of merely one finite form of the verb, even if it is the most frequently used. Turkic languages provide a possibility to reduce such phrases to the form of verbal nouns – hybrid forms lacking indices of the tense or person; nevertheless, they possess a grammatical voice marker. How are these constructions to be translated? We offer a descriptive translation:

Ярышлар узу 'to be held' – about competitions.

Lemmatizing Items Containing Polyfunctional Affixes. The Tatar language has affixes of different types:

(a) derivational affixes, expressing only lexical meaning and forming new words;
(b) inflectional affixes changing the word form (for example, case affixes);
(c) affixes serving as means of derivation as well as inflection (polyfunctional affixes).

Below we present an example of a word containing the abessive affix *сыз/-сез* – the polyfunctional affix expressing the lack/absence of the marked noun:

законсыз
law-ABESS
(1) illegal, unlawful (derivation level):
законсыз митинг 'illegal rally';
(2) lawless (morphology level):
андый законсыз булмый 'it is impossible to live without such a law'.

Currently all such items, are included in the entries with noun header word in order to unify the way of fixing them in the dictionary; for example, the item *законсыз*, regardless of its sense in an individual collocation, is included in the nest ЗАКОН 'law'. Words containing standard derivational affixes are considered as individual words.

Lemmatizing Items Containing the Equative Affix. The term Equative in Turkic languages is used to designate comparative-like constructions:

кошларча
Bird-PL, EQU
'like birds'
Әйткәнемчә
Say-POSS_1SG, EQU
'as I said'

In media texts the Equative affix is often used to refer to official words:

министр сүзләренчә
minister's word-PL, POSS_3, EQU
'according to the minister's words'.

In such constructions the Equative affix functions like a conjunction or preposition in European languages which joins parts of sentence and, if approached formally, should be cut in the process of lemmatizing.

Currently items containing the Equative affix are represented as individual collocations. For example, the collocation above *министр сүзләренчә* is not interpreted as a particular case of using the collocation *министр сүзләре* 'words of the minister' because the latter phrase is not frequent enough in the socio-political domain.

So the issue of lemmatizing collocations is not trivial in many cases, and solutions applied are based on the grammatical structure of the Tatar language as well as on the assumption of main features of socio-political texts. Most of the lemmatization problems discussed above and the solutions offered are not limited by socio-political domain; they are relevant for the Tatar language system and may be used to annotate any phrase-oriented resource for Turkic languages.

4.3   Distinguishing Synonyms and Variants of Collocations

In the course of compiling a dictionary, an important question is that of distinguishing between synonyms and collocation variants; that determines quantitative characteristics of the dictionary and peculiarities of the structure of the dictionary entry.

The question of distinguishing between synonyms and variants of word combinations at this stage of the project has rather a technical solution. At the moment we adhere to the following distinction:

(1) items formed by root words of different parts of speech are regarded as variants of the same collocation and are listed in the same line:

акционерлык җәмгыяте, акционерлар җәмгыяте 'joint-stock company'

The latter also concerns regular correspondences like ADJ + N and N + N, POSS_3; they are also interpreted as the same nominative item and are given in the same line:

икътисади кризис (ADJ + N),
икътисад кризисы (N + N, POSS_3)
'economic crisis'.

(2) Synonymous items formed by synonyms of different roots are presented in different lines:

*ислам мәдәният*е 'Islamic culture';
*мөселман мәдәният*е 'Muslim culture';
акционерлык җәмгыяте 'joint-stock company';
акционерлык оешмасы 'joint-stock company';
акционерлык ширкәте 'joint-stock company'.

The Dictionary represents numerous synonymous collocations which are used to designate topical realities of modern socio-political life. These synonymous collocations are engendered by a set of factors of different nature; this issue is quite interesting and deserves a special study.

5   Conclusion and Future Work

Compiling a dictionary of collocations for low-resource languages is a topical yet rather challenging task due to the numerous details that a lexicographer is to take into consideration, including criteria for selecting header words and collocations, lemmatizing linguistic items, distinguishing variants of collocations from synonyms. Besides such a dictionary should become a user-friendly and practically valuable new resource for its target audience, and its structure should adequately represent linguistic data and be convenient for practical use.

The compiled Tatar-Russian Socio-Political Dictionary of Collocations makes it possible: (1) to fix the real use of Tatar words, including items which are actively used in a large number of Tatar official and media texts; nevertheless those are absent in available special and bilingual Russian-Tatar dictionaries; (2) to detect and fix typical grammatical models and contexts of using items denoting socio-political realities; (3) to trace words of new mintage in the Tatar language functioning in Russian geopolitical space; (4) to keep numerous synonymous nominations used in Tatar media texts and official documents; (5) to offer Russian translations of words and collocations.

In its current state the Dictionary contains 250 header words and more than 4,000 collocations with their Russian translations and information about the grammatical structure of the collocations. In future it is planned to extend the linguistic database (adding new entries and related collocations), providing information about corpus contexts to illustrate the use of collocations.

The offered solutions for lemmatizing collocations are not limited by socio-political domain and may be used when annotating any phrase-oriented resource for Turkic languages.

6   List of Abbreviations

ABESS – Abessive, ADJ – Adjective, ATTR – Attributive, CAUS – Causative Verb, DIR – Directive Case, EQU – Equative, FUT_DEF – Future Definite, GEN – Genitive case, INT – Interrogative, N – Noun, PAST_DEF – Past Definite, PL – Plural, POSS_3 – Possessive, 3d Person, PRES – Present, VN – Verbal Noun.

Acknowledgements. The reported study was funded by Russian Science Foundation, research project № 16-18-02074.

References

Agishev, K.: Russko-tatarskiy slovar' slovosochetaniy dlya uchashchikhsia [Russian-Tatar Dictionary of Word Combinations for Students]. Fan, Kazan (2001)

Amirov, K.: Russko-tatarskij yuridicheskij slovar' [Russian-Tatar Legal Dictionary]. Tatar Publishing House, Kazan (1996)

Arslanova, G., Safiullina, F.: Uchebnyy anglo-tatarsko-russkiy slovar'-minimum [Academic English-Tatar Dictionary-Minimum]. Khater, Kazan (1999)

Beauchesne, J.: Dictionnaire des Cooccurrences. Guerin, Montréal (2001)

Benson, M., Benson, E.: Russian-English Dictionary of Verbal Collocations. John Benjamins, Amsterdam (1993)

Benson, M., Benson, E., Ilson, R.: Longman Dictionary of English Collocations (Bilingual English-Chinese). Longman, London (1997)

Benson, M., Benson, E., Ilson, R.: The BBI Combinatory Dictionary of English. John Benjamins, Amsterdam (1986)

Bosque, I. (ed.): Redes. Diccionario combinatorio del español contemporáneo. SM, Madrid (2004)

Bosque, I. (ed.): Diccionario combinatorio práctico del español contemporáneo. SM, Madrid (2006)

Borisova, Y.: Slovo v tekste. Slovar' kollokatsiy (ustoychivyih slovosochetaniy) russkogo yazyika s anglo-russkim slovarem klyuchevyih slov [A Word in the Text. Dictionary of Collocations of the Russian Language with English-Russian Dictionary of Key Words (in Russian)]. Filologia, Moscow (1995)

Corpus of Written Tatar (2017). http://corpus.tatar/

Galieva, A., Vavilova, Z., Gafarova, V.: Developing Tatar corpus-based dictionaries for educational purposes. In: INTED2017 Proceedings, 11th International Technology, Education and Development Conference Valencia, Spain, 6–8 March, 2017, pp. 9014–9022. IATED, Valencia (2017). https://doi.org/10.21125/inted.2017

Ganiev, F. (ed.): Russko-tatarskij slovar' [Russian-Tatar Dictionary]. Insan, Moscow (1997)

Garifullin, S.: English-Tatar Dictionary [Russian-Tatar Dictionary]. Magarif, Kazan (2007)

Hill, J., Lewis, M.: LTP Dictionary of Selected Collocations. Language Teaching Publications, Hove, London (1999)

Jantunen, J.H.: Corpora, phraseology and dictionaries: how does corpus research intersect language teaching and learning? In: Vilas, B.S. (ed.) Collocations Cross-linguistically: Corpora, Dictionaries and Language Teaching, pp. 97–119. Société Néophilologique, Helsinki (2016)

Klégr, A., Key, P., Hronková, N.: Česko-anglický slovník spojení: podstatné jméno a sloveso [Czech-English combinatory dictionary: noun and verb]. Karolinum, Praha (2005)

Konecny, C., Autelli, E.E.: Kollokationen Italienisch-Deutsch. Helmut Buske Verlag, Vydavateľ (2014)

Lewis, G.: Turkish Grammar. Oxford University Press, Oxford (2001)

Mcintosh, C., Francis, B., Poole, R. (eds.): Oxford Collocations Dictionary for Students of English. Oxford University Press, Oxford (2002)

Nation, P.: Learning Vocabulary in Another Language. Cambridge University Press, Cambridge (2001)

Nizamov, I.: Kratkij russko-tatarskij obshchestvenno-politicheskij slovar' [Brief Russian-Tatar Socio-Political Dictionary]. Tatar Publishing House, Kazan (1995)

Orenha-Ottaiano, A.: The compilation of a printed and online corpus-based bilingual collocations dictionary. In: Margalitadze, T., Meladze, G. (eds.) Proceedings of the XVII EURALEX International Congress. Lexicography and Linguistic Diversity, 6–10 September 2016, pp. 735–745. Ivane Javakhishvili Tbilisi State University, Tbilisi (2016)

Partington, A.: Patterns and Meanings: Using Corpora for English Language Research and Teaching. John Benjamins, Amsterdam (1998)

Rundell, M.: Macmillan Collocations Dictionary for Learners of English. Macmillan Publishers Ltd., Oxford (2010)

Safiullina, F.: Karmannyy tatarsko-russkiy i russko-tatarskiy slovar' [Tatar-Russian and Russian-Tatar Pocket Dictionary]. Tatar Publishing House, Kazan (1996)

Safiullina, F.: Tatarsko-russkiy slovar' sostavnykh clov [Tatar-Russian Dictionary of Compound Words]. Tatar Publishing House, Kazan (2002)

Safiullina, G.: Bilingual lexicography in the Republic of Tatarstan in 1990–2010. In: Margalitadze, T., Meladze, G. (eds.) Proceedings of the XVII EURALEX International Congress. Lexicography and Linguistic Diversity, 6–10 September, 2016, pp. 475–479. Ivane Javakhishvili Tbilisi State University, Tbilisi (2016)

Safiullina, G.: English-Tatar Dictionary. Kazan Federal University Publishing House, Kazan (2014)

Socio-Political Subcorpus of the Tatar language (2017). http://tugantel.tatar/corpus/op?lang=en

Stubbs, M.: Words and Phrases: Corpus Studies of Lexical Semantics. Blackwell, Oxford (2002)

Tatar-Russian Socio-Political Dictionary of Collocations (2001b). http://spdict.turklang.tatar/

Tatar National Corpus (2017). http://tugantel.tatar/

Computer Ontology of Tibetan
for Morphosyntactic Disambiguation

Aleksei Dobrov[1], Anastasia Dobrova[2], Pavel Grokhovskiy[1(✉)],
Maria Smirnova[1], and Nikolay Soms[2]

[1] Saint-Petersburg State University, Saint Petersburg, Russia
{a.dobrov, p.grokhovskiy, m.o.smirnova}@spbu.ru
[2] LLC AIIRE, Saint-Petersburg, Russia
{adobrova, nsoms}@aiire.org

Abstract. The article presents the experience of developing computer ontology as one of the tools for automatic natural language processing. A computer ontology that contains a consistent specification of meanings of lexical units with different relations between them represents a model of lexical semantics and both syntactic and semantic valencies, reflecting the Tibetan linguistic picture of the world. The article describes the approach of using computer ontology as a means of introducing semantic restrictions for morphosyntactic disambiguation on the basis of the corpus of indigenous grammatical treatises.

Keywords: Computer ontology · Tibetan corpus
Linguistic picture of the world · Automatic natural language processing

1 Introduction

The research introduced by this paper is a continuation of several research projects ("The Basic corpus of the Tibetan Classical Language with Russian translation and lexical database", "The Corpus of Indigenous Tibetan Grammar Treatises"), aimed at the development of methods for creation of a parallel Tibetan-Russian corpus [1, p. 183].

The Basic Corpus of the Tibetan Classical Language includes texts and the Corpus of Indigenous Tibetan Grammar Treatises comprise 34,000 and 48,000 tokens, respectively. Tibetan texts are represented both in a Tibetan Unicode script and in a standard Latin (Wylie) transliteration [1].

The ultimate goal of the current project is to create a formal model (a grammar and a linguistic ontology) of the Tibetan language, including morphosyntax, syntax of phrases and hyperphrase unities, and semantics, that can produce a correct morphosyntactic, syntactic, and semantic annotation of the corpora without any manual corrections.

This model should, ideally, cover both corpora; the corpus of classical Tibetan formed the basis of formal grammar at the initial stage, and at the present stage the main work was concentrated on the corpus of grammatical treatises, with the goal of achieving 100% coverage of its material; the current version is available at http://aiire.org/corman/index.html?corpora_id=67&page=1&view=docs_list.

D. A. Alexandrov et al. (Eds.): DTGS 2018, CCIS 859, pp. 336–349, 2018.

The underlying AIIRE linguistic processor needs to recognize all the relevant linguistic units in the input text. For inflectional languages the input units are easy to identify as word forms, separated by space, punctuation marks etc. It is not the case with the Tibetan language, as there are no universal symbols to segment the input string into words or morphemes. The developed module for the Tibetan language performs the segmentation of the input string into elementary units (morphs and punctuation marks - atoms) by using the Aho-Corasick algorithm [2], that allows to find all possible substrings of the input string according with a given dictionary. Thus, any part of a morpheme which is itself a morpheme with a shorter exponent, is also taken into account in the morphosyntactical analysis. The algorithm builds a tree, describing a finite state machine with terminal nodes corresponding to completed character strings of elements (in this case, morphemes) from the input dictionary. The language module contains a dictionary of morphemes, which allows the machine to create this tree in advance at the build stage of the language module, while in the runtime of the linguistic processor the tree is being loaded as a component of an executable module which brings its initialization time to minimum. Several special files were created in order to analyze Tibetan morphosyntactic structures: the grammarDefines.py file determines types of atoms (atomic units), their properties and restrictions, while the atoms.txt file (the allomorphs' dictionary) specifies the morpheme, the token type, and properties for each allomorph, also in accordance with grammarDefines.py file.

The system aims at multi-variant analysis, unlike some other systems of morphosyntactic analysis. That sometimes causes combinatorial explosions[1] in the analysis versions. In most combinatorial explosions, idioms were present, therefore, one of the strategies for eliminating morphosyntactic ambiguity was the processing of Tibetan idioms using computer ontology.

In this article, the current experience of developing computer ontology through the analysis of the issues of automatic syntactic and semantic annotation of the Corpus of Indigenous Tibetan Grammatical Treatises will be presented, and the current results of disambiguation by means of this ontology will be discussed.

2   Related Work

Tibetan can reasonably be considered as one of the less-resourced, or even under-resourced languages, in the sense in which this term was introduced in [4] and is widely used now: the presence of Tibetan on the web is limited, it lacks not only electronic resources for language processing, such as corpora, electronic dictionaries, vocabulary lists, etc., but also even such linguistic descriptions as grammars, which would be characterized by at least minimal consistency and validity of linguistic material. The only relatively complete linguistic description of the Tibetan language is the

---

[1] Following the definition of K. Krippendorf, combinatorial explosion is understood here as a situation 'when a huge number of possible combinations are created by increasing the number of entities which can be combined' [3]. As applied to parsing, these are cases of exponential growth in the number of parsing versions as the length of the parsed text and, thus, the amount of its parsed ambiguous fragments increase.

monograph by Beyer [5], but this work contains many assumptions that have not been confirmed by any language material at all. The only relatively large corpus of Tibetan texts is 'The Annotated Corpora of Classical Tibetan' (ACTib, [6]), published by Marieke Meelen, Nathan Hill and Christopher Handy in 2017 and containing, as it is stated, "80 million words", automatically produced by a segmentation and POS-tagging annotation tool (trained on a smaller corpus of 1000000 million syllables, which also includes the St. Petersburg Basic Corpora of Classical Tibetan Texts, marked up semi-manually by several participants of the team of authors of this article) without manual verification of this automatic markup. The Tibetan language has no word delimiters like spaces and generally accepted principles for dividing the text into word forms, therefore, at least in those subcorpora of this corpus that were prepared by the team of authors of this article, the division into word forms was in many cases, at least, controversial. Several publications [7, 8] have shown that Tibetan morphology and syntax are intermixed so strongly, that any division of the text into word forms is questionable, and much more grounded linguistic representation of Tibetan text is not morphological, but morphosyntactic annotation, which implies creating not a usual morphologically annotated corpus, but a treebank for Tibetan texts, the atomic units of the syntactic structures being not word forms, but morphemes. Greater accuracy of such structures from the linguistic point of view was noted not only for Tibetan: in the work [9] it was shown for Russian, and [10] advocated for total elimination of the 'word' notion as a cross-language one from the viewpoint of theoretical and comparative linguistics.

Creating a treebank involves the development of a formal grammar for the Tibetan language: even if the syntactic annotation is done manually, it is necessary to ensure the uniformity of the syntactic structures, which only a grammar can do, but having a grammar suggests creating a parser that automates the annotation procedure. The process of creating a grammar on the material of the corpus and for the automatic annotation of this corpus reveals a lot of the shortcomings of existing descriptions of the Tibetan language: many constructions are not described, for many constructions, permissible and inadmissible word order is unknown; many descriptions lead to significant ambiguity of parsing and to combinatorial explosion, i.e., following the definition of K. Krippendorf, a situation 'when a huge number of possible combinations are created by increasing the number of entities which can be combined' [3]. As applied to parsing, these are cases of exponential growth in the number of parsing versions as the length of the parsed text and, thus, the amount of its parsed ambiguous fragments increase.

Numerous studies have been devoted to solving the problem of syntactic ambiguity and combinatorial explosion, among which, in the context of this article, it is worthwhile to mention the work of Tomita [11], which proposes the idea of packing ambiguous fragments of syntactic structures into single entities, thus preventing combinatorial explosion; this idea was further developed in the works by Popov [12, 13]; and it seems also worthwhile to mention the interlevel interaction method, proposed by Tseitin in [14] and further developed in the project of the AIIRE linguistic processor [9, 15], used in this study. The packing method proposed by Tomita and redefined by Popov, in view of its latest refinements, allows to overcome combinatorial explosion during parsing, converting the multiplication of sets of versions into their addition, but this method in no way eliminates ambiguity and, accordingly, does not solve the problem in the case of corpus annotation. As for the method of inter-layer interaction, it allows to significantly reduce the syntactic ambiguity, but it involves the simultaneous semantic analysis and, correspondingly, semantic annotation of the corpus, which needs not only a grammar, but a semantic dictionary, or, as it is implemented in AIIRE, a linguistic ontology for Tibetan.

Ontologies in NLU systems are used as semantic dictionaries analogues that had been used before (cf. [16, 17], etc.); the main difference between an ontology and a semantic dictionary being that semantic valencies in semantic dictionaries were defined as lists of possible collocates and, in fact, postulated, whereas in ontologies valencies are automatically computed from the general knowledge about the world by inference engine subsystems; semantic restrictions being defined not as lists, but as base classes of ontology (that is the idea behind the mechanism of word-sense disambiguation in AIIRE [7, 8], SUMO [15], OpenCyc [18], InTez [19] etc.).

As far as the authors of this article know, at the moment, AIIRE is the only system that actually implements not only word-sense, but also syntactic disambiguation by means of the linguistic ontology without use of any statistical heuristics [9]. This article is devoted to the study of the possibilities of eliminating morphosyntactic ambiguity in the analysis of the Tibetan text by means of AIIRE and of the Tibetan ontology created within the framework of this project.

3 The Structure of the Ontology

Within modern computer linguistics, it is necessary to distinguish between linguistic thesauri and universal ontologies. The crucial difference between these concepts is that thesauri reflect more or less specified semantic relations between lexical units (words): synonymy, hyponymy, hypernymy, antonymy, meronymy, holonymy, associative relation etc.; while ontologies model strictly specified relations between concepts. Some of these concepts represent meanings of different lexical units, other have no representation in vocabulary, but are necessary for its modeling [15, p. 151]. This difference between thesauri and ontologies becomes obvious in attempts to create inference systems: ontologies are built on the basis of logical formalisms and corresponding inference rules, while thesauri generally don't provide any native mechanisms for logical inference.

The most famous and widely cited general definition of the term ontology is 'an explicit specification of a conceptualization' by Gruber [20]. Many different attempts were made to refine it for particular purposes. Without claiming for any changes to this de-facto standard, we have to clarify that, as the majority of researchers in natural language understanding, we mean not just any 'specification of a conceptualization' by this term, but rather a computer ontology, which we define as a database that consists of concepts and relations between them. Ontological concepts have attributes. Attributes and relations are interconnected: participation of a concept in a relation may be interpreted as its attribute, and vice versa. Relations between concepts are binary and directed. They can be represented as logical formulae, defined in terms of a calculus, which provides the rules of inference. Relations themselves can be modeled by concepts.

There is a special type of ontologies – so called linguistic ontologies (e.g., [21–23], etc.), which are designed for automatic processing of unstructured texts. Units of linguistic ontologies are based on meanings of real natural language expressions. Ontologies of this kind actually model linguistic picture of the world, that stands for language semantics. Ontologies, created for different languages, are not the same and are not language-independent. Differences between ontologies show differences between linguistic pictures of the world.

Ontologies, that are designed for natural language processing, are supposed to include relations that allow to perform semantic analysis of texts and to perform lexical and syntactic disambiguation.

The ontology, used for this research, was developed according with the above mentioned principles. It is a united consistent classification of concepts behind the meanings of Tibetan linguistic units, including morphemes and idiomatic morphemic complexes. The concepts are interconnected with different semantic relations. In order to create a new concept, it is compulsory to incorporate this concept into the general classification hierarchy according with the class-superclass relations (hypo/hypernymy), therefore the whole ontology has the one common superclasss. The ontology models the meanings of atomic linguistic units (morphemes) and of idiomatic combinations of these units: nominal and verbal compounds, idiomatic nominal groups, adjectival and adverbial groups. In all these cases, in addition to the meanings of an idiomatic expression, meanings of its components are also modeled in the ontology, so that they could be interpreted in their literal meanings too. This is necessary, because AIIRE natural language processor is designed to perform natural language understanding according with the compositionality principle [24], and idiomaticity is treated not merely as a property of a linguistic unit, but rather as a property of its meaning, namely, as a conventional substitution of a complex (literal) meaning with a single holistic (idiomatic) concept.

Each modeled concept is provided with a description (interpretation) intended to facilitate understanding of the decisions made by the participants of the process of editing the ontology. Each expression, the meaning of which is modeled in the ontology, is also provided with a full-scale interpretation in Tibetan from Tibetan "Great definition dictionary" [25]. If, according to the dictionary, the expression has several meanings, then the appropriate meaning (i.e. the one which is used in a particular context in the corpus) is translated into Russian (except when the expression is defined in the dictionary through synonyms).

Different relations are established between concepts. The relation of synonymy is always absolute (complete coincidence of referents with possible differences in significations). Concepts form synonymic sets (not to be confused with Wordnet synsets [26], which are sets of words). Each element of the set has the same attributes, i.e. the same relations and objects of these relations. Variance of significations within a synonymic set is compensated by automated logic rules: if Y is the synonym of X, then X has the same attributes as Y; and Y has the same characteristics as X.

Hypo-hypernymy is established between classes and subclasses or between classes and instances in situation, when one concept (hyponym) is a variety of another (hypernym).

If there is lacuna in Tibetan, it is possible to use Russian hypernym. Each concept must have at least one hypernym, except for the ontology root concept. It is necessary to create a concept hierarchy, which has basic classes (the inner classes which determine semantic valencies for example, the class "person"). Basic classes usually have a large number of relations, which appear in genitive constructions, verb valencies etc.

Relations like class-superclass provide inheritance of attributes between concepts. This mechanism allows to model semantic valencies as specific relations between some basic classes of the ontology (see below).

Totally within the framework of this research 3110 concepts that are meanings of 2906 Tibetan expressions were modelled in the ontology, 705 of them being idioms.

4 Restrictions for Morphosyntactic Disambiguation

In order to resolve the morphosyntactic ambiguity in phrases that had combinatorial explosions, four types of restrictions were established in the ontology: restrictions on genitive relations, restrictions on adjuncts (on the equivalence relation), restrictions on subjects and direct objects of verbs.

4.1 Restrictions on Genitive Relations

Restrictions on general genitive relation 'to have any object or process (about any object or process)' are imposed by establishing specific relation subclasses between basic classes in the ontology.

To exclude the possibility of first version of parsing in the example (1) the concept lus 'physical body' was connected with a genitive relation with the concept 'any creature'. This allowed to exclude version in which fame can have a body.

(1) གྲགས་པའི་ལུས

grags-pa 'i lus

be_well-known-NMLZ GEN body

(1.1) 'body of fame'

(1.2.) 'body of a famous [person]'

As a result of the use of Aho-Corasick algorithm, in the example (2) the definite pronoun so so 'every' was recognized not only as expected, but also as a possible combination of two noun roots so 'tooth', the second one together with its right context incorrectly forming the following word group:

(2) སོའི་ལུང་སྟོན་པའི་སྒྲ་སྐད

so 'i lung ston-pa 'i sgra-skad

tooth GEN grammar GEN sound

'sound of grammar of tooth'

To exclude the possibility of version in the example (2), the basic class *skad* 'language' was connected with a genitive relation 'to have a grammar' with the concept *lung ston-pa* 'grammar'. This allowed to exclude the version, in which a tooth can have grammar. As a result of the work with combinatorial explosions, the following sub-classes of the general genitive relation were introduced in the ontology: to have a grammar (about any language); to have someone (about someone); to have a place (about any creature); to have a place (about any process); to have a body (about any creature); to be part of group as a typical representative; to have meaning (about any text); to possess someone (about any object or process); to have a group (about any object or process); to have a relative (about any creature); to be like a phoneme (about a person).

4.2   Restriction on Adjuncts

Tibetan adjunct joins the noun phrase on the right side, and due to nonexistence of word delimiters (spaces) in Tibetan writing system, adjuncts can not be graphically distinguished from parts of compounds. Thus, compound (3) may be misinterpreted both as 'father-mother' ('a father, who is also a mother', ma 'mother' being interpreted as an adjunct), and as 'father's mother' (noun phrase with genitive composite), whereas the only correct interpretation is 'father and mother' (noun root group composite). While the second interpretation (which is, moreover, logically possible) can be elim-inated by just setting the correct token type in the ontology (because both the second and the third interpretations imply that the phrase is a compound (see paragraph 5), the first interpretation can not be just eliminated this way.

(3) ཕ་དང་མ

pha dang ma

father CONJ mother

'father and mother'

Thus, only semantic restrictions can reduce the number of versions and eliminate the versions of adjuncts, which are semantically incorrect. The reduction of versions with adjuncts was achieved by limiting the equivalence relation ('to be equivalent to an object or process'). Basic classes were connected with themselves with this relation so that only concepts that inherit these classes could be interpreted as adjuncts for each other.

In the example (4) 17 wrong versions (4.5–22) of parsing were built because of incorrect combinations of adjuncts: *gos hrul* 'ragged clothing' was interpreted as an adjunct to *lam kha* 'road' (4.5–6), to *lam* 'road' (4.7–10), to *lam* 'method' (4.15–18) or *kha* 'mouth' was interpreted as an adjunct to *lam* 'road' (4.11–14) or to *lam* 'method' (4.19–22).

(4) ལམ་ཁ་བོར་བ་ཡི།། གོས་ཧྲུལ་ངན་པ་སྐྱུག་བ

lam-kha bor-ba yi//gos-hrul ngan-pa skyug-ba

road lose-PST-NMLZ Gen ragged_clothing be_bad-NMLZ vomit-NMLZ

(4.1) 'bad vomiting ragged clothing, which lost the road'

(4.2) 'bad vomiting ragged clothing of [someone who] lost the road'

(4.3) 'vomiting of bad ragged clothing, which lost the road'

(4.4) 'vomiting of bad ragged clothing of [someone who] lost the road'

(4.5) 'vomiting of bad ragged clothing - the ways which lost'

(4.6) 'vomiting of bad ragged clothing - the ways of [someone who] lost [something]'

lam kha bor-ba yi//gos-hrul ngan-pa skyug-ba

road mouth lose-PST-NMLZ Gen ragged_clothing be_bad-NMLZ vomit-NMLZ

(4.7) 'bad vomiting ragged clothing - roads, which lost the mouth'

(4.8) 'bad vomiting ragged clothing - roads of [someone who] lost the mouth'

(4.9) 'vomiting of bad ragged clothing-roads, which lost the mouth'

(4.10) 'vomiting of bad ragged clothing-roads of [someone who] lost the mouth'

(4.11) 'bad vomiting ragged clothing, which lost the mouth - the road'

(4.12) 'bad vomiting ragged clothing of [someone who] lost the mouth - the road'

(4.13) 'vomiting of bad ragged clothing, which lost the mouth-the road'

(4.14) 'vomiting of bad ragged clothing of [someone who] lost the mouth-the road'

lam kha bor-ba yi//gos-hrul ngan-pa skyug-ba

method mouth lose-PST-NMLZ Gen ragged_clothing be_bad-NMLZ vomit-NMLZ

(4.15) 'bad vomiting ragged clothing - methods, which lost the mouth'

(4.16) 'bad vomiting ragged clothing - methods of [someone who] lost the mouth'

(4.17) 'vomiting of bad ragged clothing-methods, which lost the mouth'

(4.18) 'vomiting of bad ragged clothing-methods of [someone who] lost the mouth'

(4.19) 'bad vomiting ragged clothing, which lost the mouth - the method'

(4.20) 'bad vomiting ragged clothing of [someone who] lost the mouth - the method'

(4.21) 'vomiting of bad ragged clothing, which lost the mouth-the method'

(4.22) 'vomiting of bad ragged clothing of [someone who] lost the mouth-the method'

To eliminate these wrong versions of parsing the basic classes *lam* 'road', *lam* 'method' and *gos* 'clothes' were connected with themselves with the relation 'to be equivalent to an object'.

4.3   Restrictions on Subjects and Direct Objects of Verbs

Restrictions on subjects and direct objects of verbs were necessary for the correct analysis of compounds and idioms, as well as for eliminating unnecessary syntactic versions of syntactic parsing.

To exclude the versions (4.1–2, 7–8, 11–12, 15–16, 19–20) in the example (4) the restrictions were applied to the subject of the verb *skyug* 'vomit'. The subject valency of the verb was limited to the basic class 'any creature', so that only creatures can vomit. To exclude versions (4.1–3, 5, 7, 9, 11, 13, 15, 17, 19, 20) the same subject was indicated for the verb *'bor* 'lose'.

In example (5), the restrictions were also applied to the subject of the verb:

(5) སྲིད་པའི་སྒྲོན་མེ་བསུས་པའི་དྲིན་ཆེ

srid-pa 'i sgron-me bsus-pa 'i drin-che

exist-NMLZ GEN lamp invite-PST-NMLZ GEN kindness-be_great

(5.1) 'the kindness that invited the lamp of existence is great'
(5.2) 'the kindness of [someone who] exists, that invited the lamp, is great'
(5.3) 'the kindness of [someone who] invited the lamp of existence is great'[2]

In most cases, the work with a combinatorial explosion within one phrase required simultaneous use of several types of restrictions, like in (6).

(6) མིང་གཞི་རྣམས་དང་ཕྲད་པའི་ཚེ་ན་སྔོ

ming gzhi-rnams dang phrad-pa 'i tshe-na sngo
word base-PL ASS meet-PST-NMLZ GEN time-LOC become_green
(6.1) 'the word turns green in the time of meeting with the bases'
(6.2) 'in the time of meeting with the bases – words [something] turns green'
(6.3) 'in the time – word of meeting with the bases [something] turns green'
word base-PL CONJ meet-PST-NMLZ GEN time-LOC become_green
(6.4) 'in the time of bases – words and meetings [something] turns green'
(6.5) '[something] turns green in the word – the time of bases and meetings'
ming-gzhi-rnams dang phrad-pa 'i tshe-na sngo
basic_phoneme-PL ASS meet-PST-NMLZ GEN time-LOC become_green
(6.6) 'in the time of meeting with basic phonemes [something] turns green'
basic_phoneme-PL CONJ meet-PST-NMLZ-GEN time-LOC become_green
(6.7) 'in the time of basic phonemes and meeting [something] turns green'[3]

The only correct interpretation among those was (6.6).

To exclude the version (6.1), which was nonsense because it implied that words can turn green, the subject valency of the verb *sngo* 'to become green' was limited to the 'physical object' basic class (only physical objects can have colors and, thus, change them). Versions (6.3) and (6.5) were also nonsense, because they implied possibility of words being equivalent to time, so *ming* 'name, word' and *tshe* 'time' were disengaged to the basic classes 'language sign' and 'period of time', respectively.

To exclude 'time of basic phonemes' in version (6.7) the restriction on genitive relations between the basic classes 'time period' and *yi ge* 'phoneme' was established. For this purpose, the basic class *yi ge* 'phoneme' and the 'class' basic class (Tib. phreng, sde) were linked with a special subclass of the genitive relation, since phonemes can only have genitive relations with phoneme groups, where they are main representatives: *ka phreng* 'row *k*' (all consonants of the Tibetan alphabet, starting with *k*); *ca sde* '*ca* row' (the second row of the Tibetan alphabet, consisting of four graphemes, starting with *ca* - *ca*, *cha*, *ja*, *nya*).

Two wrong versions of parsing (6.2) and (6.4) were caused by incorrect interpretations of the compound *ming-gzhi* as a compound of a name root noun phrase with

---

[2] Interpretations (5.1–2) are grammatically possible, but semantically nonsense, because they imply that kindness can invite. The subject valency of the verb *bsu* 'invite' was limited to the basic class 'any creature', so that only creatures can invite. This allowed to exclude versions (5.1–2). In the version (5.3) the subject was determined correctly.

[3] These seven interpretations represent seven groups of versions that had been built before semantic restrictions described below were introduced; the total amount of parsing versions without semantic restrictions is 28.

attribute and as a noun phrase with adjunct (6.4). In order to exclude these versions, the corresponding type of the token was indicated in the ontology (see paragraph 5).

5   The System of New Types of Tokens in the Ontology

Initially, the ontology allowed marking the expression as an idiom and establishing a separate type of token, common for nominal compounds. Since a large number of combinatorial explosions were caused by the incorrect versions of compounds parsing (the same sequence of morphemes can be parsed as compounds of different types) and their interpretation as noun phrases of different types, it was decided to expand the number of token types in the ontology according to identified types of nominal and verbal compounds.

Depending on the syntactic model of the composite formation, the following types were distinguished for nominal compounds: composite noun root group (CompositeNRootGroup, (7)); noun phrase with genitive composite (NPGenComposite, (8), formed from (9)); composite class noun phrase (CompositeClassNP, (10)); named entity composite (NamedEntityComposite, (11)); adjunct composite (AdjunctComposite, (12)); and for verbal compounds: composite transitive verb phrase (CompositeTransitiveVP, (13)); verb coordination composite (VerbCoordComposite, (14)).

(7) ཨིང་ཚིག
ming-tshig
word_phrase
'words and word phrases'

(8) ཁ་རྒྱན
kha-rgyan
mouth_ornament
'mustache'

(9) ཁའི་རྒྱན
kha 'i rgyan
mouth GEN ornament
'ornament of mouth'

(10) སྐལ་བཟང
skal-bzang
fortune_be_good
'good fortune'

(11) ས་ཡིག
sa-yig
s_phoneme
'phoneme s'

(12) ཚེ་སྐབས
tshe-skabs
time_time
'time'

(13) ཕན་འདོགས
phan-'dogs
help_fasten
'assist'

(14) སངས་རྒྱས
sangs-rgyas
be_purified_be_broad
'awaken and broaden'

For all compounds the setting 'only_idiom = True' was also made. According to this setting any non-idiomatic interpretations of a compound are excluded.

Thus, in example (15) there is one composite – (16), the wrong interpretation of which formed 9 versions of parsing.

(15) ས་ཕག་བཤིབས
sa phag bshibs
earth pig arrange_in_rows-PST

(15.1) '[someone] arranged the earth - pig'
(15.2) 'earth arranged a pig'
(15.3) 'the earth - pig arranged [something]'
brick arrange_in_rows-PST
(15.4) '[someone] arranged bricks'
(15.5) 'brick arranged [something]'[4]
(16) ས་ཕག
sa-phag
earth_pig
'brick'

In the versions (15.1, 3) of the example (15), the noun-compound *sa-phag* 'brick' is parsed as a noun group with an adjunct. Indicating the correct type of token for a composite excludes these versions. The subject valency of the verb *gshib* 'arrange_in_rows' was limited to the basic class 'any creature', so that only creatures can arrange something. This allowed to exclude the remaining incorrect versions (15.2, 3, 5). Compound (16) has four versions: as an adjunct composite, as a composite noun root group, as a named entity composite and the correct one as noun phrase with genitive composite (Fig. 1).

It should be noted that specifying the correct type of token for compounds in the ontology does not always completely eliminate the ambiguity of their parsing, since the same Tibetan compound may have different morphosyntactic structures for different meanings. Thus, the compound sgra-don is a clip of two different phrases – (17) and (18).

(17) སྒྲའི་དོན
sgra 'i don
sound GEN meaning
'meaning of sound'
(18) སྒྲ་དང་དོན
sgra dang don
sound CONJ meaning
sound and its meaning'

Thus, the correct type of token in the first case is NPGenComposite, and in the second one - CompositeNRootGroup. These cases are represented in the ontology as different concepts of the same expression.

Depending on which language unit is idiomatized, Tibetan non-compound idioms are divided into separate derivatives and nominal, verbal, adjectival and adverbial phrases. Separate derivatives basically represent nouns and adjectives formed by various derivational suffixes (*khyim* 'house' → *khyim-pa* 'householder'; *nor* 'wealth' *nor-bu* 'precious stone') and verb nominalizations with uncommon meanings (*mkhas* 'to be learned' → *mkhas-pa* 'scholar', *rig* 'to understand' → *rig-pa* 'science').

---

[4] Versions (16.4) and (16.5) both represents groups of four versions of parsing, depending on the type of token of the compound (17).

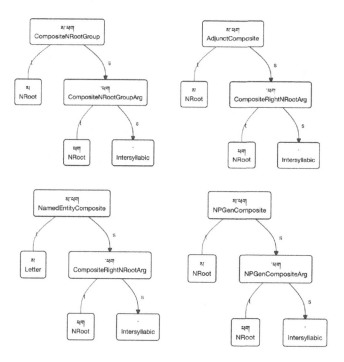

Fig. 1. *sa-phag* as CompositeNRoot, AdjunctComposite, NamedEntityComposite and NPGenComposite.

As with compounds, a list of classes of immediate constituents that can be idioms was built. E.g., on the basis of the syntactic structure of idiomatic expressions, it is possible to divide the idioms of noun phrases into instance noun phrase with quantifier, quantitative noun phrase etc. The system of token types in the ontology database has been extended with these types.

6 Conclusion

As we have seen, only semantic restrictions can delimit the variety of morphosyntactic versions while analyzing Tibetan texts due both to lack of word delimiters and to high degree of homonyny of morphemes and allomorphs.

Ontology which is necessary to introduce such restrictions will also serve as a base for build a model of Tibetan linguistic picture of the world. It would also be necessary to continue the work of disambiguation of morphosyntactic analysis of Tibetan text on the next and final level of supraphrasal units and discourse.

Acknowledgments. This work was supported by the Russian Foundation for Basic Research, Grant No. 16-06-00578 Morphosyntactycal analyser of texts in the Tibetan language.

References

1. Grokhovskii, P.L., Zakharov, V.P., Smirnova, M.O., Khokhlova, M.V.: The corpus of Tibetan grammatical works. Autom. Doc. Math. Linguist. **49**(5), 182–191 (2015). https://doi.org/10.3103/S0005105515050064

2. Aho, A.V., Corasick, M.J.: Efficient string matching: an aid to bibliographic search. Commun. ACM **18**(6), 333–340 (1975)

3. Klaus, K.: Combinatorial explosion. Web Dictionary of Cybernetics and Systems. PRINCIPIA CYBERNETICA WEB. http://pespmc1.vub.ac.be/ASC/Combin_explo.html

4. Berment, V.: Méthodes pour informatiser des langues et des groupes de langues peu dotées. Ph.D. Thesis, J. Fourier University – Grenoble I (2004)

5. Beyer, S.V.: The Classical Tibetan Language, New York (1992)

6. Meelen, M., Hill, N., Handy, C.: The Annotated Corpus of Classical Tibetan (ACTib), Part II – POS-tagged version, based on the BDRC digitised text collection, tagged with the Memory-Based Tagger from TiMBL (2017). https://doi.org/10.5281/zenodo.822537

7. Dobrov, A., Dobrova, A., Grokhovskiy, P., Soms, N., Zakharov, V.: Morphosyntactic analyzer for the Tibetan language: aspects of structural ambiguity. In: Sojka, P., Horák, A., Kopeček, I., Pala, K. (eds.) TSD 2016. LNCS (LNAI), vol. 9924, pp. 215–222. Springer, Cham (2016). https://doi.org/10.1007/978-3-319-45510-5_25

8. Dobrov, A., Dobrova, A., Grokhovskiy, P., Soms, N.: Morphosyntactic parser and textual corpora: processing uncommon phenomena of Tibetan language. In: Proceedings of the International Conference IMS-2017, pp. 143–153 (2017). https://doi.org/10.1145/3143699.3143719

9. Dobrov, A.V.: Automatic classification of news by means of syntactic semantics [Avtomaticheskaja rubrikacija novostnyh soobshhenij sredstvami sintaksicheskoj semantiki], Doctoral Thesis. Saint-Petersburg State University (2014)

10. Haspelmath, M.: The indeterminacy of word segmentation and the nature of morphology and syntax. Folia Linguistica **45**(1), 31–80 (2011)

11. Tomita, M.: An efficient augmented context-free parsing algorithm. Comput. Linguist. **13**, 31–46 (1987)

12. Popov, A., Protopopova, E., Bukiya, G.: Once more on ways to overcome structural homonymy: selection of the only structure in the Hurma parcer [Esche raz o sposobah snyatiya strukturnoy omonimii: vybor edinstvennoi struktury v parsere Hurma. In: Collection of Scientific Articles of the 19th Joint Conference "Internet and modern society" [Sbornik nauchnih statei 19 Obyedinennoy konferenzii "Internet i sovremennoe obscchestvo] IMS-2015, Saint-Petersburg, 22–24 June (2016)

13. Popov, A., Enikeeva, E.: Template search algorithm for multiple syntactic parses. In: Proceedings of the International Conference IMS-2017 (IMS2017), pp. 164–170. ACM, New York (2017). https://doi.org/10.1145/3143699.3143732

14. Tseitin, G.S.: Programming in associative networks [Programmirovanie na associativnyh setjah]. Computers in designing and manufacturing [EVM v proektirovanii i proizvodstve] (2). Mashinostroenie, Leningrad, pp. 16–48 (1985)

15. Dobrov, A.V.: Semantic and ontological relations in AIIRE natural language processor. In: Computational Models for Business and Engineering Domains, pp. 147–157. ITHEA, Rzeszow-Sofia (2014)

16. Mel'čuk, I.A., Zholkovsky, A.: Explanatory Combinatorial Dictionary of Modern Russian (1984)

17. Ideographic dictionary RUSLAN Leont'eva N

18. Matuszek, C., Cabral, J., Witbrock, M.J., DeOliveira, J.: An introduction to the syntax and content of Cyc. In: AAAI Spring Symposium: Formalizing and Compiling Background Knowledge and Its Applications to Knowledge Representation and Question Answering, pp. 44–49 (2006)

19. Rubashkin, V.Sh., Fadeeva, M.V., Chuprin, B.Y.: The technology of importing fragments from OWL and KIF-ontologies [Tekhnologiya importa fragmentov iz OWL i KIF-ontologij]. In: Proceedings of the Conference "Internet and Modern Society [Materialy nauchnoj konferencii "Internet i sovremennoe obshchestvo"]", pp. 217–230 (2012)

20. Gruber, T.R.: A translation approach to portable ontology specifications. Knowl. Acquis. **5**(2), 199–220 (1993). https://doi.org/10.1006/knac.1993.1008

21. Bateman, J.A., Hois, J., Ross, R., Tenbrink, T.: A linguistic ontology of space for natural language processing. Artif. Intell. **174**(14), 1027–1071 (2010)

22. Dahlgren, K.: A linguistic ontology. Int. J. Hum. Comput. Stud. **43**(5–6), 809–818 (1995)

23. Borgo, S., Guarino, N., Masolo, C., Vetere, G.: Using a large linguistic ontology for internet-based retrieval of object-oriented components. In: The Ninth International Conference on Software Engineering and Knowledge Engineering, pp. 528–534 (1997)

24. Pelletier, F.J.: The principle of semantic compositionality. Topoi **13**, 11 (1994). https://doi.org/10.1007/BF00763644

25. Great Tibetan-Chinese dictionary [bod rgya tshig mdzod chen mo]. Beijing (1985)

26. Miller, G.A., Beckwith, R., Fellbaum, C.D., Gross, D., Miller, K.: WordNet: an online lexical database. Int. J. Lexicograph. **3**(4), 235–244 (1990)

Using Explicit Semantic Analysis and Word2Vec in Measuring Semantic Relatedness of Russian Paraphrases

Anna Kriukova[1(✉)], Olga Mitrofanova[1], Kirill Sukharev[2],
and Natalia Roschina[1]

[1] Saint Petersburg State University,
Universitetskaya emb 11, St. Petersburg 199034, Russia
krukova.ann@gmail.com, oa-mitrofanova@yandex.ru,
tasharoshchina@gmail.com
[2] Saint Petersburg Electrotechnical University «LETI»,
Popova str., 5, St. Petersburg 197376, Russia
sukharevkirill@gmail.com

Abstract. In this study we compare two semantic relatedness algorithms, namely, Explicit Semantic Analysis (ESA) and Word2Vec. ESA represents text meaning in a high-dimensional space of concepts derived from Wikipedia. Word2Vec generates distributed vector representations from large text corpora). Experiments were carried out on the Russian paraphrase corpus of news titles and Russian ParaPlag paraphrase corpus. The paper contains thorough analysis of results and evaluation procedure.

Keywords: Text relatedness · Explicit semantic analysis · Word2vec
Russian

1 Introduction

Measuring semantic similarity and relatedness is one of the important procedures in contemporary computational linguistics, it defines how similar two words or texts are in terms of their meaning. Attempts to calculate semantic distance between words or texts have been carried out in certain research projects which gave rise to a variety of approaches leading to theoretically consistent and practically implementable solutions, cf. review [1]. Importance of research in this field is due to the fact that its results are applicable in many tasks of computational linguistics, e.g. clustering and classification of words and documents in a corpus, development of text paraphrasing and summarization systems, lexical disambiguation, etc.

There are plenty of algorithms and metrics used in measuring text similarity [2, 3]. These techniques have been implemented in certain toolkits which are in current use[1]. Most of them are discussed within a SemEval Competition (Semantic Textual Similarity task)[2]. Measures of text similarity and relatedness can be roughly divided into

---

[1] https://metacpan.org/pod/Text::Similarity.
[2] https://en.wikipedia.org/wiki/SemEval.

D. A. Alexandrov et al. (Eds.): DTGS 2018, CCIS 859, pp. 350–360, 2018.

two groups: string metrics (taking into account graphical representation of texts) and knowledge-based metrics (relying upon external knowledge databases describing word senses and their relations) [4].

String metrics can be applied to Russian data practically with no adjustments [5]: as a rule they are based on character sequences in two texts and, as a consequence, are completely independent of the language. However, results in that case cannot be particularly high, because such methods are not capable of identifying semantic relations between words (synonymy, hyponymy, etc.).

Semantic algorithms are based on additional data which allow them to overcome drawbacks of string measures. External knowledge sources can be either lexical-semantic tools or text corpora. Widely used computational solutions allowing to measure semantic distances in pairs of words can be found in knowledge databases plugins WordNet::Similarity[3], Roget's thesaurus electronic database[4], etc., cf. a survey in [6].

Vector space models provide powerful techniques of measuring text relatedness. In Latent Semantic Analysis (LSA) [7] texts are transformed to vectors by means of a dimensionality reduction operated by a singular value decomposition (SVD) on the term-by-document matrix representing the corpus. Recently developed distributional semantic models based on embeddings (Word2Vec, Doc2 Vec) allow to build vectors for words, sentences and documents [8, 9]. Within vector-based models the degree of text semantic relatedness may be computed by a conventional cosine metric.

In this paper we focus our attention on a particular algorithm, namely Explicit Semantic Analysis (ESA) [10] which inherits the advantages of knowledge-based and distributional approaches to measuring text similarity and relatedness. ESA turns out to be a hybrid technique allowing to calculate semantic distance as a cosine measure for vectors representing texts in a high-dimensional space model build for concepts described in Wikipedia.

We are the first to use ESA in performing experiments for computing semantic relatedness of Russian texts. The aim of the present study is to perform comparative tests for ESA and Word2Vec distributional embedding model in processing sets of paraphrases differing in source and transformation type.

2 Models and Tools

2.1 Explicit Semantic Analysis and Its Implementation

Explicit Semantic Analysis (ESA) is a variety of vector space models for texts or words proposed by [10]. The algorithm requires a big document collection as an external knowledge source – as a rule, Wikipedia is used for this purpose. This was one of the reasons why we have chosen ESA in our research: Wikipedia is an open and constantly growing source of tens of thousands texts in the Russian language (Russian version of Wikipedia contains now almost 1,5 mln articles). Wikipedia turns out to be a rich

---

[3] http://wn-similarity.sourceforge.net/.

[4] http://community.nzdl.org/ELKB/.

linguistic resource which is edited by users, it guaranties that Wikipedia is continuously expanding and becoming deeper, thus it contains relevant information about the world.

Within ESA approach Wikipedia articles are identified with concepts due to the fact that each article is concerned with a specific topic described in detail. The concepts are created and edited by people, so that results provided by the algorithm can be easily interpreted. Each concept is represented by a vector reflecting words which co-occur in the corresponding article. Words are assigned with weights (according to TF-IDF scheme) which represent association strength. ESA describes the meaning of a text in terms of a weighted vector of concepts sorted according to their relevance to the text (the so-called "interpretation vector"). In order to speed up semantic interpretation of texts, an inverted index is then created. An inverted index links each word with a set of articles (concepts) in which it occurs. At this point, concepts whose weights are too small are deleted, that allows to eliminate insignificant links between words and concepts.

Summing up, ESA represents text meaning in a high-dimensional space of concepts derived from Wikipedia. In this case computing semantic relatedness of texts consists in comparison of the corresponding vectors by means of conventional metrics (e.g. the cosine metric).

In order to adjust ESA algorithm to Russian text processing we decided to make use of an available implementation (instead of making it from scratch) and to process Russian Wikipedia which contains the input data for ESA.

For this reason we had to choose from a number of different resources: DKPro Similarity[5] Wikiprep-ESA[6], EasyESA[7], Research-ESA[8], ESAlib[9], Wikipedia-based Explicit Semantic Analysis[10], S-Space Package[11].

In our research we made use of S-Space Package because this library is free, it has a relatively easy programming interface and can be run out-of-the-box. It contains a considerable number of text processing instruments, including those for building semantic spaces and computing distances within them; algorithms for matrix transformation; means of data serialization. The package contains the ESA code on java[12]. One of the drawbacks is, however, a relatively old documentation on GitHub, which obliges a user to address the source code.

In order to work with data from Wikipedia, it was necessary to download a dump with articles which have been included by a certain time. We've used a xml-dump from 20.02.17 that contains almost 1 375 000 articles, and converted the obtained data into json format.

---

[5] https://github.com/dkpro/dkpro-similarity.

[6] https://github.com/faraday/wikiprep-esa.

[7] http://treo.deri.ie/easyesa/.

[8] https://code.google.com/archive/p/research-esa/.

[9] https://github.com/ticcky/esalib.

[10] https://github.com/pvoosten/explicit-semantic-analysis.

[11] https://github.com/fozziethebeat/S-Space/.

[12] https://github.com/fozziethebeat/S-Space/blob/master/src/main/java/edu/ucla/sspace/esa/ExplicitSemanticAnalysis.java .

2.2 Word2Vec and Its Implementation

Word2Vec includes a particular class of semantic models which allow to generate distributed vector representations for words (word embeddings) from large text corpora, that is, to project word contexts extracted from a corpus into a high-dimensional vector space, so that vectors for words with similar context neighbours are placed close to each other. Measuring semantic relatedness within Word2Vec is based on calculating cosine values for context vectors. Word2Vec follows the distributional hypothesis which explains semantic similarity and relatedness via common context features shared by linguistic units.

Word embeddings can be produced by two neural network models, namely, Continuous Bag-of-Words (CBOW) and Continuous Skip-Gram. CBOW model predicts words in a given context, while Skip-Gram model predicts a set of contexts for a given word.

Word2Vec was developed by a research group from Google [8], further it was expanded and elaborated in various toolkits, e.g. GenSim[13], TensorFlow[14], DeepLearning4j[15], etc. Vector representations generated by Word2Vec are used for training natural language processing tools and nowadays it is hardly possible to dispense with such data in solving most of the tasks within computational linguistics (semantic relations extraction, word and text clustering, etc.).

Word2Vec allows to work with distributed representations both for words and for phrases. Given word vectors, it is possible to calculate phrase vectors as a concatenation of word vectors. Alongside with Word2Vec one can use Doc2 Vec in this case [9].

We carried out experiments with Word2Vec Skip-Gram model being the most suitable one: it performes better than Word2Vec CBOW model in semantic and syntactic tasks as it is aimed at predicting syntagmatic relations [8]. A model pre-trained on news corpus (news_mystem_skipgram_1000_20_2015) was taken from the RusVectōrēs[16] project for Russian.

2.3 Linguistic Data

Retrieving paraphrases from texts or evaluating the degree of relatedness between them is now one of the most discussed tasks in the field of computer linguistics, for it is employed in many fields, e.g. it is closely connected with text reuse detection.

A number of competitions concerned with the task can illustrate its importance. First of all, a part of international competitions PAN[17] is dedicated to plagiarism detection and consists of six subtasks: Source Retrieval, Text Alignment, Intrinsic Plagiarism Detection, Cross-language Text Reuse Detection, External Plagiarism

---

[13] https://radimrehurek.com/gensim/models/word2vec.html.

[14] https://github.com/tensorflow/tensorflow/blob/r1.1/tensorflow/examples/tutorials/word2vec/word2vec_basic.py.

[15] https://github.com/deeplearning4j/deeplearning4j.

[16] http://rusvectores.org/ru/.

[17] http://pan.webis.de/tasks.html.

Detection, and Source Code Reuse Detection. The competitions took place since 2009 and up to 2015 and was concerned only with the English language.

However, there are some competitions for Russian as well. A seminar PlagE-valRus[18] dedicated to evaluation of plagiarism detection algorithms took place in 2016-2017. The results were announced and discussed during the Dialogue-2017 conference on Computational Linguistics. Participants could choose one of the following tracks: source retrieval, verbatim borrowed fragments (text alignment), or paraphrased borrowed fragments (text alignment).

The next stage of this seminar took place during the AINL conference in September 2017. The hackathon[19] was dedicated to retrieving paraphrased borrowings from texts and the task consisted in finding their exact bounds. One of the corpora we used in our experiments was taken from test data for this task, namely, the Paraplag corpus.

Paraplag corpus contains data for paraphrased plagiarism detection. It includes academic texts with borrowings, automatically generated paraphrases and essays with manually paraphrased texts. The size of the whole corpus is 5,7 mln tokens. We have made use of two Paraplag sub-corpora with manually generated paraphrases of different complexity. The texts included into our dataset are annotated as regards the type of paraphrasing techniques [11].

We used datasets for the "text alignment" task which consists of searching for the "original" of all borrowed sentences in a document. Two subcorpora which we employed have "medium" and "hard" difficulty level (manually_paraphrased and manually_paraphrased2). The former corpus contains sentences modified by a single paraphrasing technique (in contrast to many techniques, as in the latter corpus). Both corpora have golden standards providing information on each borrowed fragment in each document with reference to a source document and details about offset and length of the "original" fragment. We randomly chose 15 documents from the golden standards and extracted 314 and 367 paraphrases for the first and the second corpus respectively (a document usually contains several paraphrases).

It's worth mentioning, that paraphrases in the Paraplag corpus were manually created using different paraphrasing techniques:

DEL – delete some words of the original sentence.

ADD – add some words into the original sentence.

LPR (Light Paraphrase) – for texts of medium difficulty level: replace some words or phrases of the original sentence with synonyms, reorder clauses, add new words. For texts of hard difficulty level: change word forms (number, case, form and verb tense, etc.) for some words in the original sentence.

SHF (shuffling) – change the order of words or clauses in the original sentence.

CCT (concatenation) – concatenate two or more original sentences into one sentence.

SEP or SSP (sentence splitting/separation) – split the original sentence into two or more sentences (possibly with a change in the order they appear in the text).

---

[18] https://plagevalrus.github.io/.

[19] http://ru-eval.ru/plageval/rules.html.

SYN (synonymizing) – replace some words or phrases of the original sentence with synonyms, replace abbreviations to their full transcripts, and vice versa, replace the person's name with the name initial, etc.

HPR (Heavy Paraphrase) – complex rewrite of the original sentence, which combines 3–5 or even more aforementioned techniques.

CPY – copy the sentence from source and paste it into essay almost with no changes.

For all the texts from our dataset preprocessing was performed: punctuation marks and figures were deleted, lemmatization was done with the help of PyMorphy2 morphological analyzer [12].

Another source we used for collecting data for our experiments is Paraphraser.ru corpus which is a Russian corpus of paraphrases containing Russian sentence pairs extracted from news headlines.

We used the corpus version containing 6281 sentence pairs (1482 precise, 3247 loose and 2209 non-paraphrases) [13]. The corpus is manually annotated so that each pair of paraphrases gets a mark as regards semantic similarity: complete semantic coincidence ("1"), slight semantic divergence ("0") or considerable semantic difference ("-1"). For our experiments we took 120 pairs of sentences: 30 from each group "1", "0", "-1", as well as 30 pairs with sentences from different groups (we marked this group "-2").

3   Experimental Results

3.1   Measuring Relatedness of Paraphrases from Paraphraser.Ru Corpus

For the task of splitting paraphrases into 4 groups resulting cosine values were insufficient. However, when splitting them into 2 groups ("1" and "0" vs "-1" and "-2"), the results both in ESA and W2V turned out to be rather high: given the cosine value threshold 0,5 for ESA and 0,8 for W2V, precision equaled to 0,79 and 0,86 respectively. The threshold 0,8 for W2V was established empirically on the basis of error rates for different thresholds.

The Spearmen correlation coefficient between ESA and W2V cosine values is very high, namely, 0,86. For this reason while analyzing the results we have focused on cases when one of the algorithms provides a correct answer and the other one makes a mistake. Out of nine such examples W2V turned out to be better in seven and ESA in two. Table 1 gives data about some of these examples.

A possible explanation of mistakes is provided below, after the description of the next set of experiments.

3.2   Measuring Relatedness of Paraphrases from Paraplag Corpus

First of all, it's worth noticing that Pearson correlation between ESA and W2V cosine values in these experiments is also high and equals to $\sim 0,77$ for both "hard" and "medium" corpora. However, there are considerable differences in the results in terms of precision.

Table 1. Some results from the Paraphraser.ru corpus. The threshold values are 0,5 for ESA and 0,8 for W2V.

ESA	W2V	GOLD	SENT1	SENT2
0,421	0,821	1	Новые комиксы о пророке Мухаммеде появились во Франции	Во Франции издали комикс по мотивам жизни пророка Мухаммада
0,664	0,767	−1	Россия даст Бангладеш $500 млн на строительство первой в стране АЭС	Росатом назвал дату строительства АЭС в Бангладеш
0,784	0,755	1	В Ватикане перестали принимать платежи банковскими картами	ЦБ Италии запретил платежи банковскими картами в Ватикане
0,539	0,752	−1	МВД: Похищенный в Приморье металл не связан с саммитом АТЭС	При строительства моста на саммит АТЭС похитили металл на 96 миллионов

For the "hard" corpus W2V works much better than ESA: with the threshold cosine value 0,8 for W2V and 0,5 for ESA the precision reaches 0,67 and 0,58 respectively. For the "medium" corpus, on the contrary, ESA shows a bit better results, the precision namely being P = 0,93 for ESA and P = 0,91 for W2V. Some results are represented in Tables 2 and 3.

Table 2. Some results from the "medium" Paraplag corpus

KIND	ESA	W2V	Susp	Source
HPR	0,466	0,9017	Ориентирами называются особенности рельефа, отличительные местные предметы, относительно чего можно определить местоположение	Местные предметы и формы рельефа, относительно которых определяют свое местоположение, положение целей (объектов) и указывают направление движения, называются ориентирами
CCT	0,443	0,8261	В июле 2006 года Tesla решила поведать миру о своих планах, для этого инженеры компании построили красный прототип — EP2. Оба автомобиля были выставлены на мероприятии в Санта-Кларе	В июле 2006 года Tesla решила поведать миру о своих планах

(continued)

Table 2. (*continued*)

KIND	ESA	W2V	Susp	Source
HPR	0,833	0,7468	В скором времени Фостер ушел и его не на долгое время меняет Дональд Джозеф Мэскис	Вскоре Фостер отчалил и его ненадолго заменил Дональд Джозеф Мэскис (Donald Joseph Mascis из DINOSAUR Jr.)
HPR	0,650	0,7804	У Курта с детства была склонность к искусству	Курт был застенчивым мальчиком с художественными наклонностями

Table 3. Some results from the "hard" Paraplag corpus

KIND	ESA	W2V	Susp	Source
CCT, DEL	0,922	0,782	Но эти кислоты могут быть и природными, они содержаться в молоке и мясе	Такие кислоты могут быть природными
HPR	0,783	0,787	Среди известных художников начала XIX века можно назвать имена Кипренского, Брюллова, Иванова	Выдающиеся художники первой половины XIX века стали Кипренский, Брюллов, Иванов (« Явление Христа народу »)
ADD, DEL, HPR, SYN, SHF	0,606	0,786	В прошлом он был спортсменом и знал, что требуют спортсмены от их экипировки	Он тоже был спортсменом и отлично понимал, какой нужен подход к людям из спортивного мира
ADD, DEL, SYN	0,220	0,842	Занятия спортом в зрелом возрасте могут увеличить КПД работы от 20 до 25%	Считается, что занимаясь физическими нагрузками в зрелом возрасте можно тем самым увеличить коэффициент полезного действия работы от 20 до 25 процентов
CCT, SHF	0,075	0,827	Мы привыкли воспринимать такую еду как бутерброды, гамбургеры и сэндвичи как легкие перекусы на бегу, поэтому фастфуд не дает нам четкого понимания, что у нас был полноценный прием пищи	Сам принцип быстрого питания не дает возможности осознать, что был съеден полноценный обед

(*continued*)

Table 3. (*continued*)

KIND	ESA	W2V	Susp	Source
ССТ, DEL	0,152	0,801	В 1934 году была провозглашена доктрина соцреализма, а другие направления были вне закона, группировки художников были упразднены	В 1934 году в СССР была официально-провозглашена доктрина социалистического реализма

As in the previous experiment described above, in the analysis we deal with cases, where only one of the algorithms yields an incorrect result.

3.3   Error Analysis

As can be seen, W2V works better in cases where the difference between sentences is quite sharp, and ESA yields better results where they are more similar. That is why sentences that differ mainly via synonyms are better dealt with by ESA, the algorithm being capable of identifying synonymy more precisely.

In experiments with Paraphraser corpus pair (1) differing in verbs ("появляться – издавать") obtains a correct high cosine value for Word2Vec and a low value for ESA, at the same time, pair (4) differing in the lexical content as well as in syntactic structure gets an overrated value for Word2Vec and a correct low value for ESA. In tests performed with Paraplag pairs (2) ("medium" corpus) and (5) ("hard corpus") with the coinciding core phrases but differing in length get low values for ESA and high values for Word2Vec.

This may be connected with the fact that for each word Word2Vec retains many words with similar meaning, but that are both in syntagmatic and paradigmatic relations with the target word [14].

4   Summary

In course of research work on measuring semantic relatedness of Russian paraphrases by means of Explicit Semantic Analysis and Word2Vec distributional word embedding model we studied these algorithms and found specific aspects of their work on different datasets.

We carried out experiments on three corpora, one of them containing paraphrases out of news articles' titles and two other consisting of sentences from essays, manually written using different paraphrasing techniques, as well as their sources. We compared the work of both algorithms, namely, ESA based on representation of texts in a high-dimensional space model built for concepts from Wikipedia, and W2V Skip-Gram model relying on word embedding and trained on news articles.

ESA, being only recently introduced for the Russian language, has not yet been compared to any of other algorithms for computing semantic relatedness of texts, and our study has allowed us to find out its advantages and drawbacks. It turned out that W2V works better for texts that have more significant differences, while ESA is beneficial for texts that are distinct mainly in synonymous words.

Future work can deal with adjustment of some features that can be changed while preparing Wikipedia data for ESA, e.g. the minimum number of words in an article for it to be indexed or a threshold for rare words to be deleted, etc. What is more, a possible line of research is exploring whether ESA works better if we take a new and bigger version of Wikipedia containing more articles.

Acknowledgements. The authors express their gratitude to anonymous reviewers for their careful reading of the paper, for their critical comments and for giving useful suggestions that helped to improve the work.

The research discussed in the paper is supported by the RFBR grant № 16-06-00529 «Development of a linguistic toolkit for semantic analysis of Russian text corpora by statistical techniques».

References

1. Mitrofanova, O.A.: Measuring semantic distances as a problem of applied linguistics. In: Structural and Applied Linguistics (in Russian), vol. 7. St.-Petersburg (2008)
2. Mihalcea, R., Corley, C., Strapparava, C.: Corpus-based and knowledge-based measures of text semantic similarity. In: Proceedings of the 21st National Conference on Artificial Intelligence, vol. 1, pp. 775–780 (2006)
3. Šarić, F., Glavaš, G., Karan, M., Šnajder, J., Bašić, B.D.: TakeLab: systems for measuring semantic text similarity. In: SemEval 2012 Proceedings of the First Joint Conference on Lexical and Computational Semantics, vol. 1–2, pp. 441–448 (2012)
4. Bär, D., Biemann, C., Gurevich, I., Zesch, T.: UKP: computing semantic textual similarity by combining multiple content similarity measures. In: SemEval 2012 Proceedings of the First Joint Conference on Lexical and Computational Semantics, vol. 1–2, pp. 435–440 (2012)
5. Kriukova, A.: Computing semantic similarity of Russian texts by means of DKPro similarity tool (in Russian). In: IMS 2017 Proceedings, St.-Petersburg (2017)
6. Rohde, D.L.T., Gonnerman, L.M., Plaut, D.C.: An Improved Model of Semantic Similarity Based on Lexical Co-Occurrence (2005). https://github.com/hbrouwer/coals
7. Landauer, T.K., Foltz, P., Laham, D.: Introduction to latent semantic analysis. Discourse Process. **25** (1998). 10.1080/01638539809545028
8. Mikolov, T., Sutskever, I., Chen, K., Corrado, G., Dean, J.: Distributed representations of words and phrases and their compositionality. In: Proceedings of NIPS (2013)
9. Le, Q., Mikolov, T.: Distributed representations of sentences and documents (2014). http://arxiv.org/pdf/1405.4053v2.pdf
10. Gabrilovich, E., Markovitch, S.: Computing semantic relatedness using wikipedia-based explicit semantic analysis. In: Proceedings of The 20th International Joint Conference on Artificial Intelligence (IJCAI), pp. 1606–1611 (2007)

11. Sochenkov, I.V., Zubarev, D.V., Smirnov, I.V.: The ParaPlag: Russian dataset for paraphrased plagiarism detection. In: Computational Linguistics and Intellectual Technologies: Papers from the Annual International Conference "Dialogue", vol. 1, pp. 284–297 (2017)
12. Korobov, M.: Morphological analyzer and generator for Russian and Ukrainian languages. Commun. Comput. Inf. Sci. **542**, 320–332 (2015). https://doi.org/10.1007/978-3-319-26123-2_31
13. Pronoza, E., Yagunova, E.: Comparison of sentence similarity measures for Russian paraphrase identification. In: Artificial Intelligence and Natural Language and Information Extraction, Social Media and Web Search FRUCT Conference (AINL-ISMW FRUCT), pp. 74–82 (2015). 10.1109/AINL-ISMW-FRUCT.2015.7382973
14. Enikeeva, E., Mitrofanova, O.: Russian collocation extraction based on word embeddings. In: Computational Linguistics and Intellectual Technologies: Papers from the Annual International Conference "Dialogue", vol. 1, pp. 52–64 (2017)

Mapping Texts to Multidimensional Emotional Space: Challenges for Dataset Acquisition in Sentiment Analysis

Alexander Kalinin[✉] ⬚, Anastasia Kolmogorova[✉] ⬚, Galina Nikolaeva ⬚, and Alina Malikova ⬚

Siberian Federal University, 79 Svobodny pr., 660041 Krasnoyarsk, Russian Federation
{aakalinin,avkolmogorova,gnikolaeva,avmalikova}@sfu-kras.ru

Abstract. The cornerstone for any sentiment analysis research is labeled data and its acquisition. Canonical corpuses for this task contain different reviews (movies, restaurants) where sentiment can be derived from reviewer's explicit rating of a reviewed item. Ratings go with supplied comments, which are used as text samples and ratings are converted into labels. Usually emotion labels come in binary form like "negative\positive".

This simplistic approach works well when we are dealing with binary emotional model, but it turns to fail when we are dealing with more complex emotional models like "Pleasure-Arousal-Dominance (PAD)" or Lövheim's Cube, when we collect data from various sources and of different types (fiction books, social networks conversations, blog posts etc.) or when we delegate labeling to external assessors.

In the article, we describe which methodological problems we faced while collecting dataset for sentiment analysis backed by Lövheim's Cube - emotional model that represents an emotion as a point in three-dimensional space of balance of three monoamines (Dopamine, Serotonin and Noradrenaline).

These problems include the choice of necessary metadata to be collected along with text and labels, choice of tools used for labeling and survey design.

Keywords: Sentiment analysis · Emotion label · Lövheim's Cube

1 Lövheim's Cube Emotional Model

The goal of the research is to investigate linguistic representations of emotions in context of Lövheim's Cube of Emotions in Russian language and their practical aspects for example in use of sentiment analysis related tasks.

We have chosen this emotional model because it is neurologically funded. Major psychological theories of emotions provide phenomenological views of emotions - usually emotions are defined subjectively depending of what a person feels. Such approaches are vague from formalist point of view and can be hardly used computer-related fields. Otherwise, Lövheim's approach provides formal representation of emotions as functions of neurotransmitters, which are responsible for neuro-somatic response towards some stimulus. This function takes objective parameters to be the

© Springer Nature Switzerland AG 2018
D. A. Alexandrov et al. (Eds.): DTGS 2018, CCIS 859, pp. 361–367, 2018.
https://doi.org/10.1007/978-3-030-02846-6_29

arguments [2] and these parameters are mixtures or proportions of three monoamine neurotransmitters: serotonin, dopamine and noradrenaline. The balance of these monoamines forms a three-dimensional space represented by the following visual mode (Fig. 1):

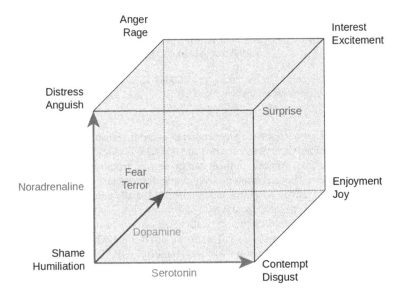

Fig. 1. Lövheim's cube of emotions

Such approach for modelling emotions numerically is also used outside neurology. It is used for creating intellectual emotional agents in robotics or computer games.

2 Data Acquisition for Sentiment Analysis

For this particular research or other research that deals with sentiment analysis, we need labeled or annotated data. A big amount of randomized samples from different sources makes it feasible for conduction of deep and reliable research in sphere of relations between a text and emotional state that stand behind it on big dataset.

Canonical corpora which are available for English and Russian mostly contain datasets with binary labels - "positive" or "negative". Such corpora can be used for building models for text classification, like reviews, forums and social networks comments, but can't be used for deep research in emotion-text interconnection as they are bound to binary scale, and emotions are certainly can't be reduced to positive\negative ones.

Unfortunately, there are few large corpora for Russian language with labels beyond binary scale, and there are no corpora labeled according to Lövheim's model neither for Russian nor for English. So in our research we faced the need of collecting such a corpus. Certainly such data should be collected from different sources with help of assessor with different social and demographic background. For this task the solution appeared to be crowdsourcing platforms, which can facilitate speed of label with help of large community.

If problem with assessors can be considered to be solved, the problem upon how to label data remains open - two main questions arise

- How to map texts with Lövheim's Cube emotional model inventory
- Which tools to provide for such a mapping

The straightforward way is to use labels - there are 8 vertices of the cube that can be used as corresponding labels. When can add a "neutral" label to this set and suggest an assessor to choose the label that he thinks is the most suitable for the text sample.

But in such approach we potentially lose information for cases with intermediate emotions. The advantage of the Lövheim's Cube emotional model is that it provides non-discrete continuous estimation of emotion states, as emotions themselves are not discrete - particular emotion can be a mix of anger and disgust, or excitement and surprise on so on, and each component in emotion can be expressed higher or lower. In discrete approach, we have to invent names for such "intermediate" emotions, but Lövheim's model allows us to use numeric coordinates. For example: what could be the name for an emotion between surprise and disgust? No need to name this complex emotion, in terms of Lövheim's Cube model it can be referenced as (0.5, 0, 1) where first coordinate is for noradrenaline level, second is for dopamine and the third is for serotonin. That problem is which tools we should provide to map text to this coordinates.

3   Mapping Emotions to Point in Space

Major feature of Lövheim's cube are coordinate axes, which correspond to degree of presence of three monoamines - noradrenaline, serotonin and dopamine. Minimal degree of all monoamines corresponds to Shame\Humiliation; the maximum degree corresponds to Interest\Excitement. It's quite straightforward to ask an assessor to set coordinates of emotion in three coordinates, for example with help of scroll bars (Fig. 2):

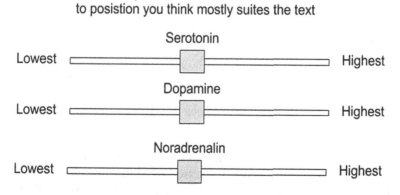

Fig. 2. Scroll-bars for assessing text with monoamines expressions

With help of scrollbars we can map their positions to the point to explicitly setting its coordinates. But here we face with problem that an ordinary assessor is not aware for such unknown thing as monoamines expression. Moreover neurological details and necessary pre-teaching of those concepts for assessor can be a serious distraction while data acquisition and so may lead to low quality of assessments.

The workaround here maybe to interpret monoamines degree of expression in terms familiar to an assessor. For example, noradrenalin is responsible for arousal, dopamine is for reinforcement, and serotonin is for self-confidence [7]. But nevertheless such terms are too abstract and quite neurology-domain specific. An ordinary assessor can hardly evaluate "reinforcement" or "arousal" of a sentiment in text. We should step out from verbalizing monoamines functions in this task and search for other variants that should definitely engage terms familiar to an assessor to operate.

3.1 Mapping Emotions Using Supporting Cube Diagonals

For representing emotional space in terms familiar for an assessor we can use oppositions between emotions situated on cube's vertices. For such oppositions we can use cube diagonals shown on Fig. 3:

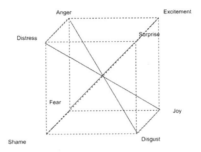

Fig. 3. Supporting diagonals of Lövheim's Cube

In such case we have four oppositions

- Distress - Enjoyment
- Rage - Disgust
- Shame - Excitement
- Fear - Surprise

Here we can see that these emotion pairs forms four dichotomies and at their extreme points monoamines' balance is inverted - fear, for example, is a negative form of surprise, rage is aggressive antonym of disgust etc. These terms are familiar for assessors, and they can reliably estimate their subjective impact of a text sample. Moreover with such approach respondents may scale the emotional expression to differentiate for example "very aggressive" text from "moderately aggressive" one.

With these oppositions we may ask assessors to adjust each slide bar to positions they think mostly corresponds to. If one of oppositions is not expressed in text the scrollbar should remain in central (neutral) position. Possible interface for this kind of task is presented on Fig. 4:

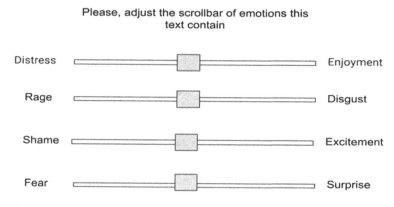

Fig. 4. Scrollbars with diagonal oppositions

Such an approach also provides opportunity to point "secondary emotions", in case when text sample is mostly about fear sentiment, but an assessor want to mention slight impact of rage. In case of discrete labels when only one answer is allowed it's impossible to do so.

To process such responds with supporting diagonals we should treat each scrollbar position as vector. For this task we need to shift the center of axes to the center of cube that can be view a neutral state. The distance from cube's center to its vertices is one. This will allow us to treat each opposition as $(-1,1)$ period and to map a slide-bar position to a vector which angle coincides with corresponding diagonal and the magnate that is equal to value chosen for a given emotional oppositions. Scrollbar that remains in neutral position are considered to be null vectors with null magnitude.

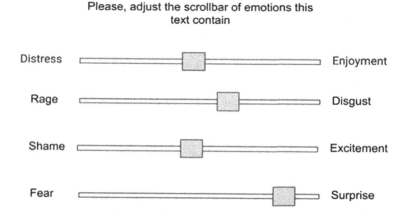

Fig. 5. Possible configuration of scrollbar

To get the final estimation of the text we sum up the vectors. For example scrollbars position on Fig. 5 tells us that an assessor considered estimated text sample to express much surprise with a bit of disgust. His estimates from scrollbar can be converted to vectors on Fig. 6 the resulting vector points on Surprise domain.

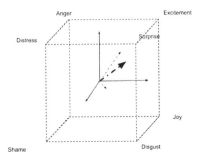

Fig. 6. Vector representation of response in Fig. 5

4   Conclusion and Discussion

At the time of writing, this approach towards data acquisition in such a continuous way is under development. This concept is believed to be very promising as it suggests stepping out from viewing text sentiment as discrete objects and moving towards more complex continuous way of accessing and representing text sentiments. Of course there several problems:

The first one is validation of Lövheim's theory itself - because of novelty and relation with neurology it has to be validated by neurological tools like tomogram or medical tests. By now, the status of his theory does not enable wide campaigns for biological studies. His Cube remains only a possible model.

The second problem for our approach is handling corner cases when an assessor decides to toggle scrollbars to random extreme position. To prevent this we might add some vector metrics including angular deviations in assessor's response.

The third problem is "user-friendliness" of our approach. In spite of provide complex answers, toggling four different sidebar is still more difficult than just choosing one label, and here we have to engage UX\UI competences. In addition, a question of correct instructions remains unresolved.

References

1. Bermingham, A., Smeaton A.: Classifying sentiment in microblogs: is brevity an advantage? In: Proceedings of the 19th ACM International Conference on Information and Knowledge Management. ACM (2010). https://doi.org/10.1145/1871437.1871741
2. Galitsky, B., Chen, H., Du, S.: Inversion of forum content based on authors' sentiments on product usability. In: AAAI Spring Symposium: Social Semantic Web: Where Web 2.0 Meets Web 3.0 (2009)

3. Garrett, J.: The Elements of User Experience: User-Centered Design for the Web and Beyond. 2nd edn. New Riders, USA (2011). https://doi.org/10.1524/icom.2.1.44.19040
4. Hu, M., Bing, L.: Mining opinion features in customer reviews. AAAI **4**(4), 755–760 (2004)
5. Jitviriya, W., Koike, M., Hayashi, E.: Emotional model for robotic system using a self-organizing map combined with Markovian model. J. Robotic. Mechatron. **27**(5), 563–570 (2015). https://doi.org/10.20965/jrm.2015.p0563
6. Larue, O.: The emergence of (artificial) emotions from cognitive and neurological processes. Biol. Inspired Cognit. Archit. **4**, 55–68 (2013). https://doi.org/10.1016/j.bica.2013.01.001
7. Lövheim, H.: A new three-dimensional model for emotions and monoamine neurotransmitters. Med. Hypotheses **78**, 341–348 (2012). https://doi.org/10.1016/j.mehy.2011.11.016
8. Marcus, A.: Design, user experience, and usability: design discourse. In: Marcus, A. (ed.) Conference 2015, vol. 9186, p. 672. Springer, Heidelberg (2015). https://doi.org/10.1007/978-3-319-20886-2

On Modelling Domain Ontology Knowledge for Processing Multilingual Texts of Terroristic Content

Svetlana Sheremetyeva[✉] and Anastasia Zinovyeva

South Ural State University, 76 Pr. Lenina, Chelyabinsk 454080, Russia
sheremetyevaso@susu.ru, bihcwd@bk.ru

Abstract. The paper reports on an ongoing research whose main objective is to address problems in ontology conceptualizing and techniques for building domain ontologies suitable for processing texts in multiple languages. Such ontologies are useful in different NLP tasks from creating semantically annotated multilingual resources to multilingual information retrieval, extraction and machine translation. Another research objective is to contribute to the pool of ontological resources for the terrorist domain as the analysis of terrorism has now been in focus as a matter of national security for more than a decade. The emphasis is made on the linguistic issues of ontology development as the main prerequisite of ontology computer realization. Our approach is a mixed top-down and bottom-up technique adjusted to the domain specificity in the multilingual context. The paper argues for a clear division between the language-dependent lexical knowledge and language-independent conceptual knowledge that, nevertheless, should be represented so as to provide as many direct mappings "lexeme-ontological concept" as possible. The approach is illustrated with an ontology prototype to process texts of terroristic content in the English, French, and Russian languages.

Keywords: Domain ontology · Conceptualization · Multilingualism
Terrorism

1 Introduction

In language engineering, ontology as means for disambiguation and a meta-language of semantic structures for natural language understanding has been in focus of research and development activities for nearly three decades. Understanding of natural language provided by ontologies is useful for different types of applications, such as information indexing and extraction, machine translation, question answering, etc. The number of works on ontologies has drastically increased since 2001, when the Semantic Web was popularized [1] with its promise of data interoperability at the semantic level. Currently, the scope of R&D on ontologies ranges from linguistic and methodological issues to tools and actual knowledge bases meant to cover general or domain knowledge. Quite a number of ontology libraries [2] make these developments publicly accessible.

© Springer Nature Switzerland AG 2018
D. A. Alexandrov et al. (Eds.): DTGS 2018, CCIS 859, pp. 368–379, 2018.

In recent years, due to globalization, multilingual exchangeability and, hence, multilingualism as related to ontologies has attracted much attention. The term "multilingual ontology" was coined and, though, it is not understood exactly in the same way by different researches, all works on ontology multilingualism share the same goal, – to make it possible to use ontological knowledge in tools that can process texts in different languages. Current efforts in this direction cover only a small set of widely used languages such as English, Spanish, French, etc. [3].

This paper reports on an ongoing research whose main objective is to address problems in conceptualization and offer techniques for building domain ontologies suitable for processing texts in multiple languages. The emphasis is made on the linguistic issues of ontology development as the main prerequisite of an ontology computer realization. The paper argues for a clear division between the language-dependent lexical knowledge and language-independent conceptual knowledge that, nevertheless, should be represented in the ontology so as to provide as many direct mappings "lexeme-ontological concept" as possible. Our approach is a mixed top-down and bottom-up technique adjusted to the domain specificity in the multilingual context. It is illustrated with an ontology prototype that could be useful for processing texts of terroristic content in the English, French and Russian languages.

Our motivation for selecting terrorism as the domain of research is that counterterrorist activity, a matter of national security, requires, among other important measures, operative analysis of text information. The studies currently conducted to analyze terrorist activity on the basis of text flows are focused mainly on two tasks:

- *tracking texts* of terrorist content, and
- *detecting and analyzing factors,* by means of which such texts affect certain targeted groups.

Within the scope of the former task, software tools are developed for searching terrorism-related text fragments on the Web [4]. The latter task, in turn, is aimed at developing methodologies and tools for analytical examination of terrorist texts in order to get ideas of both a generalized psychological profile of people involved in terrorist activities [5] and a profile of a particular terrorist [6]. Among other directions of studies on terrorism are efforts on the (a) identification of the ways certain targeted groups of population respond to the media discourse on terrorism [7] and (d) prevention of the population radicalization [8]. Up to now, there are no satisfactory solutions to these problems and though the work has already started on developing ontological resources for certain aspects of terrorist domain, they are so far meant for processing information in English [9–11], which makes a multilingual terrorist ontology a useful contribution to the field.

The paper is structured as follows. Section 2 discusses the understanding of the term "multilingual ontology" in the context of different types and interpretations of the term "ontology" and gives an overview of the currently developed terrorist domain ontologies. Section 3 describes our development process. We conclude with the summary and future work.

2   Background

2.1   Ontologies and Multilinguality

Since most of ontologies are developed based on the lexica of one language (often English) and use the concept labels formulated in the same language (again, often English), multilingualism in ontologies is generally understood in two major senses –as the capability of one ontology to be applied to processing texts in different languages and as the adaptation (or understandability) of the ontology labels for the users in different languages. The way these issues are dealt with in R&D on ontological multilingualism depends greatly on how the term "ontology" is interpreted.

The classical definition of ontology as "an explicit specification of a conceptualization.", where conceptualization is "an abstract, simplified view of the world that we wish to represent for some purpose" [12, p. 199] is accepted by all ontology developers and is generally understood as a structural framework for organizing and representing knowledge with a set of concepts and the relationships between them. However, although the definition above underlies all works on ontology development, its specific interpretations vary greatly. The most important difference in interpreting the term "ontology" that has a major impact on approaches to ontological multilingualism concerns the way the developers understand the relation between the natural language lexica inventory and ontological concepts; the opinions here are quite polar.

In the framework of the first approach, ontologies are required to be language-independent. Such ontologies are not a priori linked to any natural language like, for example, the commercial ontology Mikrokosmos originally created for machine translation [13], SUMO (Suggested Upper Merged Ontology) for information processing [14] and BFO (Basic Formal Ontology), originally designed for the purposes of information retrieval [15]. In domestic linguistics, the most promising research on creating language-independent ontological resources for text understanding is described in [16].

Within the second approach, ontologies are treated as thesaurus-like structures whose elements are defined by the properties of lexica in a specific language. A well-known example of such language-dependent resources, often called ontological, is the famous WordNet thesaurus [17].

The language-independent ontologies allow multilinguality in the first sense (applicability to texts in multiple languages) per definition, provided every unit in the vocabulary of a particular language is referred (according to special rules) to an ontology concept. The fact that in most of such ontologies concepts are labeled with words in a certain language (most often English) is explained by the convenience convention only. The wordings of concept labels in this case have nothing to do with the meaning of the concept, which is only specified by the slot fillers in a concept frame. To raise the degree of applicability of language-independent ontologies to processing multilingual texts involves creating for every new language explicit linking of its lexica to ontology concepts (which of course is not trivial and might require refining the set of ontological concepts and/or their properties). Here the goal of the developers is to provide for the interfacing tools that could reduce developers' efforts on mapping lexica to ontological concepts. A success example of such work on multilingualism is the Mikrokosmos acquisition interface, through which the senses of English, Spanish, and Chinese lexical units are described by means of ontological concepts and a set of restriction rules. In Russia, a multilingual language-independent ontology based on SUMO concepts is being developed in Laboratory of Computational Linguistics of the Institute for Information Transmission Problems of the Russian Academy of Science [16]. There are also efforts to provide for the multilingualism in the second sense, – the adaptation, or understandability of the ontology for the users in different languages. The most frequent proposition is to localize the labels of ontology concepts, rather than modifying the ontological conceptualization. In [18] it is suggested to be done by the association of word senses in different languages to ontology concepts through a special linguistic model, while in [19] a tool for a semi-automatic translation of ontology labels is suggested as a means of ontology localization. One more localization technique is to manually annotate ontological concepts with labels in different languages [20]. Obviously, translating or annotating concept labels in multiple languages could only be applied to very restricted domain ontologies, where one might hope for similar cross-linguistic conceptualizations.

Within the framework of language-dependent ontological research, the ontological multilingualism is approached in a different way. Here one can find tools proposed for a semi-automatic procedure to create separate ontologies for different natural languages [3] or methodologies on how to relate different ontologies that have been initially grounded in each language of interest by mapping both, the data and the metadata, among the language-specific ontologies [21]. A similar, though not exactly the same technique is proposed in [22], where one language-dependent ontology is used as a seed resource to build another language-dependent ontology.

2.2   Domain Ontologies on Terrorism

An ontology focused on the terrorism domain should ideally contain all knowledge about terrorist events, which is a very complex phenomenon and includes a huge number of situations and relationships that are next to impossible to structure in the form of a unified system. In practice, the ontological R&D either focus on certain aspects of terrorism or are limited to constructing basic domain ontologies that do not reflect the specificity of each type of terrorist activity. Depending on their purpose and targeted user, ontologies vary in content, depth, breadth, and set of concepts, acquisition techniques, and representation formalisms posing a lot of problems to their developers. Therefore, up to now terrorist ontologies are mostly research projects under development, rather than products. We further illustrate the above with a brief overview of the following terrorism domain ontologies:

- PiT (ontology for the *Profiles in Terror* web portal) [9],
- AIT (Adversary–Intent–Target) [11],
- Ontology of Terrorism [10, 23].

The first two ontologies are being developed by American researchers, while the third resource is a Russian project. Bearing in mind the purpose of our research, these ontologies are examined along the following parameters: ontology scope and the end user, link to a top-level ontology, acquisition technique, basic concepts, instances, and representation formalism.

The PiT ontology is developed for the strategic purposes of the U.S. intelligence counterterrorism services. It is designed to track terrorist activities and represents knowledge about the terrorist network, comprising a set of organizations and individual terrorists, as well as their numerous relations. The conceptual knowledge is extracted from English databases containing information about terrorist attacks and terrorists. The ontology labels are worded in English. The resource is not linked to any upper level ontology and develops its own structuring of the reality using a mixed acquisition technique where to relate more detailed concepts to broader ones is often achieved with the Russian stacking doll solution. For example, to such broad concepts as EVENT (to which the actual act of terrorism refers), PERSON (participants in the terrorist act) and ORGANIZATION (terrorist groups), the developers add concepts describing the CIVIL STATUS OF THE TERRORIST, his CONTACTS with other people, various METHODS OF SECRET COMMUNICATION and so on. PiT is a fairly extensive ontology: it contains 70 concepts and 173 properties. In addition to concepts, the ontology contains instances, which are a must, because a large part of the ontology is to cover various features of individual terrorists and relations between them. The ontology is written in OWL.

The AIT ontology is developed for the U.S. intelligence counterterrorism services as well. It is designed to predict terrorist attacks based on English language data on terrorist organizations, their intentions, and available weapons. AIT should become part of a comprehensive BOOT ontology, which stands for *Basic Ontology of Terrorism*. AIT is linked to the upper-level ontology BFO [15] and follows its division of the reality into MATERIAL OBJECTS, QUALITIES, and PROCESSES. AIT splits the top-level concepts of BFO into concepts specific to the terrorism domain. The acquisition approach is a mixed technique where the ontology development starts with a model statement: "A terrorist attack occurs when an adversary, with intent and capability, uses a weapon against a target" [11, p. 13]. All keywords from this statement are analyzed by the terrorist domain experts in order to identify concepts related to the keywords. This results into a set of basic terms of the domain and relations between them. The ontology labels are worded in English. AIT comprises only 11 basic relations and 4 inverse ones, but it has many concepts obtained by specifying those from the model statement. The role of instances in AIT is less significant than that in the PiT ontology, because, as claimed by the developers, specific terrorist attacks, especially those that happened in the distant past, are no help in solving counter-terrorism problems of today. The ontology language is OWL.

The Ontology of Terrorism of the Russian project is developed to study the concept of terrorism in the English and Russian discourses of international news on terrorism. This ontology does not have a strictly formulated applied orientation and is meant for further research in linguistics, philosophy, and social science. The authors conduct the content analysis of the concept TERRORISM based on its definitions found in different Russian and English dictionaries with the use of component semantic analysis and thus build taxonomy of terrorism-related concepts. The main concept of the ontology is

TERRORISM, which is then specified into SUBJECT, OBJECT, RESULTS, and CONSEQUENCES OF TERRORIST ACTIVITY. Each concept is represented as a frame with slots filled with concept properties. The labels of this ontology have Russian wording. The authors then extract lexical units from Russian and English comparable corpora and link them to the ontology concepts.

3   Ontology Building

3.1   Preliminary Remarks

In our project, we follow three basic methodological assumptions. The first is that ontology is a reusable language-independent resource; the second is that "domain-specific knowledge is not isolated from general world knowledge" [24, p. 233] and we, therefore, link our ontological resource to the upper-level Mikrokosmos ontology [13] to reuse the knowledge that is already there. We follow the initial Mikrokosmos division of the reality into OBJECTS, EVENTS, and PROPERTIES, and use its formalism. We also keep concept labels worded in English.

There are three major approaches to ontology building that rely on top-down, bottom-up or mixed techniques. In the top-down approach ontology building goes from the most universal concepts to more specific ones; the bottom-up approach, on the contrary, proposes to get universal concepts by generalizing the narrower ones. In the mixed approach, first key concepts are defined, which are further generalized and detailed [23]. Ontology concepts can be derived from different sources, such as encyclopedias, thesauri, dictionaries, glossaries, databases, corpora, etc., using one or a combination of formal (e.g., statistical) or informal (e.g., intuitive/heuristic) methods.

Our third assumption is that conceptual information on terrorism can be extracted from multilingual comparable domain corpora using mixed acquisition techniques.

3.2   Defining Concepts

We started our work with the acquisition of comparable corpora on terrorism as found in the Internet in three languages, – Russian, English and French, and dated no earlier than 2016. The resulting corpora are of about half a million words each and are mostly composed of news articles reporting on terrorist attacks.

It is well known that it is noun and verb phrases that are the closest to the text content. We therefore concentrated on extracting these types of phrases from every corpus. The extraction was done in two steps. First, the seed set of phrases up to four components long were automatically extracted from the English, French, and Russian corpora with the universal extractor of typed phrases and key words described in [25]. The extractor was preliminary trained for the terrorist domain in all the three languages. We further searched the corpora for longer lexical items with the regular "find" functionality using the seed set of 4-component phrases. The resulting set was grouped into semantic fields according to the sense closeness within the same language, and across languages. We are fully aware that words and phrases identified as translation equivalents are rarely completely identical in sense but, following [18], we use this approach for the practical purpose of providing for multilingualism.

Attributing some of the lexemes to a certain semantic field can be purely corpus based, which is the case, e.g., with the English items *militant* and *fighter*. In general use these lexemes are defined as people who act militant or fight, respectively, without any reference to terrorism. In the domain corpus, they are used in the same contexts as the word *terrorist* when military actions are described and we could not find the use of *militant* and *fighter* meaning "good guys" of the military conflict.

An example of grouping of multilingual lexical items in conceptual areas is shown in Table 1 for the semantic field 'Terrorist'. All lexical items in this table, when used in the corpora, share the same semantic component – 'a person who employs methods of terror to achieve their own goals'. The numbers in brackets show the absolute frequency (F) of a lexical item in the corresponding corpus.

Table 1. Fragments of the lexical lists grouped into the semantic field 'Terrorist'

English (F)	French (F)	Russian (F)
terrorist (1374)	terroriste (1823)	террорист (1949)
militant (437)	kamikaze (168)	боевик (1793)
fighter (401)	combattant (164)	смертник (237)
gunman (357)	femme kamikaze (6)	террорист-смертник (133)
suicide bomber (131)	djihadiste (4)	террористка-смертница (16)
jihadi (7)	loup solitaire (3)	террористка (6)
female suicide bomber (5)	terroriste de l'EI (3)	джихадист (4)
female terrorist (4)	combattant terroriste (2)	игиловец (4)
lone-wolf terrorist (4)	femme terroriste (2)	террорист-одиночка (4)
ISIS terrorist (3)	recruteur terroriste (2)	террорист-вербовщик (4)

This stage of corpora analysis resulted in the identification of the following main semantic fields covering multilingual lexicons in the terrorist corpora: terrorism and its types, terrorist attacks (events), countries or cities the attacks occur in, specific places the attacks occur in, temporal parameters of the attacks, attack means, attack strategies, organizations behind the attacks, attitude towards attacks, terrorist organizations, terrorists, terrorism support, terrorist goals, results and consequences of the attacks, targets, counter-terrorism activities, war and politics, religion.

Based on this semantic classification of lexical units, we first defined TERRORISM, TERRORIST ATTACK, TERRORIST, TERRORIST ORGANIZATION, and WEAPON as basic terrorist domain concepts and then, by analyzing particular semantic properties of the lexemes in the lexical semantic fields, searched for more domain-specific concepts related to them. For example, the TERRORIST lexical semantic field gave us such (PROPERTY ATTRIBUTE) concept as GENDER and (PROPERTY RELATION) concepts as AGENT-OF MILITARY-ACTIVITY for *fighter* or AGENT-OF SUICIDE-BOMB-ATTACK for *suicide bomber*.

When the initial set of concepts and relations between them were defined, we further augmented and refined the pool of concepts and relations by using the text-template technique. For example, such RELATION concepts as IS-A and INSTANCE-OF can be linked (though not exclusively) to the following parallel text templates in English, French and Russian as

A *is /est /это* B
B *such as /comme /такие как* A
A *and other /et autres /и другие* B,

wherein B is a more generalized concept, while A is a more specific one or an instance; in cursive are parallel English, French and Russian textual fragments.

Another example of our multilingual acquisition templates is

attack /attaque /атака with /using /involving /avec /au moyen de /c использованием /c применением A,

wherein the words in cursive are textual manifestations of the RELATION concept IN-STRUMENT in the three languages.

By processing the corresponding corpora with these templates, we identified such subconcepts of the WEAPON concept as BLADE WEAPON, FIREARM, CHEMICAL WEAPON, and EXPLOSIVE DEVICE. With the same templates we identified a number of concepts that are actually not weapons, but can be used as such by terrorists: CAR, TRUCK, and PLANE, i.e. vehicles. These concepts have no connection to WEAPON in the ontology, but are linked to some specific attack types by means of the relation INSTRUMENT. Up to now our ontology contains 92 concepts, 20 relations, 7 attributes.

Fragments of the ontology for the terrorist domain as linked to the second-level Mikrokosmos concepts OBJECT and EVENT and the third-level concept RELATION (the child of PROPERTY) by the IS-A relation are shown in Figs. 1, 2 and 3, respectively. Figure 4 shows an ontological fragment with different types of relations.

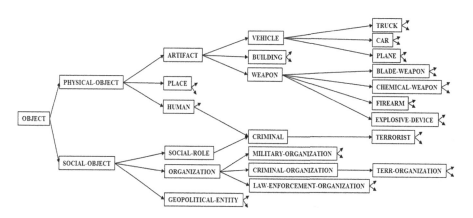

Fig. 1. A fragment of the OBJECT branch of the ontological tree for the Terrorism domain

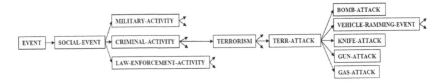

Fig. 2. A fragment of the EVENT branch of the ontological tree for the terrorism domain

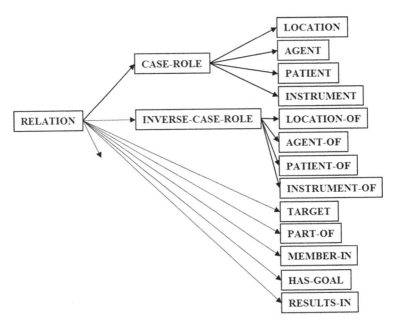

Fig. 3. A fragment of the RELATION branch of the ontological tree for the terrorism domain

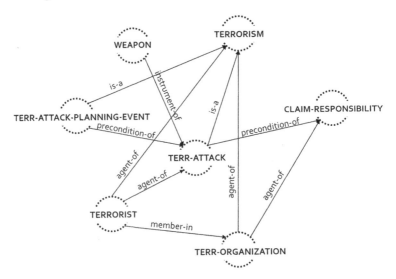

Fig. 4. A fragment of the ontology for the terrorism domain

Following the Mikrokosmos formalism, the structure of CONCEPTS in the terrorism ontology is represented by frames, though of a simplified structure. A frame has a name (of a CONCEPT) and slots. The name is simply a label to index CONCEPTS. English words are used as names of CONCEPTS by pure convention. The meaning of a CONCEPT is conveyed by its properties, represented as slots. Slots, in turn, have facets describing some finer distinctions between the possible fillers of the slot: VALUE, SEM, and DEFAULT. The fillers of a VALUE facet are actual value(s) (if any) for a given slot of the given concept. The fillers of the SEM facet are semantic constraints (selectional restrictions), which refer to ontological concepts, from which the fillers of the VALUE or DEFAULT facets should be chosen. Facets can be filled with CONCEPTS, INSTANCES of concepts, literal symbols, a scalar range and, a number. All CONCEPT frames have non-special and special slots. Special slots for all kinds of concepts are DEFINITION, IS-A, SUBCLASSES and INSTANCES, their values are not inherited, while the SEM facets are inherited in the ontology. The DEFINITION slot has only the VALUE facet whose filler is a definition of the concept in English intended only for human consumption. The DEFINITION slot is mandatory. Figures 5 and 6 give examples of OBJECT and EVENT concept frames.

LABEL	EXPLOSIVE-DEVICE	VALUE
DEFINITION	"an object filled with a bursting charge and exploded by means of a fuse, by impact, or otherwise"	VALUE
IS-A	WEAPON	VALUE
SUBCLASSES	IMPROVESED-EXPLOSIVE-DEVICE	VALUE
HAS-PART	EXPLOSIVE-SUBSTANCE	SEM
INSTRUMENT-OF	BOMB-ATTACK	DEFAULT
	CRIMINAL-ACTIVITY, MILITARY-ACTIVITY	SEM

Fig. 5. The frame of the OBJECT concept EXPLOSIVE-DEVICE

LABEL	BOMB-ATTACK	VALUE
DEFINITION	"an attack performed by a terrorist or a group of terrorists using an explosive device as weapon"	VALUE
IS-A	TERR-ATTACK	VALUE
SUBCLASSES	SUICIDE-BOMB-ATTACK	VALUE
LEGALITY-ATTRIBUTE	NO	VALUE
NUMBER-OF-EXPLOSIONS	"≥ 1"	VALUE
AGENT	TERRORIST, TERR-ORGANIZATION-PART, TERR-ORGANIZATION	SEM
THEME	CROWD	DEFAULT
	HUMAN	SEM
INSTRUMENT	EXPLOSIVE-DEVICE	SEM
LOCATION	BUILDING, PUBLIC-TRANSPORTATION-LOCATION, PLACE	SEM

Fig. 6. The frame of the EVENT concept BOMB-ATTACK

4 Conclusions

We have presented an ongoing research aimed at the development of terrorism ontology for multilingual text processing on the example of the English, French, and Russian languages. The ontology is linked to the upper-level language-independent Mikrokosmos ontology and reuses its formalism in a simplified version. A methodology for the acquisition of ontological concepts based on extracting corpus-based translation equivalents and grouping them into semantic fields with the subsequent refinement by textual templates has been proposed. The methodology can most likely be used on the material of a broader set of languages that would of course include the development of corresponding language-dependent textual templates. Our future work is aimed at enlarging both the depth and the breadth of the ontology. We are also planning to develop rules for mapping language expressions to ontology concepts.

References

1. Berners-Lee, T., Hendler, J., Lassila, O.: The semantic web. Sci. Am. **284**(5), 34–43 (2001). https://doi.org/10.1038/scientificamerican0501-34
2. D'Aquin, M., Noy, N.F.: Where to publish and find ontologies? A survey of ontology libraries. Web Semant.: Sci., Serv. Agents World Wide Web **11**, 96–111 (2012). https://doi.org/10.1016/j.websem.2011.08.005
3. Alatrish, E.A., Tošić, D., Milenkov, N.: Building ontologies for different natural languages. 10.2298/CSIS130429023A, https://www.researchgate.net/publication/270471292_Building_ontologies_for_di-fferent_natural_languages. Accessed 28 Jan 2018
4. Choi, D., Ko, B., Kim, H., Kim, P.: Text analysis for detecting terrorism-related articles on the web. J. Netw. Comput. Appl. **38**, 16–21 (2014). https://doi.org/10.1016/j.jnca.2013.05.007
5. Behavioral Assessments Based on Automated Text Analyses. https://goo.gl/7RXXeN. Accessed 28 Jan 2018
6. Brynielsson, J., Horndahl, A., Johansson, F., Kaati, L., Mårtenson, C.: Harvesting and analysis of weak signals for detecting lone wolf terrorists. Secur. Inform. **2**(1), 1–15 (2013). https://doi.org/10.1186/2190-8532-2-11
7. Aly, A.: The terrorists' audience: a model of internet radicalisation. J. Aust. Prof. Intell. Off. **17**, 3–19 (2009)
8. Weimann, G., von Knop, K.: Applying the notion of noise to countering online-terrorism. Stud. Confl. Terror. **31**, 883–902 (2008). https://doi.org/10.1080/10576100802342601
9. Mannes, A., Golbeck, J.: Ontology building: a terrorism specialist's perspective. In: Proceedings of Aerospace Conference, EEEI, pp. 1–5 (2007). https://doi.org/10.1109/AERO.2007.352794
10. Zhdanova, S.Yu., Mishlanova, S.L., Polyakov, V.B.: The concept of terrorism in the discourse of international news online resources (in Russian). Vector Sci. Togliatti State University. Ser.: Pedagog., Psychol. **4**(11), 100–103 (2012)
11. Turner, M., Turner, J., Weinberg, D.: A Simple Ontology for the Analysis of Terrorist Attacks. https://goo.gl/tqyTRG. Accessed 30 Jan 2018
12. Gruber, T.R.: A translation approach to portable ontology specifications. Knowl. Acquis. **5**(2), 199–220 (1993). https://doi.org/10.1006/knac.1993.1008
13. Nirenburg, S., Raskin, V.: Ontological Semantics. MIT Press, Cambridge (2004)

14. Niles, I., Pease, A.: Linking lexicons and ontologies: mapping wordnet to the suggested upper merged ontology. In: Proceedings of the 2003 International Conference on Information and Knowledge Engineering (IKE 2003), pp. 412–416 (2003)
15. Arp, R., Smith, B., Spear, A.D.: Building Ontologies with Basic Formal Ontology. MIT Press, Cambridge (2015). https://doi.org/10.7551/mitpress/9780262527811.001.0001
16. Boguslavskii, I.: Semantic analysis based on linguistic and ontological resources. In: Proceedings of the 5th International Conference on the Meaning-Text Representations. Barcelona, 8–9 September, pp. 25–36 (2011)
17. Miller, G.A., Beckwith, R., Fellbaum, C., Gross, D., Miller, K.J.: Introduction to WordNet: an on-line lexical database. Int. J. Lexicogr. 3(4), 235–244 (1990). https://doi.org/10.1093/ijl/3.4.235
18. Montiel-Ponsoda, E., Aguado de Cea, G., Gómez-Pérez, A., Peters, W.: Modelling multilinguality in ontologies. In: Proceedings of COLING 2008, Companion volume – Posters and Demonstrations, pp. 67–70. Manchester (2008)
19. Espinoza, Mauricio, Gómez-Pérez, Asunción, Mena, Eduardo: Enriching an ontology with multilingual information. In: Bechhofer, Sean, Hauswirth, Manfred, Hoffmann, Jörg, Koubarakis, Manolis (eds.) ESWC 2008. LNCS, vol. 5021, pp. 333–347. Springer, Heidelberg (2008). https://doi.org/10.1007/978-3-540-68234-9_26
20. Chaves, M., Trojahn, C.: Towards a Multilingual Ontology for Ontology-driven Content Mining in Social Web Sites. https://goo.gl/sZKmS2. Accessed 30 Jan 2018
21. Embley, D.W., Liddle, S.W., Lonsdale, D.W., Tijerino, Y.: Multilingual Ontologies for Cross-Language Information Extraction and Semantic Search. https://pdfs.semanticscholar.org/6884/41a96b6da61295c7df39b70db2f28531370a.pdf, last accessed 2018/01/30. https://doi.org/10.1007/978-3-642-24606-7_12
22. Cherkassky, N.C.: How to Construct Multilingual Domain Ontologies. https://goo.gl/CXZ6R6. Accessed 30 Jan 2018
23. Mishlanova, S.L., Kupriyanycheva, E.A.: Osobennosti reprezentatsii kontsepta terrorizm v angloyazychnom diskurse novostei internet-resursov (in Russian) [Peculiarities of the Terrorism Concept Representation in the English-Language Internet News Discourse]. Filologicheskiye zametki [Philological Studies], pp. 265–276 (2012)
24. Moreno, A., Pérez, Ch.: From text to ontology extraction and representation of conceptual information. In: Actes de quatrièmes rencontres «Terminologie et Intelligence Artificielle», pp. 233–242 (2011)
25. Sheremetyeva, S.: Automatic extraction of linguistic resources in multiple languages. In: Proceedings of NLPCS 2012, 9th International Workshop on Natural Language Processing and Cognitive Science in conjunction with ICEIS 2012, pp. 44–52. Wroclaw, Poland (2012)

Cross-Tagset Parsing Evaluation
for Russian

Kira Droganova[1] and Olga Lyashevskaya[2,3(✉)]

[1] Institute of Formal and Applied Linguistics, Faculty of Mathematics and Physics,
Charles University in Prague, Prague, Czech Republic
`droganova@ufal.mff.cuni.cz`
[2] School of Linguistics, National Research University
Higher School of Economics, Moscow, Russia
[3] Vinogradov Institute of the Russian Language RAS, Moscow, Russia
`olesar@yandex.ru`

Abstract. Cross-tagset parsing is based on the substitution of one annotation layer for another while processing data within one language. As often as not, either the native tagger or the dependency parser used in (pre-)annotation of the Gold treebank is not available. The cross-tagset approach allows one to annotate new texts using freely available tools or tools optimized to user's needs. We evaluate the robustness of Russian dependency parsing using different morphological and syntactic tagsets in input and output. Qualitative analysis of errors shows that the cross-substitution of three morphological tagsets and two syntactic tagsets causes only a mild drop in performance.

Keywords: Dependency parsing · Cross-tagset parsing
Parser evaluation · Russian language treebanks
Universal dependencies · SynTagRus

1 Introduction

It is not uncommon that a computational linguist faces the following challenges:

- annotation tools or data are not available;
- a larger corpus exists only in different annotation standards;
- the linguistic theory or IT practice moves on changing the inventory of corpus tags.

The cross-tagset parsing is an approach that attempts to develop and extend linguistic corpora by the substitution of one annotation layer for another in the

The article was prepared within the framework of the Academic Fund Program at the National Research University Higher School of Economics (HSE) in 2018 (grant 18-05-0047) and by the Russian Academic Excellence Project «5-100». The work by Kira Droganova was partially supported by the GA UK grant 794417 and by SVV project number 260 453.

D. A. Alexandrov et al. (Eds.): DTGS 2018, CCIS 859, pp. 380–390, 2018.
https://doi.org/10.1007/978-3-030-02846-6_31

data within one language. Whereas certain approaches, such as using coarse-grained POS tags or delexicalized feature set, is widely used, the more practical strategy would be adapting alternative annotations, especially if a morphological or syntactic parser involved in the annotation of the gold collection is not freely available or lacks a disambiguation module or a new word guesser.

The cross-tagset parsing can be seen as a spin-off of the cross-lingual parsing task [12,21], with the notable difference that in the cross-tagset parsing the direct projection of POS and dependencies is unreliable by default, even though the lexical layer (and associated distributional probabilities) remain the same.

In this paper, we present the cross-tagset parsing approach and evaluate the robustness of the cross-tagset annotation for Russian.

At present, several existing corpora are annotated in different schemes. Many popular morphological taggers developed for Russian (see [10,18]) are based on different tagsets. As a result, the morphological tagsets of the SynTagRus tree-bank [1,2,6], Russian National Corpus (RNC) core collection [8], RuTenTen and other large corpora [9] are not consistent with each other. As for syntactic annotation, the most common representation is dependency trees which are implemented considerably differently in a number of language specific parsers [22] and treebanks, e.g. SynTagRus and UD-Russian (universaldependencies.org, [16,17], cf. also Fig. 1).

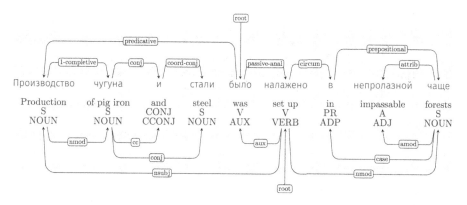

Fig. 1. Annotation of a sample sentence in SynTagRus (above) and UD-Russian (below).

We evaluate the dependency parsing models, which take as an input three different morphological tagsets, namely, the original SynTagRus tagset, RNC tagset, and Russian UD tagset (v. 1.4). We expect roughly comparable accuracy of the models although there is no perfect matching among POS classes, and the borders of inflectional categories also overlap across tagsets. As a next step, we experiment with two full tagsets (morphology and dependency relations), namely SynTagRus and Russian UD tagsets [5,11]. We expect that the overall score of the UD model will improve compared to the scores of the SynTagRus

Table 1. POS tags used in the datasets.

Dataset 1		Dataset 2		Dataset 3		Dataset 4	
A	100104	A	107806	ADJ	91065	ADJ	91021
ADV	39371	ADV	43049	ADP	76040	ADP	75675
CONJ	44128	CONJ	44128	ADV	43049	ADV	43414
INTJ	74	INTJ	74	CONJ	31848	AUX	6139
NUM	11486	NUM	10018	DET	16746	CONJ	31848
PART	29341	PART	25663	INTJ	74	DET	16790
PR	76044	PARTCP	13660	NOUN	237569	INTJ	74
S	267683	PR	76044	NUM	10018	NOUN	237569
V	98897	S	241875	PART	25663	NUM	10018
-	-	SPRO	19574	PRON	23185	PART	25663
-	-	V	85237	SCONJ	12280	PRON	23185
-	-	-	-	SYM	694	SCONJ	12280
-	-	-	-	VERB	98897	SYM	694
-	-	-	-	-	-	VERB	92758

model since the classes of syntactic relations are more coarse-grained and they are less lexically dependent.

As a standard tool, we use MaltParser [15], the most cited tool for parsing Russian data as of 2017, which is superior in terms of replicability. The crucial point is that we are not trying to achieve state-of-the-art results in dependency parsing, rather, the aim of our study is to evaluate the robustness of MaltParser models trained on Russian data and explore the influence of various tagsets on the syntactic annotation quality.

The paper is structured as follows. Section 2 briefly describes tagsets and tools in use. Section 3 reports on the experiments and results. Section 4 is focused on the analysis of typical cross-parsing errors.

2   Tagsets and Tools

Table 1 summarizes the key features of the tagsets: an inventory of POS tags and their raw frequencies. For full documentation of dependency labels, see [3,11]. The datasets are described below, see Sect. 3.1. Note that the dependency annotation schemes differ not only in the amount of labels but also in the topology of syntactic trees, see Fig. 2. The SynTagRus sentence structures looks like a linegraph and meaningful nodes connect through the functional parts of speech. In contrast, the UD sentence structures tend to star-graph, the edges connecting the meaningful nodes such as verbs and nouns. The functional parts of speech such as articles and prepositions are moved to the periphery.

Compared to the UD structure, the average path length from the root node to the functional node is almost twice as long in the SynTagRus structure (Table 2).

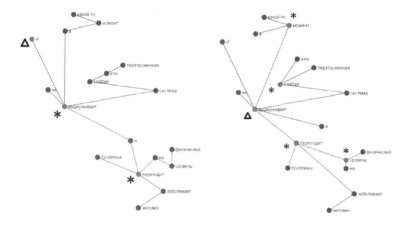

Fig. 2. Sentence structures in SynTagRus (left) and UD (right).

Table 2. Average path length.

	Average path length to a meaningful node	Average path length to a functional node
The SynTagRus sentence structure	3.75	2.8
The UD sentence structure	2.0	1.8

The models are trained and tuned using POS tags and morphological features. Identical settings are applied to the datasets. The training algorithm is nivreeager [14], a Library for Large Linear Classification [7] is used as a machine learning package.

The models are available at https://github.com/Kira-D/Russian-SynTagRus-Parser/tree/master/DTGS2018.

Evaluation was performed with MaltEval [13] and TedEval [23].

3   Experiments

Several tagset-specific datasets were created and evaluated. The first experiment involved Datasets 1—3. Three models which have different morphology layers were trained to test how the morphological tagsets affect the parsing quality.

The second experiment employed pairwise experiment evaluation for Datasets 1 and 4. Two models were tested to assess the potential losses in parsing quality caused by switching to automatically converted UD treebank, where among other issues the sentence structure and morphological annotation principle were completely different from SynTagRus.

Table 3. LAS & UAS.

Model	LAS	UAS
Dataset 1. SynTagRus POS + SynTagRus deps	$84.3\%^{-1.5}$	**89.7%**
Dataset 2. Mystem POS + SynTagRus deps	$83.6\%^{-2.2}$	$89.4\%^{-0.3}$
Dataset 3. UD POS + SynTagRus deps	$83.7\%^{-2.1}$	$89.4\%^{-0.3}$
Dataset 4. UD POS + UD deps	**85.8%**	$87.9\%^{-1.8}$

Table 4. Error distribution.

Mistaken label	I completive	circumstantial	root	quasi-agentive	II completive
Dataset 1. SynTagRus POS + SynTagRus deps					
Error rate	18.00%	28.57%	21.06%	26.45%	37.48%
Overall error rate	1.83%	1.80%	1.20%	0.86%	0.83%
Dataset 2. Mystem POS + SynTagRus deps					
Error rate	19.00%	29.99%	21.70%	39.18%	25.89%
Overall error rate	1.93%	1.89%	1.24%	0.86%	0.85%
Dataset 3. UD POS + SynTagRus deps					
Error rate	19.00%	29.50%	21.06%	37.89%	25.39%
Overall error rate	1.93%	1.86%	1.20%	0.84%	0.83%
Dataset 4. UD POS + UD deps					
Mistaken label	conj	root	nmod	parataxis	nsubj
Error rate	26.84%	22.10%	6.86%	33.06%	7.24%
Overall error rate	1.42%	1.26%	1.11%	0.67%	0.52%

3.1 Datasets

The four datasets are identical with respect to the underlying texts.

Dataset 1: includes an original SynTagRus morphological and syntactic annotation, comprising 67 dependency relations (deps), 11 POS tags and 14 feature categories (feats).

Dataset 2: includes the morphological layer converted from SynTagRus to Mystem tags [19] and SynTagRus dependency labels (67 deps, 12 POS tags and 14 feats).

Dataset 3: includes the morphological layer annotated in UD tags and original SynTagRus sentence structures and dependency labels (67 deps, 13 POS tags and 12 feats).

Dataset 4: includes the UD morphological and syntactic annotation (34 deps, 14 POS tags and 12 feats).

The data have been split into three parts: the training set (80%), the development set (10%) and the testing set (10%). The overall size of the datasets is over 1,000,000 tokens (ca. 59,000 sentences).

Table 5. TED Accuracy and TedEval LAS and UAS (projective trees only).

Model	TED accuracy	LAS	UAS
Dataset 1. SynTagRus POS + SynTagRus deps	92.7%	86.2%	91.8%
Dataset 2. Mystem POS + SynTagRus deps	92/39%	85.6%	91.4%
Dataset 3. UD POS + SynTagRus deps	92.34%	85.6%	91.4%
Dataset 4. UD POS + UD deps	**94.63%**	87.5%	89.6%

Table 6. TED Accuracy: pairwise evaluation (projective trees only).

Pair	TED accuracy
Dataset 1. SynTagRus POS + SynTagRus deps	96.19%
Dataset 4. UD POS + UD deps	**97.39%**

3.2   MaltEval: Results

Table 3 shows the attachment scores. The highest LAS (85.8%) is achieved on the UD tagset whereas the highest UAS (89.7%) is obtained on the SynTagRus tagset.

Table 4 demonstrates the top-5 most frequently occurring label errors for each dataset. Even though the attachment scores are identical, error structure varies from tagset to tagset. The Error rate column shows the ratio of specific mismatch among all labels of this type. The Overall error rate shows the ratio of specific mismatch among all mismatches.

3.3   TedEval: Results

Table 5 shows TED accuracy scores. The highest score (94.63%) is achieved on the UD tagset. The drawback of TedEval is that the tool does not support non-projective trees, which are typical for Russian. For the purpose of the experiment we extracted projective trees from the datasets and measured TED accuracy on the samples of approximately 3000 sentences (for each dataset).

Table 6 provides the pairwise experiment evaluation for Dataset 1 and Dataset 4. The UD model shows moderately better results (97.39%).

4   Analysis

The results prove that the method allows one to substitute one morphological layer and sentence structure for alternative ones and still preserve the same level of parsing quality (97.39% TED accuracy is achieved on the UD tagset and 96.19% on SynTagRus tagset).

The model trained on the UD full tagset and sentence structures provides better quality on dependency labels parsing than the model trained on the SynTagRus original sentence structures, dependency labels and UD POS tags (85.8%

vs. 84.3% LAS). This result can be explained by reduced number of dependency relations: 34 UD dependency labels versus 67 SynTagRus dependency labels.

The TED accuracy scores seems excessively high. However, there can be seen some typical mismatches in the output.

The most frequent mismatch which the UD model produces is "*nmod*" mixed up with "*dobj*" (see Figs. 3 and 4) and vice versa. The second most frequent mismatch is "*parataxis*" incorrectly labeled as "*conj*" (see Figs. 5 and 6). These relations cannot be distinguished by using only the morphological data, and not semantics. The root node errors are triggered by the training options that force the model to generate additional root nodes.

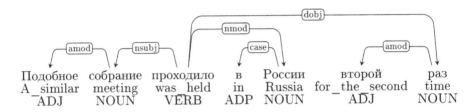

Fig. 3. An example of "*nmod*" mixed up with "*dobj*".

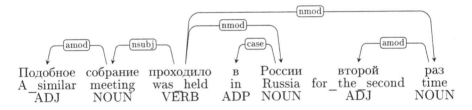

Fig. 4. The sentence from Fig. 3 (gold standard).

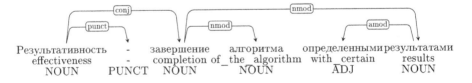

Fig. 5. An example of "*parataxis*" mixed up with "*conj*".

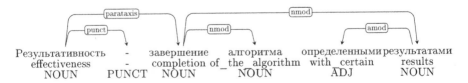

Fig. 6. The sentence from Fig. 5 (gold standard).

The most common confusion pairs in the SynTagRus parsing are "*circumstantial*" and "*I completive*" (usually the 2nd argument of the verb), "*circumstantial*" and "*II completive*" (usually the 3rd argument of the verb), "*quasi-agentive*" (1st argument of the predicate noun) and "*I completive*", "*I completive*" and "*II completive*".

The label "*quasi-agentive*" appears instead of "*I completive*" mostly when the head of this phrase is a verb noun (see Figs. 7 and 8).

Two other frequent cases are "*I completive*" attached instead of "*II completive*" (see Figs. 9 and 10) and "*circumstantial*" attached instead of "*I completive*" (see Figs. 11 and 12). A valency dictionary and lexical-semantic selection knowledge is required to distinguish between these labels. Without such data, any attempt to identify the relations automatically is tantamount to guesswork. This mismatch occurs frequently since I and II completives and circumstantial deps are among the top-frequent dependency relation in the testing set (Table 4).

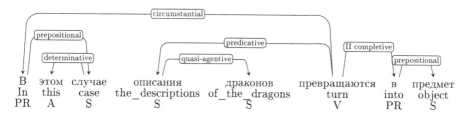

Fig. 7. An example of "*I completive*" mixed up with "*quasi-agentive*".

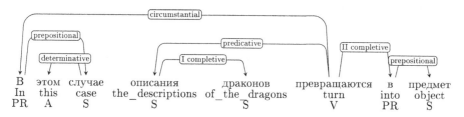

Fig. 8. The sentence from Fig. 7 (gold standard).

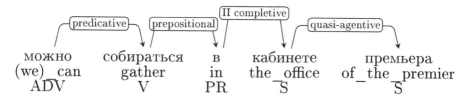

Fig. 9. An example of "*I completive*" mixed up with "*II completive*".

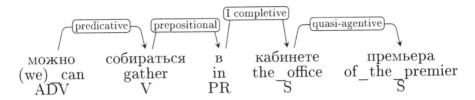

Fig. 10. The sentence from Fig. 9 (gold standard).

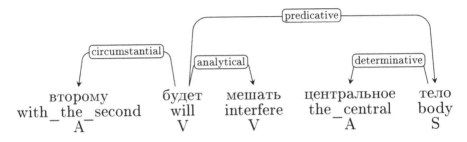

Fig. 11. An example of "*I completive*" mixed up with "*circumstantial*".

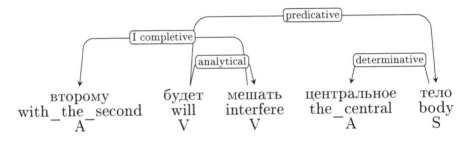

Fig. 12. The sentence from Fig. 11 (gold standard).

5   Conclusion

In this work, we explored the method of tagset substitution in syntactic dependency parsing. We investigated the robustness of the Russian dependency learning using MaltParser if (a) the genuine POS and morphological features annotation and/or (b) syntactic dependency annotation is substituted with an alternative tagset.

In general, our results (84.3% LAS, 89.7% UAS on the SynTagRus dataset, 85.8% LAS, 87.9% UAS on the UD dataset) over-perform the best MaltParser model for Russian reported in [20] (83.4% LAS, 89.4% UAS) but not the results of [4] who use the joint morphological and syntactic approach with hard lexical constraints (88.0% LAS, 93.0% UAS). Substitution of the SynTagRus tags with UD and Mystem tags results in 0.6–0.7% drop in LAS and 0.3% drop in UAS. Interestingly, [20] report 0.6% drop in both LAS and UAS when they use

MTE tags instead of SynTagRus tags. Thus, we are optimistic about the use of alternative taggers in Russian statistic dependency analysis.

In the future, more work should be done to tune the models based on UD tags and deploy the methods of models' cross-evaluation in order to compare the parsing of elliptic and other complex trees.

References

1. Apresian, Ju., Boguslavsky, I., Iomdin, L., Lazursky, A., Sannikov, V., Sizov, V., Tsinman L.: ETAP-3 linguistic processor: a full-fledged NLP implementation of the MTT. In: Proceedings of the First International Conference on Meaning-Text Theory, pp. 279–288 (2003)
2. Boguslavsky, I., Iomdin, L., Frolova, T., Timoshenko, S.: Development of a Russian tagged corpus with lexical and functional annotation. In: Proceedings of the MONDILEX Third Open Workshop. Bratislava, Slovakia, 15–16 April 2009 (2009)
3. Boguslavsky, I., Iomdin, L., Sizov, V., Tsinman, L., Petrochenkov, V.: Rule-based dependency parser refined by empirical and corpus statistics. Proc. DepLing **2011**, 318–327 (2011)
4. Bohnet, B., Nivre, J., Boguslavsky, I., Farkas, R., Ginter, F., Hajič, J.: Joint morphological and syntactic analysis for richly inflected languages. Trans. Assoc. Comput. Linguist. **1**, 415–428 (2013)
5. Droganova, K., Zeman, D.: Conversion of SynTagRus (the Russian dependency treebank) to Universal Dependencies. ÚFAL Technical Report TR-2016-60, ISSN 1214-5521 (2016)
6. Dyachenko, P., et al.: Sovremennoe sostojanie gluboko annotirovannogo korpusa tekstov russkogo jazyka (SynTagRus) [SynTagRus, a deeply annotated corpus of Russian texts: present state of the art (in Russian)]. In: Russian National Corpus: 10 years. Trudy Instituta russkogo jazyka im. V. V. Vinogradova. Moscow, vol. 6, pp. 272–299 (2015)
7. Fan, R.-E., Chang, K.-W., Hsieh, C.-J., Wang, X.-R., Lin, C.-J.: LIBLINEAR: a library for large linear classification. J. Mach. Learn. Res. **9**, 1871–1874 (2008)
8. Grishina, E., Rakhilina, E.: Russian National Corpus (RNC): an overview and perspectives. In: Proceedings of AATSEEL-2005. Washington, 27–30 December (2005)
9. Khokhlova, M.: Comparison of high-frequency nouns from the perspective of large corpora. In: RASLAN 2016 Recent Advances in Slavonic Natural Language Processing, pp. 9–17 (2016)
10. Lyashevskaya, O., et al.: Ocenka metodov avtomaticheskogo analiza teksta: morfologicheskije parsery russkogo jazyka [NLP evaluation: Russian morphological parsers (in Russian)]. Computational Linguistics and Intellectual Technologies, vol. 9 (16), pp. 318–326 (2010)
11. Lyashevskaya, O., et al.: Universal dependencies for Russian: a new syntactic dependencies tagset. In: Series: Linguistics, WP BRP 44/LNG/2016 (2016)
12. McDonald, R., Petrov, S., Hall, K.: Multi-source transfer of delexicalized dependency parsers. In: Proceedings of EMNLP, pp. 62–72 (2011)
13. Nilsson, J.: User Guide for MaltEval 1.0 (beta) (2014). http://www.maltparser.org/malteval.html
14. Nivre, J.: An efficient algorithm for projective dependency parsing. In: Proceedings of the IWPT 2003, pp. 149–160 (2003)

15. Nivre, J.: MaltParser: a language independent system for data-driven dependency parsing. Nat. Lang. Eng. **13**, 95–135 (2007)
16. Nivre, J., et al.: Universal dependencies v1: a multilingual treebank collection. In: Proceedings of LREC-10 (2016)
17. Nivre, J., Agić, Ž., et al.: Universal Dependencies 2.1, LINDAT/CLARIN digital library at the Institute of Formal and Applied Linguistics (ÚFAL), Faculty of Mathematics and Physics, Charles University (2017). http://hdl.handle.net/11234/1-2515
18. NLPub: Russian NLP tools page (in Russian) https://nlpub.ru/
19. Segalovich, I.: A fast morphological algorithm with unknown word guessing induced by a dictionary for a web search engine. In: Proceedings of MLMTA-2003 (2003). https://tech.yandex.ru/mystem/
20. Sharoff, S., Nivre, J.: The proper place of men and machines in language technology. In: Processing Russian without any Linguistic Knowledge. Computational Linguistics and Intelligent Technologies, vol. 10 (17), pp. 657–670 (2011)
21. Tiedemann, J.: Cross-lingual dependency parsing with universal dependencies and predicted PoS labels. In: Proceedings of the Third International Conference on Dependency Linguistics (Depling 2015), pp. 340–349 (2015)
22. Toldova, S., et al.: Otsenka metodov avtomaticheskogo analiza teksta 2011–2012: sintaksicheskie parsery russkogo jazyka [NLP evaluation 2011–2012: Russian syntactic parsers (in Russian)]. Computational Linguistics and Intelligent Technologies, vol. 11(18), pp. 797–809 (2012)
23. Tsarfaty, R., Nivre, J., Andersson, E.: Cross-framework evaluation for statistical parsing. In: Proceedings of EACL 12, France, 2012 (2012)

Active Processes in Modern Spoken Russian Language (Evidence from Russian)

Natalia Bogdanova-Beglarian[1] and Yulia Filyasova[2(✉)]

[1] Saint-Petersburg State University, Saint-Petersburg 199034, Russia
nvbogdanova_2005@mail.ru
[2] Saint-Petersburg Mining University, Saint-Petersburg 199106, Russia
phill.yield@gmail.com

Abstract. Various application fields of linguistics including automatic recognition and speech processing, teaching foreign languages and interpreting colloquial speech, characterization of sociolinguistic speech diversity, linguistic "portrayal" of a certain community (social dialect) and a particular persona (idiolect), linguistic examination (for example, for counter-terrorism efforts), among others, require not only extensive lexical and grammatical resources, but also the description of speech production mechanisms, mainly those, typical of spontaneous speech. Regrettably, the latter are almost always neglected by traditional dictionaries and grammar books, being out of scope of linguistic analysis. The knowledge of such mechanisms is necessary for *colloquial studies (colloquialistics)* per se, a branch of linguistics which studies everyday spoken language. The authors of this article make an attempt to systematize processes proceeding in modern colloquial language through the reliance on domestic and foreign professional academic literature and research results obtained from the ORD-corpus (everyday Russian spoken language) analysis.

The research was fulfilled with the support of RFBR (Russian Foundation for Basic Research) No. 17-29-09175 "Diagnostic features of sociolinguistic variation in Russian everyday spoken language (evidence from a corpus)".

Keywords: Spoken language · Corpus linguistics · Grammaticalization
Pragmaticalization · Hesitation · Metacommunication · Reduction
Reduplication · Semantic change · Desemantization · Parceling
Self-correction · Idiomatization

1 Introduction

There is a variety of processes which characterize speech in any spoken language, such as hesitation and metacommunication, reduction and reduplication, semantic change and desemantization, parceling and self-correction, pragmaticalization, idiomatization and grammaticalization. These processes are quite active since they generate new lexicogrammatical and pragmatic units and, reversely, eliminate well-known elements, which used to be traditional, from common lexicon (being pragmaticalized), or transform them into new elements.

All phenomena of this kind are only typical of spoken language; they constitute its specific nature and need to be systematically described both in terms of the processes

© Springer Nature Switzerland AG 2018
D. A. Alexandrov et al. (Eds.): DTGS 2018, CCIS 859, pp. 391–400, 2018.

mentioned above, and those linguistic units which emerge as a result of their active influence.

2 Material

All processes, considered in the present article, are illustrated, with rare exceptions, by contexts, excerpted from the ORD-corpus of Russian everyday speech, which is currently one of the most extensive linguistic resources, comprising recordings of 130 informants and more than 1000 interlocutors, whose participation realistically represents social stratification of the modern Russian-speaking society. The corpus contains approximately 1250 h of spoken language, more than 2800 communication episodes and one million transcribed word forms [2, 7–12, 25].

3 Hesitation

Hesitation means the presence of interruptions or pauses in speech, caused not by its syntactic structure, but by the fact that speakers, generating speech hic et nunc, have to operate two speech production processes simultaneously, in the face of time deficit: selecting language units and developing their ideas. Hesitation phenomena are inherent features of spontaneous speech, and their functional value can be considered in terms of cognitive, affective-state, and social interaction variables [23, p. 51]. For this reason, hesitation phenomena have become the basis of a new academic field, both interesting and prospective, which appeared at the interface between linguistics, psychology, medicine and the theory of probabilistic processes [22].

Hesitation fulfills the functions of speech planning, memory activities, signals clause boundaries, changes of mood or topic, aiding intelligibility for listeners [18, 21].

Despite the fact that there are different classifications of hesitation phenomena in professional linguistic literature [1, 3, 4, 26], all authors admit that hesitation pauses can be both physical (silent) and filled with verbal or nonverbal sounds. Verbal hesitations are of great interest for application purposes as they are derived from notional linguistic units, and during speech perception and transcription processes it is quite a difficult task for experts to separate them from meaningful speech segments, not to mention non-native speakers and automated recognition systems, cf. (about special markers in the ORD-corpus transcription) [25, pp. 242–243]:

- *a vot (e-e) na... n... vot nashe vot eto vot (e-e) vot eto vot/vot tut/tut slozhnee gorazdo/da//potomu chto/znachit/ja vot vot (e-e) vot eti/nu v principe/znachit/nu/p...po moim/pon'atijam znachit/ja zhe ne otlichu tak skazhem/tadzhika ot uzbeka chto nazyvaets'a/da da da//da?*
- *a/vs'o ... a u men'a na da... d... dacha na etom/kak jego/na Dunae;*
- *tak prosto nu chtob povypendrivats'a znaete tam;*
- *vot jest'/veshchi takie/vot/nu/u l'udej khobbi naprimer/da? *P *V nu(:)/tam skazhem/*P nu/ne znaju/pajaet chto-to.*

These examples clearly show that verbal hesitatives are not only diverse, but also polyfunctional. Apart from the hesitation function, they also act as discourse markers-navigators (*nu v principe/znachit/nu*), reflexive markers (*tak skazhem, chto nazyvaet-sja*), discourse end markers and at the same time, metacommunicative markers (*znaete tam*), among others (see below and [30]).

4 Pragmaticalization

Pragmaticalization is defined as the transition of grammatical forms and certain lex-emes to the category of communicative-pragmatic units, capable of acting as independent elements, expressing speakers' reactions to the communicative situation and social environment [16, pp. 288–289; 17]. Traditional lexemes, phrases or even predicative units turn into *pragmatemes* [5], or *pragmatic markers* [6]: *eto samoe, kak skazat', (nu) (ty) znaesh', vot (etot) vot, tuda-s'uda, kak jego (jejo, ikh), kak eto, (ja) ne znaju* and others. The majority of such markers express hesitation (see Sect. 3). However, their functions are quite various – that makes it more difficult to identify them while transcribing, translating or automatic speech processing. Let us consider a few examples from the ORD-corpus and specify contextual function(s) of each unit in the discourse (the types of pragmatic markers):

- *da tam kakie-to/eti samye/i (eshcho vot)/chto-to po-moemu/ona kakie-to protokoly raznoglasija pishet//ja ne znaju* (markers of discourse beginning/hesitatives *kakie-to/eti samye* + *da tam* + search hesitative *kakie-to/eti samye* + discourse end marker/metacommunicative *ja ne znaju*);
- *v obshchem/*V *P kakoj-to marazm/takoe vpechatlenie sozdajots'a//*P i starushka eta/*P vot tak znaete (e...e) *P vz'ala/razorvala recept/i brosila tam jej* (hesitative/deictic marker *vot tak* + metacommunicative/hesitative *znaete*);
- *byli (e) kak-to vot/(e) (...) vot eti/kak ikh? l'amblii?ili kak eto?* (hesitatives *kak-to vot, vot eti/kak ikh?* +reflexive *ili kak eto?*);
- *s drugimi//*P nu (...) nespecialistami tak skazhem//*P v toj oblasti/v kotoroj ja rabotaju* (reflexive/euphemistic marker);
- *i ona prosto/u nejo tam na na urovne podsoznanija srabatyvaet/net/ne khochu/potomu chto//ja ne znaju pochemu/dumaju chto* (discourse end marker);
- *sprosila by/ja by tebe objasnila by/*V i ty by () uzhe davno by sdelala/i mne by/v poldes'atogo/nervy ne trepala by/s etoj jerundoj/durackoj! s gektarami! *P chto oni iz vas/zhivotnovodov khot'at () etikh (...) fu ty () pakharej (...) chjortovykh vyrastit'/ chto li?* (marker of self-correction/hesitative);
- *nu vot//*P i tut zvonok v dver'//stoit etot muzhik//*P tipa togo chto blin/*P *H *P davajte obshchats'a !* (discourse beginning marker *nu vot* + xenomarker *tipa togo chto*);
- *slushaj/gde-to (...) berut eti (...) vzryvnye veshchestva* (hesitative/rhythm-forming marker);
- *da on teper' tam s radostju/predstavl'aesh' tam//vs'o slomal/poly vse tam po... povylamyval//vs'o teper' sam/govorit/vs'o chto mozhno/sdelat'/*P remont nado//i t'oplye poly/i vse dela/vs'o tam sdelaju* (a series of hesitation, rhythm forming

markers   *tam* + metacommunicative   *predstavljaesh'* + marker-approximator,   a substitute for a number of enumerations *vse dela*);

- *postojali/privet-privet **tam/bla-bla-bla*** (rhythm forming marker *tam* + marker-approximator, a substitute for insertions of other people's speech *bla-bla-bla*).

The list of contextual examples, illustrating pragmatemes, can be extended. However, it is clear from the those given above that it is absolutely necessary to compile a special *dictionary of pragmatic markers*, intended for various practical purposes, containing a complete description of their discourse functions in spoken language, including indications of their polyfunctionality and "compositionality". Markers of the same type tend to occur in one and the same context *taki tak mol; to-s'o p'atoe-des'atoe; eto samoe kak jego* and others, thus intensifying the pragmatic effect, which might not be achieved by only one marker. As it is expected, the dictionary should also include some information about social and psycholinguistic characteristics of pragmatic markers, which will help to widen the scope of the research.

5   Grammaticalization

Grammaticalization is a process which results in adding certain grammatical functions to lexical units and constructions, thus generating new grammatical units; alternatively, grammatical units acquire new functions, in other words, one grammatical element is transformed into another grammatical element [19, p. 1].

According to E.C. Traugott (2003), grammaticalization involves (1) structural decategorization, (2) shift from membership in a relatively open set to membership in a relatively closed one, (3) bonding (erasure of morphological boundaries), (4) semantic and pragmatic shift from more to less referential meaning via invited inferencing, phonological attrition, which may result in the development of paradigmatic zero [27, p. 644]. *Primary* and *secondary* grammaticalization is further distinguished. The latter studies the development of grammaticalized units [28, p. 270].

There are a lot of examples illustrating different types of grammaticalization in the spoken Russian language. The process of conjunctionalization (conjunction generation) is quite active:

- *chto-to golova tak gr'aznits'a stala//znaesh'/**vrode** v shapke khodish'//*P **a** vs'o ravno* (PARTICLE/INTRODUCTORY WORD → CONJUNCTION);
- *vot samoe interesnoe/chto my () mnogoe zabyvaem/da/**kazalos'** by/[no] okazy-vaets'a net/ono nyr'aet v podsoznanie/podkorku golovnogo mozga* (PARTICLE/INTRODUCTORY WORD → CONJUNCTION);
- *on govorit vot vidimo sejchas eto naobor ot takaja propaganda/chto **jakoby** nevydajut vizu/*V [no] na samom dele vot jemu dali; eta obshchestvennaja organizacija/po idee/jest' v kazhdom rajone/uprave/vezde. No zvonka ottuda ne dozhdeshs'a* (PARTICLE/INTRODUCTORY WORD → CONJUNCTION);
- *po obshchim kriterijam/ne govor'a o konkretnykh standartakh; ni odnu gruppu ni v odnom sezone/krome dvukh grupp dekabr'a my ne mozhem prodat' po 530 jevro/uzh **ne   govor'a   o   tom   chtoby** prodat'   po   510* (PARTICLE → PREPOSITION → CONJUNCTION);

- *my obrabatyvaem informaciju/i on u nas konechno *N zverski (?) tormozit//*P vot* **uchityvaja** *(...)* **chto** *(...) Marina (...) analitik (...) zagruzhaetsja (?)// chto/mozhet/nichego *N//*P po krajnej mere* (PARTICIPLE → CONJUNCTION);
- *to jest' ja ne byla zan'ata ni kastr'ul'ami ni magazinami eto* **jedinstvenno chto** *juzhnyj rynok pokupala dl'a dushi sebe chto khotela* (ADVERB → CONJUNCTION).

The ADVERB *tuda-s'uda* can act in spoken language as a MARKER-APPROXIMATOR (*on () on mne zvonit/tipa/koroche/my vs'o gotovo/koroche/**tuda-s'uda**/a ja govor'u/ja v otpuske*); the VERB *govorit'* in its finite forms (usually – reduced) functions as a XENO-MARKER- (*grit/gyt/gr'u*, among others).

These examples clearly demonstrate that different processes occur simultaneously, and new units emerge as a result of several processes, not only one: grammaticalization can proceed along with pragmaticalization, reduction or reduplication (see above). Let us analyze a few examples below.

The pronominal adjective *takoj* is characterized by a wide range of functions:

- XENO-MARKER (*ja tak tyn-dyn-tyn-tyn-tyn/no vjekhala//on* **takoj**/*on* **takoj**/*nu(:) molodca! nu pojekhali ots'uda*);
- VERBAL HESITATIVE (*on vidish' li/on* **takoj**/*vot etot*);
- FIGURATIVE MARKER (*tam sid'at babus'ki* **takie**/*na ostanovke*);
- MARKER-INTENSIFIER (*kakoj u nikh epos! *P geroi* **takie**/*voobshche*);
- MARKER OF NEGATIVE INTENSIFICATION (*da/my* **takie** *p'janicy/Natasha*) and many others.

The particle/preposition *tipa* functions as a polyfunctional unit with the variations *tipa togo* and *tipa togo chto*:

- EXPLANATORY CONJUNCTION: *vypili tam chego-to viski ... t'fu vodku s koloj/* **tipa** *poveselilis'; on takoj/jazhe skazal/nado () golovoj vo vse storony krutit'/chtoby sheja/chto/*V nu chto/chtoby sheja slomalas'/**tipa togo chto** ne nado bojats'a*;
- FINAL MARKER: *eto uzhe kak patefon/znaesh'/**tipa togo***;
- MARKER OF AGREEMENT: *je shcho odin bonus. – Nu-nu-nu/**tipa togo**. Da*;
- XENOMARKER: *i tut z... zvonok v dver' /i Val'demar zakhodit/**tipa togo chto** kto takaja Val'demar! *P gde ty?/*P a pojezdka (e) v etu samuju (:)/(e...e) v Novosibirsk/on govorit/otmenilas'.*

The adverb *tam* has also a number of pragmatic functions:

- VERBAL HESITATIVE: *aga-ga-ga//*P stil'nuju* **tam** *vs'akuju mebel'*;
- RHYTHM FORMING PRAGMATEME: *ja znaju chto nichego ne izmenits'a/@/ ugu/@ chego* **tam** *() psikhovat'*;
- XENO-MARKER: *i ona kak na nas naletela! vot* **tam** *ty-ty-ty-ty-ty-ty/da my alkashi* **tam**/*nu chto-to tam takoe/ja ne pomn'u*;
- MARKER-APPROXIMATOR: *no ona sejchas tozhe budet golovu morochit'/adres* **tam**;
- MARKER-APPROXIMANT (along with the underlined element): *grubo govor'a* **tam**/*chetyre s polovinoj na dva vosem' des'at' gde-to.*

These examples provide evidence of a combined effect produced by speech processes in a colloquial discourse. This conclusion is related to that made by G. Diewald: pragmaticalization is argued to represent a subclass of grammaticalization, which displays essential core features of grammaticalization processes, but is distinguished from other subtypes of grammaticalization processes by specific characteristic traits (concerning function and domain as well as syntactic integration) [15]. Cf.: pragmatic functions are genuinely grammatical functions which are indispensible for the organization of spoken dialogic discourse, as well as for the coherence of written texts... no clear line can be drawn between pragmatics and grammar, since traditional 'grammatical' categories (e.g. tense, aspect, and mood expressions) may be found to have pragmatic functions, and discourse-related categories (e.g. topic and focus) and may display a grammatical dimension [14, p. 74].

Special linguistic attention should be paid to such cases when new grammatical constructions, not only grammatical units, emerge as a result of the grammaticalization process. These grammatical constructions are characterized by integrity, stability and reproducibility (construction-collocations):

- *(Q?) – P – chto li? (gde ? v kafe chto li ?; a iz chego on ? iz morkovki chto li ?) –* the construction has three modifications:

  *chto / chego – P – chto li? (tak a **chego** / malo **chto li** ?; i **chto zh** ? sejchas po-drugomu **chto li** ?) (Q? –* the invariable ***chto*** has a modification ***chego***;

  *chto / chego – P – chto li? (nu **chto zh** ja / durnoj **chto li?** (= 'ne durnoj');*
  *chego / mne zhalko **chto li** ? (= 'ne zhalko')* with the meaning 'ne P';

  P – chto li ?!(ser'jozno chto li?, s uma soshol chto li?, ne znaesh' chto li?, ne vidish' chto li?) the construction without the initial Q? turns into speech cliché – rhetorical question/bewilderment/surprise;

- *P – net? (id'osh' net budesh' net?) –* a reduced modification of the coded *P aut net?,* in which the word *net* means an action or condition, opposite to that, which is mentioned in the first part of the utterance.

The suggested list of constructions can further be extended, but this one already provides evidence that the active processes also include *idiomatization*.

6   Idiomatization

In the broad sense, idioms are those units which meet the criteria of stability, reproducibility, compositionality and meaning integrity. The meaning of an idiom is not a sum of the meaning of its constituents [13, 29]. The analysis of the ORD-corpus shows that idioms occurring in our everyday speech are not units from traditional dictionaries, but colloquial neologisms which possess a huge idiomatic potential. The following excerpts, undoubtfully, illustrate the idiomatization process:

- *vs'aka durka v boshku,* (occasional idiom);
- *bred sovershennoj sivoj kobyly* (combination of two traditional idioms);
- *s odnim polotencem na pereves* (modification of a traditional idiom);

- *bez nikakikh, po khodu* (prepositional-nominal form);
- *kak raz dl'a blondinok* (generalizing phrase, clear to native speakers);
- *eto ja udachno zashol* (precedent text);
- *vremena ne stol' otdal'onnye* (modified precedent text);
- *znaesh' kak/chto/gde…; vidish' kak/chto/gde* and others of this kind (demonstrative-rhetorical colloquial constructions);
- *takoe vpechatlenie/oshchushchenie chto…; (tut) takoe delo; (chto-nibud') v takom duhe* (construction with the word *takoj*).

As can be seen from above, idiomatization, along with pragmaticalization, is almost inseparable from grammaticalization. The suggested list contains examples illustrating this process, and can be further extended while analyzing the ORD-corpus material. It is obvious that the idiomatic potential of our everyday colloquial speech is enormous. Identification of meaning through other units can serve as idiomatization criteria (*vremena ne stol' otdaljonnye* = 'recently'; *eto ja udachno zashol* = 'I have been lucky').

7 Other Processes in Spoken Language

Giving a brief overview of other processes in spoken language, it is necessary to underline again their interrelation with each other.

As a result of **reduction**, *interjectional pragmatemes* emerge along with simply reduced forms of highly frequent words: *zdras'te!/zdras'te pozhalsta!* and *shchas!*, hesitation pragmateme *shchas-shchas (shchas)*. The process of *grammaticalization*: etiquette   interjection → interjectional   pragmateme'; and *pragmaticalization*: adverb → pragmateme, verbal search hesitative. The connection between reduction of speech units and grammaticalization is also found in the English language: the process of grammaticalization which studies simplification of units, shift from more to less complex structures, is called reduction [28, p. 270]. This definition is illustrated by the following examples: *have to – hafta, has to – hasta, want to – wanna, going to – gonna*, along with others.

Reduplication, broadly defined as the repetition of part or all of one linguistic constituent to form a new constituent with a different function (cf. *lexicalization* and *grammaticalization*) [20, p. 1]. According to F. Rozhanskiy (2015), the relation between the original and reduplicated forms can be described not only through the notion of doubling, but also through inexact similarity [24].

Reduplication, present in the Russian language, e.g. *takoj-s'akoj, tuda-s'uda, p'atoe-des'atoe*, embraces different contextual elements:

- *a ja jej vs'o prigotovila/p'atoe des'atoe tridcat' p'atoe*;
- *Znachit/my prikhodim k vyvodu/chto tak ili s'ak… ili br'ak/ili kak khochesh'… muzyka jest' product sinteticheskij* (evidence from oral sub-corpus of the National Corpus of the Russian Language UP).

Parceling and *self-correction,* along with phrasal cut-off and syntactic ellipsis, are also important processes in colloquial spontaneous speech:

- *dyrka zhe tam vot takaja// <pauza> bol'shaja*;
- *tut u nas sneg lezhit// <pauza> pr'amo voobshche*;
- *takie glu… (…) uzhasy rasskazyvajut*;
- *a jejo zva (…) nazyvat'.*

A typical combination of *pragmaticalization*, *semantic change* and *phonetic reduction* is given below as exemplified through the colloquial reduced form *shhas* (from *sejchas*):

(1) – *Poluchaets'a? – **Shchas** pokazhem. Brat'ja Kozlovy s gotovnostju zasopeli (UP)*;
(2) – ***Shchas kak pikhnus' – kostej ne sober'osh' (UP)***;
(3) *Ja? Ha! **Shchas** pr'am! Nedostojny oni etogo! (UP)*;
(4) *tak/**shchas shchas shchas** ja//aga/vot oni (ORD).*

While example (1) demonstrates *shchas* in its traditional meaning of time, described in dictionaries, contexts (2)–(3) show weakened meanings of this unit, caused by *desemantization* (adverb → interjectional pragmateme, isolated or as part of collocations). Example (4) gives evidence of almost complete loss of its meaning (*pragmaticalization*) (adverb → verbal search hesitative, usually accompanied by *reduplication*) [25].

8   Conclusion

The overview of processes, typical of colloquial spontaneous everyday speech, firstly, clearly demonstrates the opportunities of corpus approach for speech material analysis; secondly, reveals the diachronic evolution of many linguistic units, both contextual/occasional and systematic, which precedes future lexical and grammatical changes in the Russian language. Identification and linguistic, including lexicographical, description of such units can be considered as relevant objectives of colloquial studies (colloquialistics). Awareness of these processes is, as it was mentioned earlier, important for various application fields of linguistics.

References

1. Alexandrova, O.A.: Rechekommunikativnyj status pauzy kolebania. Ph.D. Thesis. Veliky Novgorod, 208 p. (2004)
2. Asinovsky, A., Bogdanova, N., Rusakova, M., Ryko, A., Stepanova, S., Sherstinova, T.: The ORD speech corpus of Russian everyday communication "one speaker's day": creation principles and annotation. In: Matoušek, V., Mautner, P. (eds.) TSD 2009. LNCS (LNAI), vol. 5729, pp. 250–257. Springer, Heidelberg (2009). https://doi.org/10.1007/978-3-642-04208-9_36
3. Belickaja, A.A.: O roli khezitacionnykh pauz v spontannoj rechi. Filologia i Literaturovedenie, February 2014, no. 2 (2014). http://philology.snauka.ru/2014/02/697

4. Blankenship, J., Kay, Ch.: Hesitation phenomena in English speech: a study in distribution. Word **20**(3), 360–372 (1964). https://doi.org/10.1080/00437956.1964.11659828

5. Bogdanova-Beglarian, N.V.: Pragmatemy v ustnoj povsednevnoj rechi: opredeleniepon'atia i obshhaja tipologia. Vestnik Permskogo universiteta. Rossijskaja i zarubezhnaja filologia **3** (27), 7–20 (2014)

6. Bogdanova-Beglarian, N., Filyasova, Yu.: Discourse vs. pragmatic markers: a contrastive terminological study. In: 5th International Multidisciplinary Scientific Conference on Social Sciences and Arts SGEM 2018, SGEM2018 Vienna ART Conference Proceedings, March 2018, vol. 5, no. 3.1, pp. 123–130 (2018). www.sgemvienna.org. https://doi.org/10.5593/sgemsocial2018h/31/S10.016

7. Bogdanova-Beglarian, N.V., Asinovsky, A.S., Blinova, O.V., Markasova, E.V., Ryko, A.I., Sherstinova, T.Yu.: Zvukovoj korpus russkogo jazyka: novaja metodologija analiza ustnoj rechi. Jazyk i metod: Russkij jazyk v lingvisticheskih issledovanijah XXI veka. Vyp. 2, Krakow, pp. 357–372 (2015). ISBN 978-83-233-4001-0, eISBN 978-83-233-9400-6

8. Bogdanova-Beglarian, N., Martynenko, G., Sherstinova, T.: The "One Day of Speech" corpus: phonetic and syntactic studies of everyday spoken Russian. In: Ronzhin, A., Potapova, R., Fakotakis, N. (eds.) SPECOM 2015. LNCS (LNAI), vol. 9319, pp. 429–437. Springer, Cham (2015). https://doi.org/10.1007/978-3-319-23132-7_53

9. Bogdanova-Beglarian, N., et al.: Sociolinguistic extension of the ORD corpus of Russian everyday speech. In: Ronzhin, A., Potapova, R., Németh, G. (eds.) SPECOM 2016. LNCS (LNAI), vol. 9811, pp. 659–666. Springer, Cham (2016). https://doi.org/10.1007/978-3-319-43958-7_80

10. Bogdanova-Beglarian, N., Sherstinova, T., Blinova, O., Martynenko, G.: An exploratory study on sociolinguistic variation of Russian everyday speech. In: Ronzhin, A., Potapova, R., Németh, G. (eds.) SPECOM 2016. LNCS (LNAI), vol. 9811, pp. 100–107. Springer, Cham (2016). https://doi.org/10.1007/978-3-319-43958-7_11

11. Bogdanova-Beglarian, N.V., Sherstinova, T.Ju., Blinova, O.V., Martynenko, G.Ja.: Korpus «Odin rechevoj den'» v issledovanijakh sociolingvisticheskoj variativnosti russkoj razgovorno jrechi. Analiz razgovornoj rechi (AR³-2017): trudy sed'mogo mezhdisciplinarnogo seminara, St. Petersburg, pp. 14–20 (2017)

12. Bogdanova-Beglarian, N., Sherstinova, T., Blinova, O., Martynenko, G.: Linguistic features and sociolinguistic variability in everyday spoken Russian. In: Karpov, A., Potapova, R., Mporas, I. (eds.) SPECOM 2017. LNCS (LNAI), vol. 10458, pp. 503–511. Springer, Cham (2017). https://doi.org/10.1007/978-3-319-66429-3_50

13. Collocational and Idiomatic Aspects of Composite Predicates in the History of English. John Benjamins, B.V. (1999). ISBN 9789027298751

14. Degand, L., Evers-Vermeul, J.: Grammaticalization or pragmaticalization of discourse markers? More than a terminological issue. J. Hist. Pragmat. **16**(1), 59–85 (2015). https://doi.org/10.1075/jhp.16.1.03deg

15. Diewald, G.: Pragmaticalization (Defined) as grammaticalization of discourse functions. Linguistics **49**(2), 365–390 (2011). https://doi.org/10.1515/ling.2011.011

16. Graf, E.: Interjections in Russian as Interactive units. Frankfurt on Main, 328 p. (2011)

17. Günther, S., Mutz, K.: Grammaticalization vs. pragmaticalization? The development of pragmatic markers in German and Italian. In: Bisang, W., Himmelmann, N.P., Wiemer, B. (eds.) What Makes Grammaticalization? A Look from its Fringes and its Components, pp. 77–107. Language Arts & Disciplines, Berlin (2004). ISBN 978-3-11-019744-0

18. Hlavac, J.: Hesitation and monitoring phenomena in bilingual speech: a consequence of code-switching or a strategy to facilitate its incorporation? J. Pragmat. **43**(15), 3793–3806 (2011). https://doi.org/10.1016/j.pragma.2011.09.008

19. Hopper, P.J., Traugott, E.C.: Grammaticalization, 2nd edn., 276 p. Cambridge University Press, Cambridge (2003). ISBN 9780521804219
20. Inkelas, S.: Non-concatenative derivation: reduplication. In: Lieber, R., Skekauer, P. (eds.) The Oxford Handbook of Derivational Morphology, 46 p. Oxford University Press, Oxford (2013). https://doi.org/10.1093/oxfordhb/9780199641642.013.0011
21. Lounsbury, F.G.: Pausal, Juncture and Hesitation Phenomena. In: Psycholinguistics, pp. 96–101. Waverly Press, Baltimore (1954)
22. Nikolaeva T.M.: Lingvistika teksta: sovremennoe sostojanie i perspektivy. In: Novoe v zarubezhnoj livgvistike, vyp. VIII. Lingvistika teksta. Moscow, pp. 5–39 (1978)
23. Rochester, S.R.: The significance of pauses in spontaneous speech. J. Psycholinguist. Res. 2 (1), 51–81 (1973). https://doi.org/10.1007/bf01067111
24. Rozhanskiy, F.: Two semantic patterns of reduplication. In: The Why and How of Total Reduplication: Current Issues and New Perspectives, Studies in Language, vol. 39, no. 4, pp. 992–1018 (2018). https://doi.org/10.1075/sl.39.4.02roz
25. Russkij jazyk povsednevnogo obshchenia: osobennosti funkcionirovania v raznykh social'nykh gruppakh. Kollektivnaja monografia. St. Petersburg, 244 p. (2016). ISBN 978-5-9906824-4-3
26. Stepanova, S.B.: Obshchee i individual'noe v khezitacijakh (na materiale russkoj spontannoj rechi). In: Proceedings of XXXV International Philological Conference, 13–18 March 2006, St. Petersburg, Section of Phonetics. Part 1, no. 20, pp. 24–32 (2006)
27. Traugott, E.C.: Constructions in grammaticalization. In: The Handbook of Historical Linguistics, pp. 624–647. Blackwell, Oxford (2003). https://doi.org/10.1002/9781405166201.ch20
28. Traugott, E.C.: Grammaticalization. In: Continuum Companion to Historical Linguistics, pp. 269–283. Continuum Press, London (2010). ISBN-10 144114465X
29. Vinogradov, V.V.: Russkij jazyk: Grammaticheskoe uchenie o slove, Moscow, 640 p. (1986)
30. Zvukovoj korpus kak material dl'a analiza russkoj rechi. Kollektivnaja monografia. Chast' 2. Teoretichskie i prakticheskie aspekty analiza. Tom 2. Zvukovoj korpus kak material dl'a novykh leksikograficheskikh proektov, St.-Petersburg, 364 p. (2015)

Author Index

Printed in the United States